MEDICAL
AND
ECONOMICAL BOTANY

MEDICAL
AND
ECONOMICAL BOTANY

BY

JOHN LINDLEY, Ph.D. F.R.S.,

VICE-SECRETARY OF THE HORTICULTURAL SOCIETY OF LONDON; PROFESSOR OF BOTANY IN UNIVERSITY COLLEGE, LONDON, THE ROYAL INSTITUTION OF GREAT BRITAIN, AND TO THE SOCIETY OF APOTHECARIES

WITH NUMEROUS ILLUSTRATIONS

GYAN

Published by

Gyan Publishing House
5, Ansari Road
Daryaganj, New Delhi-110002
Phone: 011-47034999, 9811692060
E-mail: books@gyanbooks.com

Distribution Network
gyanbooks.com
India, USA, Canada, UK, Australia, France

ISBN : 978-81-212-2261-7 (HB)
First Published, 1849

2nd Impression 2021

Printed at: Gyan Press, Delhi.

Medical and Economical Botany
Author: John Lindley

MEDICAL AND ŒCONOMICAL BOTANY.

BY

JOHN LINDLEY, PH.D., F.R.S.,

VICE-SECRETARY OF THE HORTICULTURAL SOCIETY OF LONDON;
PROFESSOR OF BOTANY IN UNIVERSITY COLLEGE, LONDON,
THE ROYAL INSTITUTION OF GREAT BRITAIN, AND TO THE SOCIETY OF APOTHECARIES.

WITH NUMEROUS ILLUSTRATIONS.

LONDON:
BRADBURY & EVANS, 11, BOUVERIE STREET.
MDCCCXLIX.

TABLE OF CONTENTS.

VII.—EXOGENS—(*continued.*)

PART III.

MEDICAL AND ŒCONOMICAL BOTANY.

PAUCIS utatur medicus remediis, iisque selectis.

Qui potest mederi simplicibus, dolosè et frustra quærit composita.

Medicus notitiâ plantae destitutus de viribus ejusdem nunquam juste judicavit.

Systemate, Qualitate, et Experientiâ, eruitur omnis usus plantarum.

Alimenta à toxicis, uti medicamenta à venenis, non natura sed dosis distinguit.

LINNÆUS, *Mat. Med. Canones.*

MEDICAL AND ŒCONOMICAL BOTANY.

THE Vegetable Kingdom contains, among a large quantity of plants of no known importance to man, various useful species employed in medicine, the arts, or in the many branches of domestic œconomy.

The principal part of those which can be brought by teachers in Europe under the notice of students, or which, from their great importance, deserve to be among the earliest subjects of study, are mentioned in the following pages, where they are arranged in the manner proposed in the "*Vegetable Kingdom*" of the author, with the sequence of matter departed from in a few instances, when it was believed that the convenience of younger students would be consulted by doing so. The author trusts that this selection will be found to have been made in such a way that all teachers who possess reasonably extensive means of illustrating their lectures, and all Botanic Gardens, may furnish the larger part of the species which are mentioned. A small selection was indispensable; firstly, because a greater work would have been beyond the reach of the majority of purchasers; and secondly, because experience shews us that those who have to study a science of observation, such as Botany, require to concentrate their attention, in the first instance, upon a limited number of objects.

In the work above referred to, the Vegetable Kingdom is, in the first place, divided into Classes; these are subdivided into Sub-classes, which are themselves broken up into Alliances; beneath the Alliances are placed the Natural Orders, under which are disposed those final subdivisions termed Genera.

Therefore, genera are groups of species; orders are groups of genera; alliances are groups of orders; sub-classes, when they are employed, are groups of alliances, and classes are assemblages of sub-classes (if present), or of alliances, or, in some instances, of orders only. Thus we have—

1. CLASSES.
2. **Sub-classes.**
3. ALLIANCES.
4. **Orders.**
5. GENERA.
6. *Species.*

In the following pages the heading of each of these subdivisions is printed in the type just employed.

The plan of the work is similar to that so happily employed by Linnæus in his Materia Medica, a book invaluable in its day, although now forgotten, and better adapted to the objects of medical men than any thing which has since appeared. Each species, placed in its due position in the classification, has its vulgar or officinal name, as well as that which it bears in science, and, in a few words, the country whence it comes, the quality it has been said to possess, and the uses to which it has been found applicable. Very short phrases are also given, for the purpose of showing how the genera or species are to be distinguished from each other.

The whole plan of the work is, in fact, to point out *distinctions:* and the fewest possible words are employed with this view. In many instances the distinctions may not be absolute, but they are sufficient for ordinary purposes. Those who wish to become better acquainted with Botany will, of course, make the "*Vegetable Kingdom*" of the author the subject of ulterior study. For the convenience of those who may desire to do so, references are made to that work throughout all the following pages.

The authorities generally taken for the quality and uses of officinal plants, are the invaluable works of Pereira, Royle, and Christison. Where plants are not officinal, those other sources of information have been consulted, from which the more extended statements in the "*Vegetable Kingdom*" have been derived.

CLASSES.

The CLASSES of plants are seven, viz :—

I. THALLOGENS; having no sexes, and no distinct separation of stem and leaves.

II. ACROGENS; having no sexes, and distinct stems and leaves.

III. RHIZOGENS; having sexes, and a mycelium.

IV. ENDOGENS; having sexes, a root and stem, endogenous wood, and parallel-veined leaves.

V. DICTYOGENS; having sexes, a root and stem, exogenous roots, endogenous stems, and netted leaves.

VI. GYMNOGENS; having sexes, a root and stem, exogenous wood, and naked seeds.

VII. EXOGENS; having sexes, a root and stem, exogenous wood, and seeds in seed-vessels.

CLASS I. THALLOGENS (V. K., p. 5.)

Of this there are three Alliances; viz.,

ALGALS. Submersed; having no mycelium.
FUNGALS. Aerial; having a mycelium.
LICHENALS. Aerial; having no mycelium.

The natural orders of these alliances need not occupy the attention of the student. Few of the numerous species are of importance, either in medicine or common life.

THE ALGAL ALLIANCE (V. K., p. 8.)

A few species are employed as food. Any medicinal properties which they have been found to possess appear to be owing to the presence of iodine, which the marine species contain in common with other oceanic productions.

PORPHYRA. *Agardh.*

Frond flat, very thin, more or less purple, leafy, not gelatinous. *Spores* roundish, arranged in fours (usually) and filling all the frond.

1. P. *laciniata* Agardh.—(LAVER. SLOKE. SLOKAUN.) Fig. 4.

Fronds clustered, deeply cleft, with broad segments variously lobed and cut at the edges, bright purple.

Habitat. On rocks and stones in the sea. Annual.
Quality. Saline, nutritious.
Uses. Employed as food, salted; eaten with pepper, vinegar, and oil. Said to be useful in scrofulous affections and glandular tumours.

2. P. *vulgaris* Agardh.—(LAVER.)

Fronds undivided, broadly lanceolate, wavy, bright lively purple.

Habitat, Quality, and *Uses,* as the last.

ULVA. *Linnæus.*

Frond membranous, green, flat, sometimes inflated. *Spores* minute, lying in fours.

1. U. *latissima* Linn.—(GREEN LAVER. GREEN SLOKE. OYSTER GREEN.)

Marine. Frond broad, oblong, roundish, wavy, full green, completely covered by fructification.

Habitat. Rocks and stones in the sea. Annual.
Quality. Bitterish, salt, mucilaginous.
Uses. Employed as food, stewed and seasoned with lemon-juice. Ordered for scrofulous patients.

LAMINARIA. *Lamouroux.*

Frond leathery, flat, without a midrib. *Spores* forming close spots, and imbedded in the thickened surface of some part of the frond.

1. L. *digitata* Lamouroux.—(TANGLE. SEA WARE. SEA-GIRDLES. SEA-WAND. RED-WARE.)

Stem woody, cylindrical, gradually expanding into a leathery roundish-oblong frond deeply cleft into many linear divisions.

Habitat. Rocks in the sea, in deep water.
Quality. Nutritious.
Uses. When young employed as food for both man and cattle.

ALARIA. *Greville.*

Frond membranous with a stout midrib; stem pinnated. *Spores* pear-shaped, vertically arranged in the thickened leaflets.

1. A. *esculenta* Greville. — (BADDERLOCKS. HENWARE. HONEYWARE. MURLINS.) Fig. 2.

Frond from 2 to 12 feet long, olive-green; stem 4 to 8 inches long, pinnated with several short flat narrow leaflets.

Habitat. Rocks in the sea, in deep water. Annual.
Quality. Nutritious.
Uses. The midrib is eaten, when the frond is stripped off.

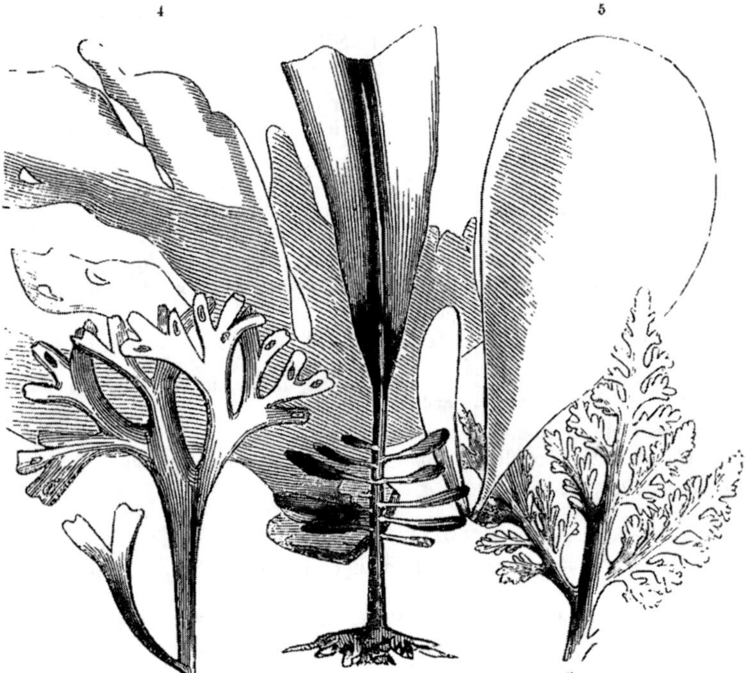

CHONDRUS. *Stackhouse.*

Frond cartilaginous, dilated upwards, flat, ribless, dichotomous, purplish or livid red. *Tetraspores* lying in round masses in the substance of the frond.

1. C. *crispus* Lyngb.—(CARRAGEEN MOSS. IRISH MOSS. PEARL MOSS.) Fig. 1.

Frond thickish, cartilaginous, dichotomous, flat or curled, segments linear wedge-shaped; masses of spores hemispherical, concave on one side.

Habitat. Rocky sea-shores of Europe.
Quality. Nutritive, emollient, demulcent.
Uses. Pulmonary complaints, chronic diarrhœa, dysentery, scrofula, rickets, enlarged mesenteric glands, irritation of bladder and kidneys.—Jellies, soups, &c.

Fig. 1.—Chondrus crispus in fruit; 2. Alaria esculenta; 3. Laurencia pinnatifida; 4. Porphyra laciniata; 5. Iridæa edulis.

FUCUS. *Linnæus.*

Frond leathery, dichotomous, flat, linear; usually furnished with large air-cells. *Spores* arranged in tubercles buried in mucus and collected in heads, through pores in which they are discharged.

1. F. *vesiculosus* Linn.—(SEA WARE. SEAWRACK. KELP WARE. BLACK TANG. SWINE TANG.) Fig. 6.

Frond quite entire, with a midrib; air-cells round, usually in pairs; heads of spores terminal, oblong, blunt, yellowish.

Habitat. Sea-shores. Perennial.
Quality. Detergent, discutient, sub-nutritious.
Uses. Glandular affections and scrofulous tumours. A dentifrice. Makes good manure. Furnishes *Kelp.*

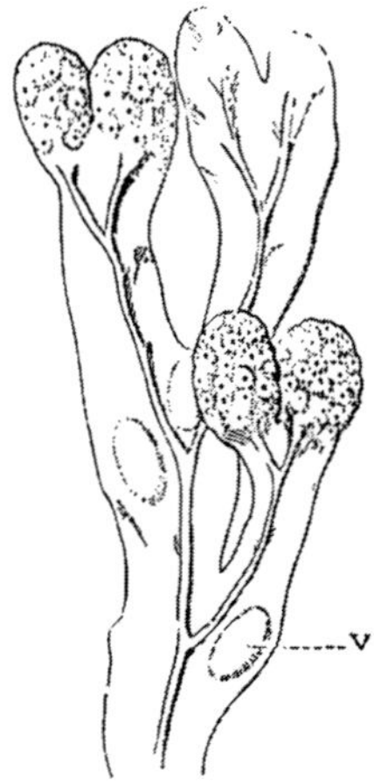

6

IRIDÆA. *Bory.*

Frond flat, cellular, expanded, between fleshy and cartilaginous, purplish-red. *Tetraspores* collected in gelatinous spheres buried in the substance of the frond.

1. I. *edulis* Bory.—(DULSE in the S.W. of England.) Fig. 5.

Frond undivided, obovate or wedge-shaped, very succulent, dull purple, tapering into a short stalk.

Habitat. Rocks in the sea. Biennial.
Quality. Nutritious.
Uses. Employed as food by fishermen, either raw, or pinched between hot irons.—*Stackhouse.*

LAURENCIA. *Lamouroux.*

Frond cylindrical or compressed, between cartilaginous and gelatinous, yellowish or purplish red. *Tetraspores* contained in ovate cystocarps, and lying imbedded in the branches.

1. L. *pinnatifida* Lamouroux.—(PEPPER DULSE.) Fig. 3.

Frond dull purple or greyish, compressed, cartilaginous, twice or thrice pinnatifid, the divisions blunt, entire or lobed. Cystocarps near the end of the branches, the size of poppy seed.

Habitat. Rocks in the sea. Annual.
Quality. Pungent, nutritious.
Uses. A condiment, when other sea-weeds are eaten.

RHODOMENIA. *Greville.*

Frond flat, membranous, pink or red, veinless, sessile, with a very short stem. *Tetraspores* in the substance of the frond or collected in superficial spheres or coccidia.

1. R. *palmata* Greville.—(DULSE. DILLISK. DELLISH. DUILLISG, or Water-leaf, among the Highlanders.—*Grev.*)

Frond purple, leathery, or somewhat membranous, broadly wedge-shaped,

Fig. 6.—Portion of the frond of Fucus vesiculosus in fructification; v, one of the vesicles by which it floats.

irregularly cut, with dichotomous segments entire at the edge or furnished with lateral leaflets ; spores distributed in cloud-like spots over the whole frond.

Habitat. On rocks in the sea.
Quality. Nutritious. Smells of violets.
Uses. Employed as food by the poor of many northern nations. Sudorific.—*Lightfoot.*

PLOCARIA. *Nees.*

Frond cartilaginous, cellular as if jointed, cylindrical or compressed, branched, dull red. *Tetraspores* immersed in the substance of the surface, or collected in superficial spheres or coccidia.

7

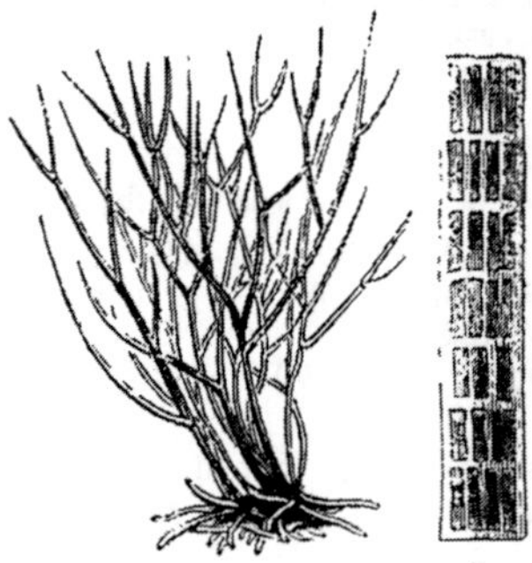

a b

1. P. *Helminthochortos* Nees.—*Gracilaria*, or *Sphærococcus*, or *Gigartina Helminthochortos* of others (CORSICAN MOSS.*) Fig. 7.

Frond cartilaginous, filiform, tufted, entangled ; the primary stem creeping ; the branches setaceous, somewhat dichotomous, striated transversely at the separations of the tiers of red cells.

Habitat. On the coast of Corsica, and elsewhere in the Mediterranean.
Quality. Anthelmintic, nutritious, saline, strong-scented.
Uses. In removing the Ascaris lumbricoides.

2. P. *tenax* Nees.—*Sphærococcus* or *Gracilaria tenax* of authors.

Frond somewhat gelatinous, slippery, filiform, dichotomous; branches spreading, the uppermost reflexed and acute ; coccidia hemispherical, sessile, scattered.

Habitat. The Chinese seas.
Quality. Gelatinous, glutinous, nutritious.
Uses. Soups and jellies among the Chinese ; also as size and gum.

Fig. 7.—Plocaria Helminthochortos ; *a* natural size ; *b* one of the branches much magnified.

* As sold in the shops, this consists of various marine productions, especially of Laurencia obtusa, with a very little Plocaria intermixed.

THE FUNGAL ALLIANCE (V. K., p. 29.)

With the single exception of Ergot, these are excluded from the modern practice of medicine. Ergot itself is a mere disease of the ovary of grasses, caused by the attacks of a parasite of this alliance (*see* Oidium.) Fungals are, however, among the more useful friends of man as food, and among his most dangerous enemies as parasites, destroying the sources of his food. The following are the most common and important examples.

AGARICUS. *Linnæus.*

Pileus bearing on one side vertical, unequal plates or gills, forming a lamellate hymenium. *Veil* single.

1. A. *campestris* Linnæus. — (COMMON MUSHROOM.) Fig. 8.

Pileus fleshy, dry, whitish, silky or scaly, fragrant when broken, and not changing colour; hymenium pink, free, becoming brown or blackish; stipe solid, white, having a ring.

Habitat. Pastures, dunghills.
Quality. Nutritious, fragrant.
Uses. As food, and as sauce (ketchup.)

8

2. A. *oreades* Bolton.—(FAIRY-RING MUSHROOM. SCOTCH BONNETS. CHAMPIGNON.) Fig. 9.

Pileus fleshy, tough, somewhat bossed, first pale-brown, and becoming whiter with age; hymenium whitish, with distinct gills; stipe solid, with no ring, round, whitish, with a skin separating into longitudinal fibres.

9

Fig. 8. Agaricus campestris, with its mycelium, or spawn; 9. Agaricus oreades in different states of growth.

Habitat. Pastures and lawns, where it grows in irregular circles.
Quality. Nutritious, stimulant, rather fragrant.
Uses. Frequently strung on thread, dried in the shade, and pounded, as an addition to sauces.

AMANITA. *Greville.*

Pileus bearing on the under side vertical, unequal plates, forming an *hymenium*. *Veil* double; the outer covering the whole plant when young.

1. A. *muscaria* Greville.—(FLY AGARIC.) Fig. 11.

Pileus bright orange red, warted, striated at the edge; gills white; stipe bulbous.

Habitat. Woods, especially of Fir and Birch.
Quality. Narcotic, poisonous.
Uses. Produces intoxication and delirium. The infusion employed to kill flies.

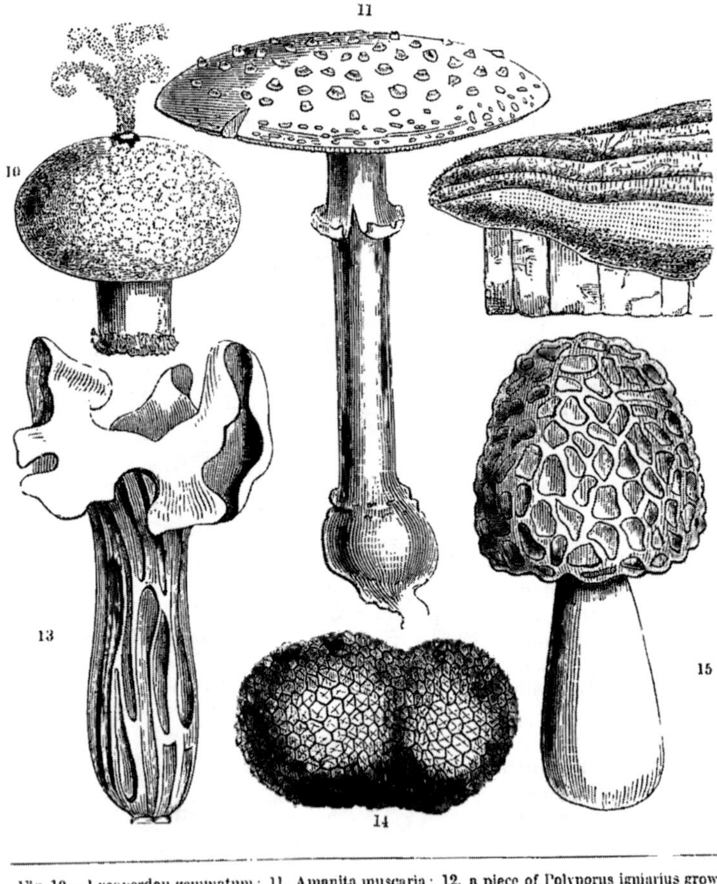

Fig. 10.—Lycoperdon gemmatum; 11. Amanita muscaria; 12. a piece of Polyporus igniarius growing on wood; 13. Helvella crispa; 14. Tuber cibarium; 15. Morchella esculenta.

POLYPORUS. *Micheli.*

Pileus fleshy ; completely blended with the hymenium, which is pierced with thin-sided, rather angular, tubular, vertical passages.

1. P. *igniarius* Fries.—(HARD AMADOU. TOUCHWOOD. SPUNK.) Fig. 12.

Hard ; pileus thick, obtuse, nearly even, ferruginous, at length cinereous: the edges and minute convex pores cinnamon.

Habitat. On old trees.
Quality. Styptic.
Uses. Beaten till soft, it staunches slight wounds. Sliced, it forms good razor-strops.

2. P. *fomentarius* Fries.—(SOFT AMADOU. GERMAN TINDER.)

Pileus somewhat triangular, smooth, brownish grey, soft within : the edge pale bluish white or yellowish, as well as the very minute pores, becoming ferruginous.

Habitat. On old trees.
Quality. Styptic, elastic.
Uses. Forms tinder ; staunches slight wounds ; forms small surgical pads.

3. P. *officinalis* Fries. *Boletus Laricis* Jacquin. —(Αγαρικον of Dioscorides.)

Pileus between corky and fleshy, warted, stalked, banded with yellow and brown ; pores yellowish.

Habitat. On Larch trees in Southern Europe.
Quality. Smells like new flour. Taste bitter, nauseous.
Uses. A drastic purgative, now rarely employed.

MORCHELLA. *Dillenius.*

Pileus convex, with a ribbed, irregularly excavated *hymenium*.

1. M. *esculenta* Linnæus.—(MORELL.) Fig. 15.

Pileus conical, ovate or globose, united to the stipe at the contracted base ; ribs of the hymenium anastomosing into distinct cells.

Habitat. Woods, orchards, cinder-walks.
Quality. Nutritious.
Uses. A delicate article of food.

HELVELLA. *Linnæus.*

Pileus turned downwards, lobed, with an even *hymenium*.

1. H. *crispa* Scopoli.—(MITRE MUSHROOM.) Fig. 13.

Pileus crisp and irregularly lobed, pale yellowish brown ; stipe white, ribbed, fistular, irregularly excavated.

Habitat. Woods.
Quality. Nutritious.
Uses. A delicate article of food.

TUBER. *Micheli.*

A fleshy, firm, roundish mass, filled with veins and minute cavities, in which stand the *spore-cases*.

1. T. *cibarium* Sibthorp.—(TRUFFLE.) Fig. 14.

Surface black, covered with angular warts.

Habitat. In the earth, beneath trees, especially Beeches.
Quality. Nutritious, fragrant, stimulating.
Uses. A common ingredient in sauces and rich dishes.

LYCOPERDON. *Tournefort.*

A globular, fleshy mass, eventually bursting and discharging a multitude of powdery spores.

1. L. *gemmatum* Batsch.—(COMMON PUFFBALL.) Fig. 10.

Round, tapering to the base, covered with little rough warts.

Habitat. Fields and meadows.
Quality. Acts mechanically as a styptic, by means of its brown spores.

2. L. *giganteum* Batsch. *Bovista giganteum* Nees.—(GIANT PUFFBALL.)

Oblong or roundish, white, very large, smooth, with the skin cracking into angular spaces, pulpy at first, becoming dry.

Habitat. Fields and plantations.
Quality. Styptic.
Uses. When dry, staunches slight wounds. The smoke stupifies bees.

16 17 18 19

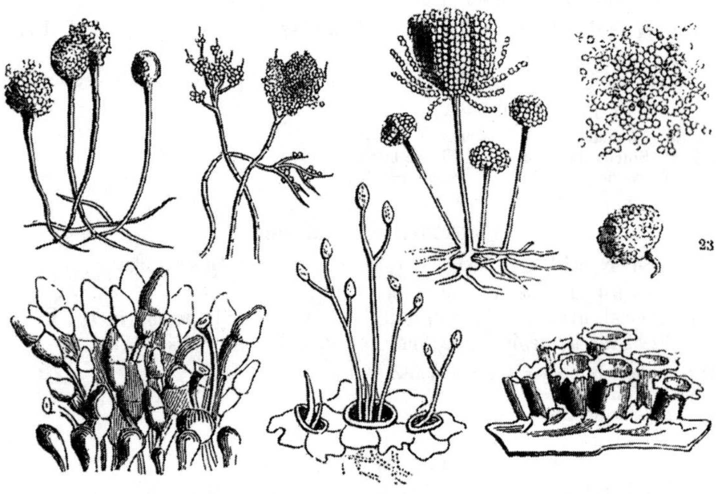

23

20 21 22

TUBURCINIA. *Fries.**

Microscopic. Hollow granular oblong balls, collected in subcutaneous patches.

1. T. *Scabies* Berkeley.—(POTATO SCAB.) Fig. 23.

Spots brown, oblong, becoming confluent; balls attached transversely to a short stipe.

Habitat. Beneath the skin of the tuber of the Potato.
Quality. Produces superficial cavities and pits, but is of no further injury.

Magnified figures of: Fig. 16. Mucor mucedo; 17. Aspergillus glaucus; 18. Penicillium glaucum; 19. Uredo Caries, as it appears when rubbed in water; 20. Puccinia Graminis; 21. Botrytis infestans; 22. Æcidium Berberidis; 23. Tuburcinia Scabies.

* A few species, neither medical nor economical, are here admitted for the sake of the student, because they produce marked effects upon the sources of our medicine or food.

Puccinia. *Persoon.*

Microscopic. Pear-shaped or oblong bodies, containing spores, having internal partitions, and furnished with a slender stalk, by which they are attached to an internal mycelium.

1. P. *graminis* Persoon.—(Corn Mildew.) Fig. 20.

Pitch brown or black, growing in irregular lines, which become occasionally confluent.

Habitat. Leaves and straw of Grasses.
Quality. Preys upon the juices of plants, especially of Corn, and prevents the grain from swelling.

Æcidium. *Persoon.*

Microscopic. Cellular membranous sacs, bursting at the side or apex, and discharging numerous spores.

1. Æc. *cancellatum* Persoon.

Forms spots, which are yellow at first, and then become red and prominent, in figure somewhat conical, splitting at the side in an irregular manner; spores brown.

Habitat. The leaves of Pear trees.
Quality. A destructive parasite, greatly injuring the trees which it attacks.

2. Æc. *Berberidis* Persoon.—(Berberry Blight.) Fig. 22.

Forms roundish bright red spots, in figure elongated, bursting irregularly at the end, and becoming cups; spores orange.

Habitat. Leaves of the common Berberry.
Quality. A troublesome parasite, erroneously supposed to blight Wheat.

3. Æc. *Urticæ* De Candolle.—(Nettle Blight.)

Spore-cases form oblong orange heaps on the under side of the leaves, each being nearly round, and finally gaping wide.

Habitat. Common on Nettles.
Quality. Like other parasites, deprives the plant on which it grows of the organisable matter intended for its own nutrition.

Erysiphe. *Hedwig.*

Microscopic. A fleshy, somewhat gelatinous mass, opening at the collapsing apex, and standing in the centre of a free floccose superficial mycelium.

24

1. E. *communis* Schlechtendahl.

Mycelium dirty white, resembling cobweb, at length forming spots; rays simple, acute.

Habitat. On the living leaves of herbaceous plants.
Quality. A surface parasite, infesting various plants, especially the Pea, which it overruns and destroys.

2. E. *bicornis* Link. Fig. 24.

Mycelium-milk-white, densely entangled into a white skin; rays very short, forked, and warted at the point.

Habitat. Very common on Maple leaves, rendering them hoary.
Quality. As in the last.

Fig. 24. Magnified figure of Erysiphe bicornis.

BOTRYTIS. *Micheli.*

Microscopic. Consisting of erect branched *threads*, at the ends of which grow clusters of spores (spore-cases.)

1. B. *infestans* Montagne.—(POTATO MILDEW.) Fig. 21.

Tufts of threads, lax, erect, white, branching at the ends; spores lateral and terminal, solitary, oblong, with a granular nucleus.

Habitat. Leaves of the common Potato.
Quality. Attacks the leaves and stems, and aggravates the disease to which this plant has become subject.

2. B. *Bassiana* Montagne.—(SILKWORM-ROT. MUSCARDINE.) Fig. 25.

Threads erect, branched in a racemose manner, with clusters of spores at the end of the short lateral divisions.

Quality. Kills silkworms in great numbers.

OIDIUM. *Link.*

Microscopic. *Threads* white or brightly coloured, simple or irregularly branched, moniliform above, and breaking up into more or less elliptic *spores.—Berkel. MSS.*

1. O. *abortifaciens* Berkel. *Ergotœtia abortifaciens* Quekett (*Ergot.*)

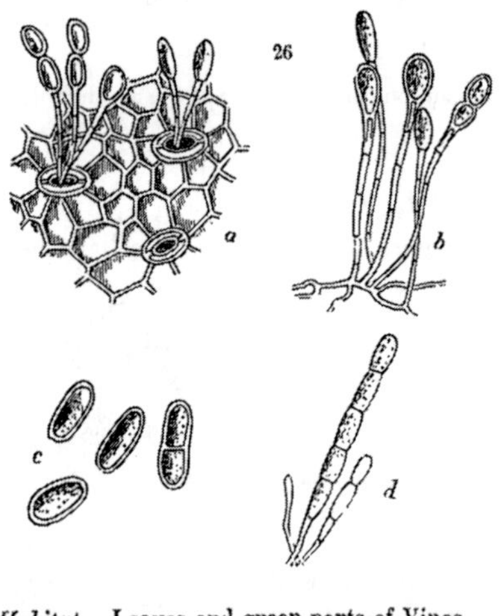

Threads white, irregularly branched; spores abundant, elliptic, containing two nuclei. —*Berkeley.*

Habitat. Ovary of Grasses and Eleocharis, causing the disease called Ergot, destroying the ordinary growth, and compelling it to assume a horned appearance.
Quality. The ovary thus affected becomes hardened and deformed.
Uses. Ergot or spurred Rye is a dangerous narcotic poison, producing gangrene. Used in medicine to increase the action of the nterus in difficult parturition.

2. O. *Tuckeri* Berkeley.—(VINE MILDEW.) Fig. 26.

Fertile threads elongated; spores large, elliptic or oblong, at length septate.—*Berkeley.*

Habitat. Leaves and green parts of Vines.
Quality. A formidable parasite, destroying the functions of the skin of the parts it attacks.

Fig. 25.—Magnified view of Botrytis Bassiana; 26. Do. of Oidium Tuckeri, at *a* making its way through the stomates of a Vine leaf; *b*, a plant, with mycelium; *c*, spores; *d*, Oidium of the Peach tree.—*M.J.B.*

GRAPHIUM. *Corda.*

Microscopic. *Stem* erect, fibrous, capitate, pencilled, floccose ; flocci continuous, breaking off into continuous, homogeneous spores, which are glued together at first, but at last separate.

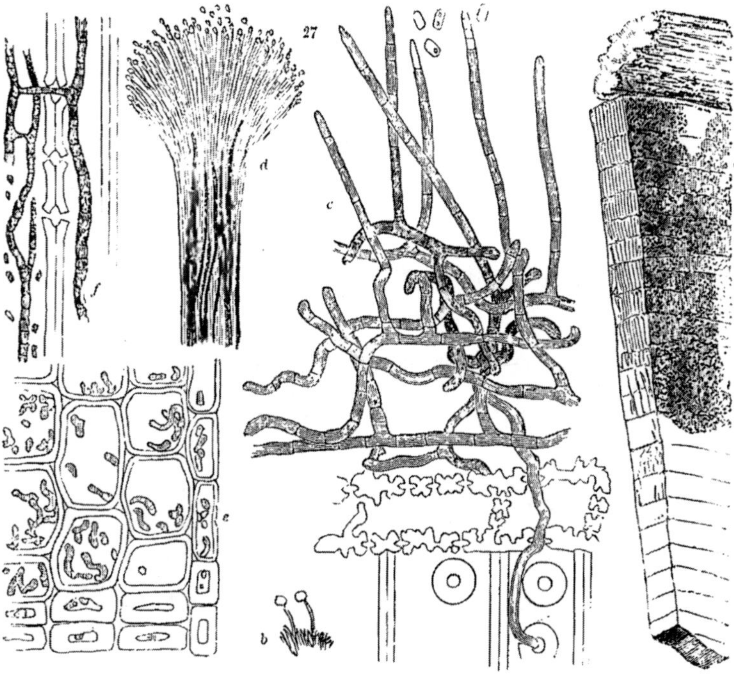

1. G. *penicilloides* Corda. Fig. 27.

Effused, black ; stem even, black, opaque, brown in the middle, above pale dirty white ; spores cylindrical, white, hyaline.

Habitat. Dead Fir wood and Poplar wood.
Uses. A destroyer of timber.

ASPERGILLUS. *Micheli.*

Microscopic. Cobweb-like strata, producing threads, at the ends of which grow spores arranged in rows, the rows themselves being collected in pencils or tufts.

1. A. *glaucus* Link.—(BLUE MOULDINESS.) Fig. 17.

Cobweb-like strata white ; the fertile threads simple, capitate ; spores loosely packed, becoming glaucous.

Habitat. Decaying substances everywhere. Cheese.
Quality. Assists in decomposition.

Fig. 27.—Various states of Graphium penicilloides ; *a*, a piece of wood covered with it, of the natural size ; *b*, a small portion, magnified 12 diameters ; *c*, a portion very highly magnified, with a piece of the fir wood out of which it grows ; *d*, top of a thread in fruit ; *e*, view of wood of Pinus infected by it ; *f*, a vertical section of the same.—*Unger*.

LANOSA. *Unger.*

Microscopic. Branched, transparent, jointed *threads*, the terminal joints of the small lateral branches of which at length acquire a reddish colour, and separate at the articulations, producing oblong spores.

28

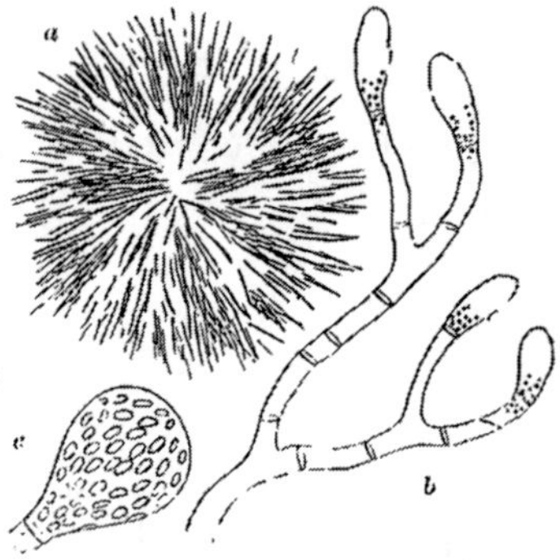

1. L. *nivalis* Unger.—(SNOW-MOULD.) Fig. 28.

White patches, a foot or more in diameter, themselves consisting of numerous entangled circular patches, finally becoming red, as if dusted with red powder.

Habitat. Beneath snow, on Grasses and cereal crops.
Quality. Commits great ravages among the plants which it attacks, sometimes destroying whole crops of corn. Especially injurious to Barley and Rye.

MUCOR. *Micheli.*

Microscopic. Cobwebby masses, consisting of tubular septate threads bearing at the end a roundish membranous spore-case, which bursts when plunged in water.

1. M. *Mucedo* Linnæus.—(COMMON MOULDINESS.) Fig. 16.

Cobweb-like; the fertile threads simple; spores and spore-cases blackish.

Habitat. Fruit, paste and preserves.
Quality. Destroys the quality of the substances it infests.

UREDO. *Persoon.*

Microscopic. Subcutaneous patches or masses of simple powdery spores.

1. U. *Caries* De Candolle—U. *fœtida* of some.—(BUNT. SMUT-BALLS. PEPPER-BRAND.) Fig. 19.

Spores black, rather large, spherical, fœtid.

Habitat. Within the ovary of corn; 4,000,000 may be contained in a grain of Wheat.
Quality. Destructive to corn, destroying all the interior.

2. U. *segetum* Persoon.—(SMUT. DUST-BRAND.)

Spores black, minute, spherical, scentless.

Habitat. Rachis and receptacle of Grasses, especially Oats and Barley.
Quality. Destructive to such corn, but less so than the last.

Fig. 28.—Magnified views of Lanosa nivalis; *a*, a patch of the plant; *b*, one of its threads; at *c*, inflated and filled with spores.

PENICILLIUM. *Link.*

Microscopic. Cobweb-like or mothery flocculent masses, producing simple globose *spores* disposed in patches about the pencil-shaped ends of septate fertile threads.

1. P. *glaucum* Greville.—(THE VINEGAR PLANT.)

Mycelium forming a close tough crust-like or leathery web; branches somewhat entangled and bifid; spores verdigris-green.

Habitat. On decaying bodies, and in fluids in a state of acetification.

Quality. Assists in the decomposition of decaying matter, and augments rapidly the acetous fermentation of saccharine fluids. A bit placed in sugar and water soon changes it to vinegar.

MYCODERMA. It is probable that the flocculent substance which forms in various infusions when they become "mothery," and which bears this name, is only the mycelium of Mucor, Penicillium, and other Fungals of a similar nature. The accompanying cuts, from a paper of Mr. Berkeley's, illustrate this. Fig. 29 is a view of the mycelium of Mucor subtilissimus as found in water, with one ball of the reproductive bodies formed when the mycelium reached the air; 30 is Penicillium candidum in the same state, the greater part of which had been formed in water; but a couple of branches reaching the air produced the true fructification of this genus of Fungals.

29

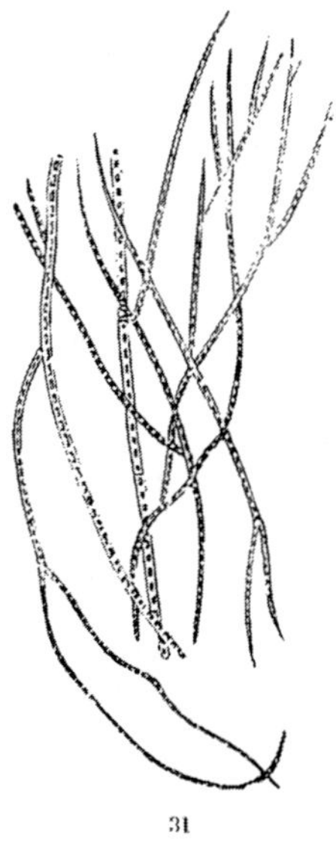

31

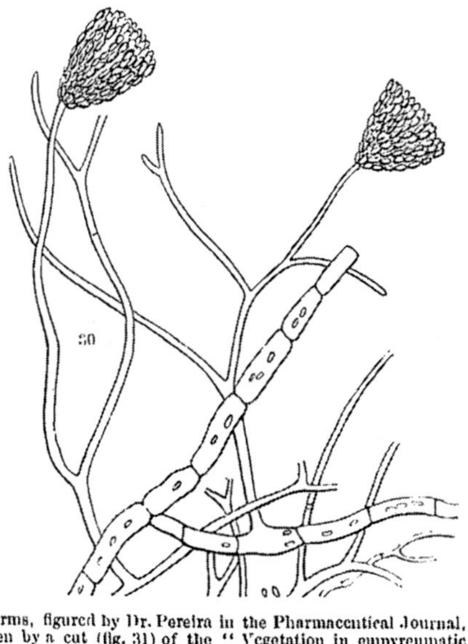

30

Their identity with some of the Mycoderms, figured by Dr. Pereira in the Pharmaceutical Journal, is sufficiently evident; as will be seen by a cut (fig. 31) of the "Vegetation in empyreumatic succinate of ammonia," borrowed from the *Pharm. Journal*, vol. vii. f. 8, p. 341.

THE LICHENAL ALLIANCE (V. K., p. 45.)

To a considerable amount of nutritious gelatinous matter is here added a variable quantity of bitterness, which renders the species tonic and stomachic. Some are among the most valuable dyeing plants; others supply food to animals, as the Cenomyce to reindeer.

GYROPHORA. *Acharius.*

Thallus leafy, horizontal, peltate. *Shields* round, sessile, adnate, covered with a black membrane, the surface marked with circles or plaits.

32

1. G. *proboscidea* Acharius.—(TRIPE DE ROCHE.*) Fig. 35.

Thallus membranous, wrinkled, with elevated netting, smoky brown, lobed and notched at the edge; shields convex, plaited.

Habitat. Mountainous and very northern regions, on rocks.
Quality. Nutritious, bitter, nauseous, purgative.
Uses. Food in arctic regions.

2. G. *erosa* Acharius.—(TRIPE DE ROCHE.)

Thallus membranous, wrinkled, almost black, irregularly lobed, perforated at the circumference, and notched; shields convex, variously plaited.

Habitat. Mountainous, and very northern regions, on rocks.
Quality. Like the last.

CENOMYCE. *Acharius.*

Thallus shrubby, perpendicular, branched, usually covered with leafy scales. *Shields* sessile, round, convex, without a border, the sides reflexed.

1. C. *rangiferina* Acharius.—(REINDEER MOSS.) Fig. 39.

Stalks of the shields erect, long, rough, cylindrical, greenish white, very much branched; the axils pierced; the branches scattered, entangled; shields roundish, brown, on small erect stalks.

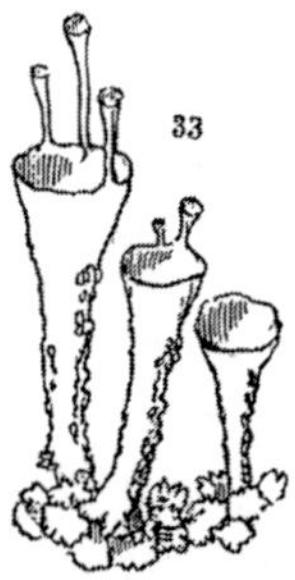

33

Habitat. Moors, heaths, especially mountainous or northern.
Quality. Nutritious.
Uses. Constitutes important food for reindeer and other animals in high northern latitudes.

2. C. *pyxidata* Acharius.—(CUP LICHEN. CUP MOSS.) Fig. 33.

Thallus leafy, erect, the lobes crenulate, forming long, granulated, rough greyish green cups, on the edge of which stand the brown convex shields.

Habitat. Moors and dry woods.
Quality. Bitter, gelatinous.
Uses. Hooping-cough; as Iceland Moss. Febrifugal.

Fig. 32.—Magnified shields of Gyrophora; 33. Cenomyce pyxidata, a little larger than natural size.

* By some mistake a figure of Peltidea aphthosa is given for this by Dr. Pereira.

PARMELIA. *Acharius.*

Thallus leafy, horizontal, lobed, and cut. *Shields* orbicular, fixed by a central point, concave, bordered by the inflexed thallus.

1. P. *parietina* Acharius.—(YELLOW WALL LICHEN.) Fig. 36.

Thallus circular, bright orange yellow; lobes radiating, appressed, rounded, crenate, and crisp, granular in the centre; shields deep orange, concave, with an entire border.

Habitat. Trees and walls. Common.
Quality. Bitter. Yields a yellow colouring matter.
Uses. In intermittent fevers.

LECANORA. *Acharius.*

Thallus crustaceous, flat, uniform. *Shields* orbicular, plano-concave, thick, sessile, bordered by a rim formed out of the crust.

Various Lichens, natural size. Fig. 34. Lecanora Parella; 35. Gyrophora proboscidea; 36. Parmelia parietina; 37. Lecanora tartarea; 38. Roccella fuciformis; 39. Cenomyce rangiferina; 40. Peltidea aphthosa; 41. Peltidea canina.

1. L. *Parella* Acharius.—(PERELLE. CRAB'S-EYE LICHEN.) Fig. 34.
Crust dirty white, determinate, plaited and warty ; shields scattered, thick, concave, whole coloured.

Habitat. Rocks in mountainous countries.
Quality. Yields a purple dye, equal to that of Archill.

2. L. *tartarea* Acharius.—(CUDBEAR.) Fig. 37.
Crust thick, granulated, and tartareous, greyish white ; shields scattered, convex, at length flat, yellow brown inclining to flesh colour, with a thick inflexed border, becoming wavy.

Habitat. Rocks, &c., in alpine countries.
Quality. Yields a rich purple colour.
Uses. Employed in dyeing yarn. The source of Litmus, used as a test for acids, when it becomes red, and for alkalies, by which the blue colour is restored.

STICTA. *Acharius.*

Thallus leafy, leathery, lobed, with numerous little cavities. *Shields* on the under side, fixed by a central point, flat, surrounded by an elevated border, formed of the thallus.

1. S. *pulmonaria* Acharius.—(LUNGWORT. OAK-LUNGS.) Fig. 42.

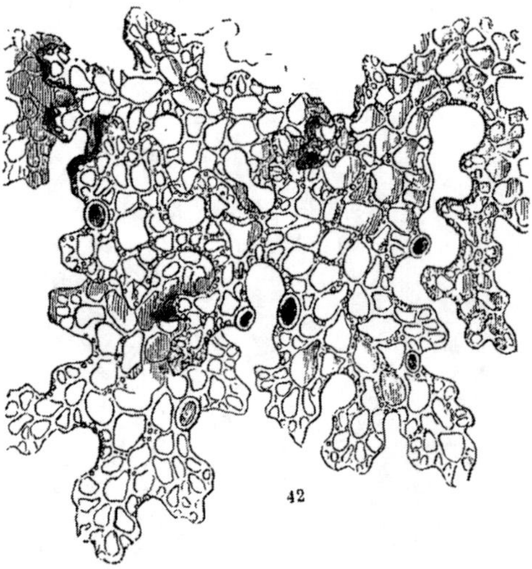

42

Thallus spreading, olive green, pale brown when dry, pitted and netted, much lacerated, with broad rounded lobes ; shields marginal, red-brown, with a thick border.

Habitat. Trunks of trees in mountainous countries.
Quality. Bitter. Furnishes a good brown dye. Nutritious.
Uses. Pulmonary affections, hæmorrhage ; a light diet ; as a substitute for hops.

PELTIDEA. *Acharius.*

Thallus leafy, horizontal, lobed, woolly beneath, bearing the shields on the

Fig. 42.—Sticta pulmonaria ; natural size.

lobes. *Shields* roundish, attached to the upper side of the lobes, and having a border formed of the thallus.

1. P. *canina* Acharius.—(Dog Lichen. Ground Liverwort.) Fig. 41.

Thallus thick, glaucous grey, greenish when moist, somewhat furrowed, with rounded lobes, beneath white, with branching veins and fibres ; shields vertical, revolute, reddish brown, with a crenulate border.

Habitat. On the ground, among moss, &c.
Uses. An imaginary remedy for hydrophobia.

2. P. *aphthosa* Acharius.—(Thrush Lichen.) Fig. 40.

Thallus light green, smooth, sprinkled with brown warts ; lobes broad, rounded, the fertile ones contracted ; shields large, red-brown, with a jagged border.

Habitat. Moist alpine rocks.
Quality. Purgative, anthelmintic. The Swedes boil it in milk as a cure for aphthæ.

Cetraria. *Acharius.*

Thallus leafy, spreading or erect, lobed and lacerated, smooth on each side. *Shields* orbicular, attached obliquely to the edge of the thallus, planoconcave, bordered by the inflexed thallus.

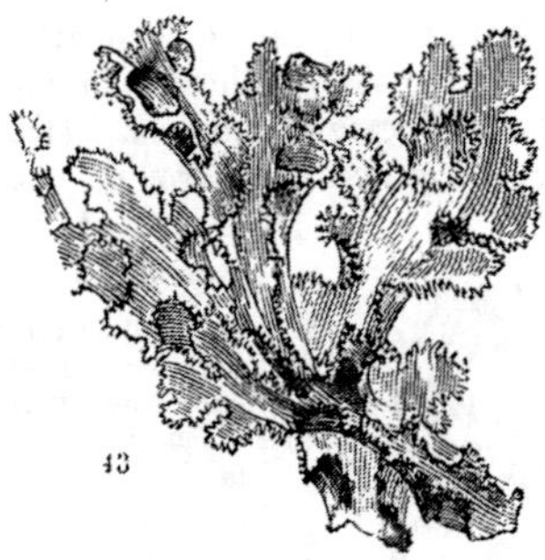

43

1. C. *islandica* Acharius.—(Iceland Moss.) Fig. 43.

Thallus erect, tufted, olive-brown, paler on one side, lacerated, channelled, and bordered with tooth-like fringes ; shields brown, with a raised border.

Habitat. On the ground, in northern regions.
Quality. Mucilaginous, demulcent, tonic, bitter.
Uses. A light diet for invalids, and a mild tonic.
Phthisis, chronic catarrh, dyspepsia, chronic diarrhœa, dysentery.—*Pereira.*

Roccella. *Acharius.*

Thallus between leathery and cartilaginous, branched and cut, erect, terete. *Shields* orbicular, adnate, plano-convex, with a thickened elevated border.

1. R. *tinctoria* De Candolle.—(Orchill. Archill. Dyer's Moss.)

Thallus suffruticose, branched, terete, erect, greyish brown, bearing powdery warts ; shields almost black and pruinose, with a scarcely raised border ; juice deep yellow.

Habitat. Rocks near the sea, especially in southern places.
Quality. Dyes purple.
Uses. Formerly used for preparing Litmus paper. Dyes various articles of manufacture.

2. R. *fuciformis* De Candolle.—(Flat Orchill.) Fig. 38.

Thallus flat, branched, nearly erect, greyish white, with powdery warts ; sap not yellow.

Habitat, Quality, &c., like the preceding, but, according to the Rev. Mr. Salway, very inferior as a dye.

Fig. 43.—Cetraria islandica ; natural size.

CLASS II. ACROGENS (V. K., p. 51.)

Here also occur three Alliances ; viz.—

MUSCALS. Spore-cases immersed or calyptrate.
LYCOPODALS. Spore-cases axillary or radical. Spores of two sorts.
FILICALS. Spore-cases marginal or dorsal. Spores of one sort.

Little of obvious importance to man occurs among the members of this alliance, which are objects of botanical interest rather than of medical or dietetical value. It is needless to detain the student with the natural orders, or with more than a very few instances of useful species.

THE MUSCAL ALLIANCE (V. K., p. 54.)

Equisetum. *Linnæus.*

(*Order.* Equisetaceæ, or Horsetails ; V. K., p. 61.)

Spore-cases growing beneath peltate scales collected in cones, splitting on one side, without operculum, and with an *elater* to every spore.

1. E. *hyemale* Linnæus. (Dutch Rush. Shave-grass.)

Stem naked, very rough, mostly branching at the base ; sheaths pallid, white at top and bottom, with deciduous teeth ; cone terminal.

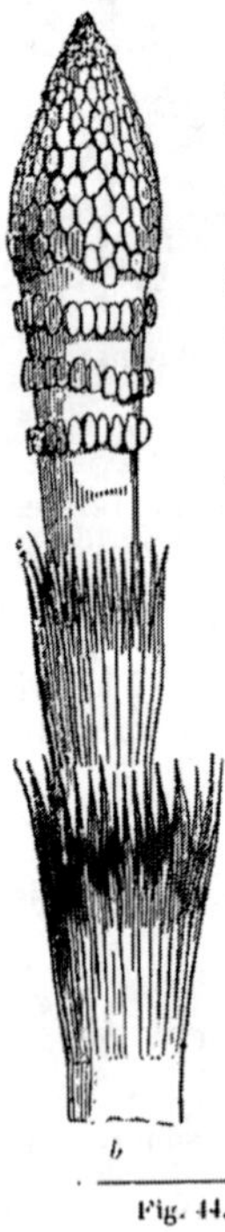

Habitat. In swampy places.
Quality. Rough with flinty points. (Said to be astringent, diuretic, emmenagogue.) The rhizomes nutritious.
Uses. Employed for polishing wood, ivory, and brass. Food in time of famine.

2. E. *fluviatile* Linnæus. (Water Horsetail.) Fig. 44.

Stems of two kinds : barren, covered with numerous rough doubly angular branches : fertile ones unbranched, pallid, with large, loose, deeply-toothed sheaths.

Habitat. Watery places.
Uses. The starch contained in the tubers of the rhizome nutritious ; said by Haller to be the plant eaten by the Romans under the name of Equisetum.

44

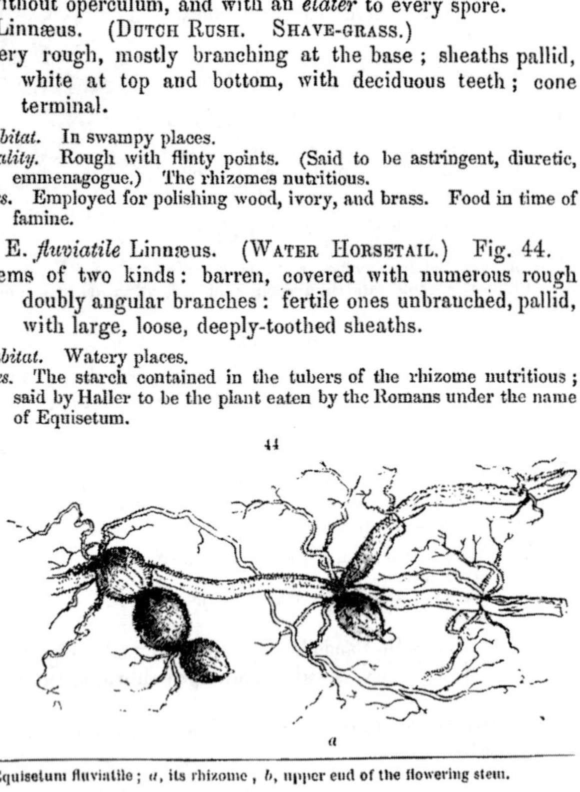

b a

Fig. 44. Equisetum fluviatile ; *a*, its rhizome , *b*, upper end of the flowering stem.

SPHAGNUM. *Linnæus.*

(*Order.* Bryaceæ, or Urn-mosses ; V. K., p. 64.)

Spore-case an urn closed by a deciduous lid, having a toothless brim, and capped by an irregularly torn *calyptra.*

1. S. *obtusifolium* Ehrhart. (BOG-MOSS.) Fig. 45.

Spongy. Leaves whitish, ovate, obtuse, closely imbricated, tumid, with large spiral-coated perforated cells.

Habitat. Wet commons, bogs and moors.
Quality. Has the property of absorbing moisture readily, and parting with it slowly. Slightly nutritive.
Uses. Employed by gardeners for covering the roots of plants and preserving them moist ; also for drainage. A wretched food in barbarous countries.

N.B.—This is the genus of Mosses spoken of in Lindley's "Introduction to Botany," vol. i. p. 52, fourth edition, as being remarkable for the spiral structure of the cells composing its leaves, and for the presence of large pores in their sides.

a

45

46

THE LYCOPODAL ALLIANCE (V. K., p. 68.)

LYCOPODIUM. *Linnæus.*

(*Order.* Lycopodiaceæ, or Club-mosses ; V. K., p. 69.)

Spore-cases 2-valved, solitary in the axils of leaves.

1. L. *clavatum* Linnæus. (COMMON CLUB-MOSS. SNAKE-MOSS.) Fig. 46.

Stem creeping ; branches ascending, often entangled ; leaves incurved with thread-like points ; spikes two or three, cylindrical, stalked, with dilated membranous bracts.

Habitat. Damp hills, low mountains, and moors.
Quality. Spores inflammable, called *Witch-meal* and *Vegetable Sulphur.* "The minute copious volatile seeds (spores) are used in Germany for artificial lightning on the stage, and are sold in the shops. When dispersed in the air, they take fire with a candle, and suddenly explode."—*Smith.* Herbage emetic.
Uses. For rolling up pills ; dusting infants ; Plica polonica ; a rude and unsafe emetic ; employed in scorbutic affections.

2. L. *Selago* Linnæus. (FIR-MOSS.)

Stems dwarf, erect, forked ; leaves in eight rows, lanceolate, pointless, slightly spreading.

Habitat. Mountain heaths, and lowland sandy wet tracts.
Quality. Astringent, emetic, drastic, narcotic.
Uses. A powerful irritant ; keeps blisters open ; a counter-irritant ; a rude purgative for very strong persons ; forms a detergent lotion against vermin.

Fig. 45.—Sphagnum obtusifolium ; *a*, a leaf magnified ; 46. A branch of Lycopodium clavatum.

3. L. *rubrum* Chamisso—L. *catharticum* Hooker. (YATUM CONDENADO.) Fig. 47.

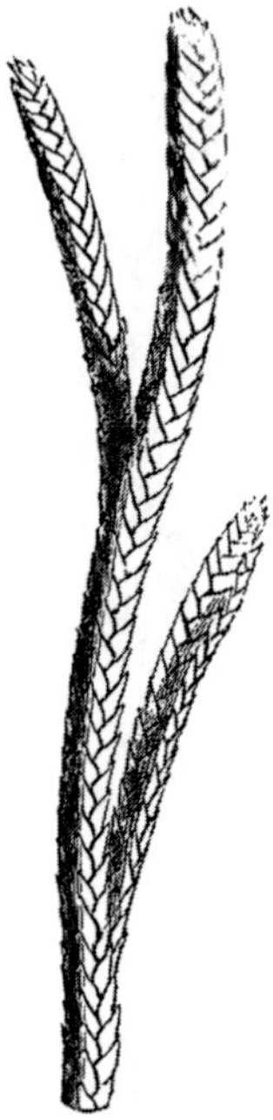

47

Stem ascending, dichotomous; branches four-cornered; leaves closely imbricated, ovate, acuminate, stiff, keeled, ciliated, with the spore-cases in their axils.

Habitat. Equatorial America, on mountains.
Quality. Hypercathartic.
Uses. In elephantiasis, leprosy. A medicine of great activity.

THE FILICAL ALLIANCE; OR, FERNS.

(V. K., p. 74.)

ADIANTUM. *Linnæus.*

Spore-cases in sori, hidden beneath rounded, reflexed, marginal, distinct, indusia with a vertical ring.

1. A. *Capillus Veneris* Linnæus. (MAIDENHAIR.) Fig. 48.

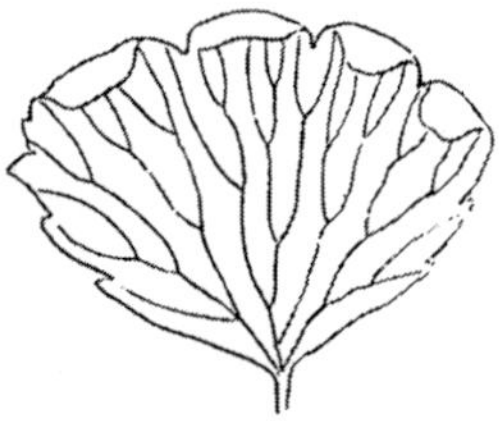

48

Leaf doubly compound; leaflets alternate, wedge-shaped on capillary stalks; indusia oblong.

Habitat. Moist rocks and old walls.
Quality. Rhizome slightly astringent, fragrant when dry; somewhat emetic.
Uses. With syrup and orange flowers, makes bad capillaire.

2. A. *pedatum* Linnæus. Fig. 49.

Leaves pedate: divisions pinnate; leaflets halved, oblong, lunate, cut at the upper edge.

Habitat. North America.
Quality. Sweet, slightly styptic, and fragrant.
Uses. With syrup and orange flowers, makes the best capillaire.

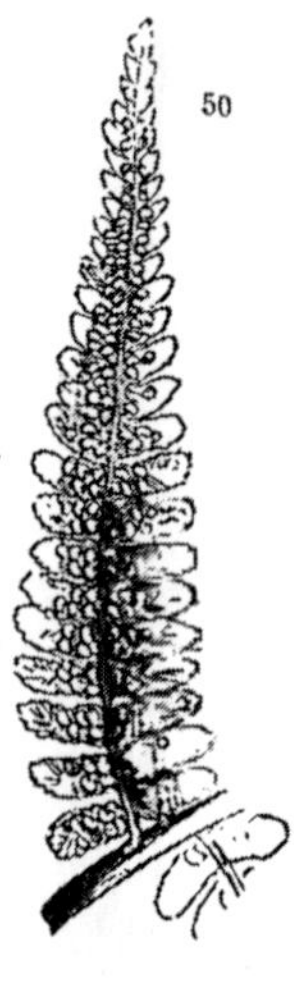

50

ASPIDIUM. *Swartz.*

Spore-cases in roundish scattered dorsal sori, covered by an orbicular or kidney-shaped indusium, with a vertical ring.

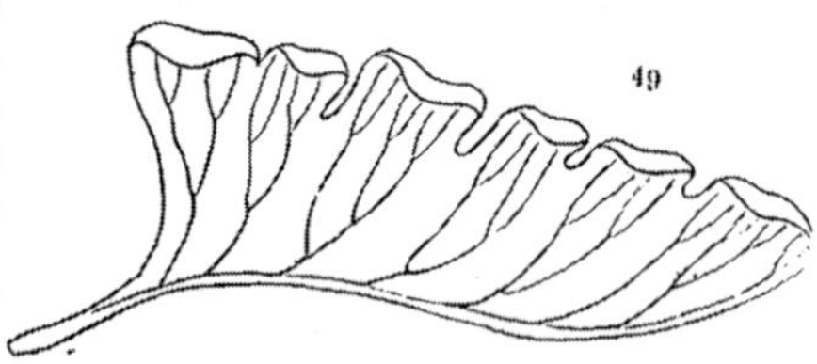

49

1. A. *Filix mas* Swartz — *Nephrodium Filix mas.* (MALE FERN.) Fig. 50.

Fig. 47.—Piece of Lycopodium rubrum; 48. Leaflet of Adiantum Capillus Veneris, a little magnified; 49. Leaflet of Adiantum pedatum, a little magnified; 50. Pinna of Aspidium Filix mas.

Leaf doubly pinnate ; leaflets obtuse, serrate, partly confluent ; stalk ramentaceous ; indusia near the midrib, kidney-shaped.

Habitat. Dry ditches and banks.
Quality. Astringent, emetic.
Uses. Anthelmintic ; against *Bothriocephalus latus.—Pereira.*

PTERIS. *Linnœus.*

Spore-cases in marginal dorsal lines covered by the inflexed edge of the leaf, with a vertical ring.

1. P *aquilina,* Linnæus.—(BRAKE, BRACKEN.) Fig. 51.

51

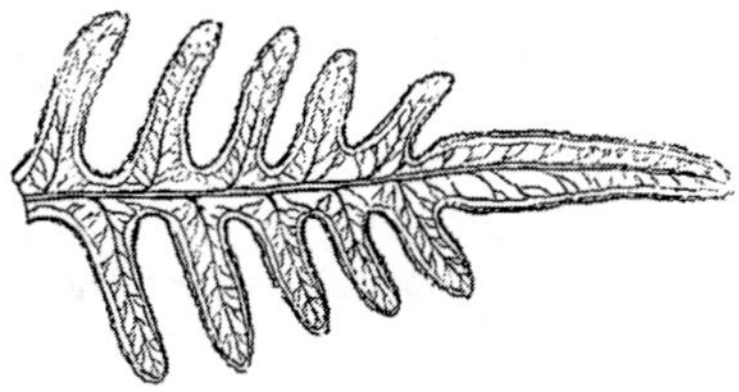

Leaf 3 times pinnate, with lanceolate bluntish segments, of which the lowest are pinnatifid, and the upper gradually smaller ; terminal lobes large, undivided.

Habitat. Heaths, parks, &c.
Quality. Astringent, anthelmintic, bitter.
Uses. The rhizome has been used as a substitute for hops ; it furnishes a wretched bread.

OSMUNDA. *Linnœus.*

Spore-cases in branched masses, stalked, with a horizontal ring.

1. O. *regalis* Linnæus.—(OSMUND ROYAL.)

Leaflets oblong, nearly entire, dilated at the base ; clusters of spore-cases panicled, terminal.

Habitat. Bogs and wet meadows.
Quality. Tonic, styptic.
Uses. Rachitis.

CLASS III. RHIZOGENS (V. K., p. 83.)

No species are of importance in this country.

Fig. 51. Pinna of Pteris aquilina.

CLASS IV. ENDOGENS (V. K., p. 105.)

The following are the chief Alliances:

GLUMALS. Flowers glumaceous.

ARALS. Flowers unisexual, naked, or scales or hairs; on a simple spadix.

PALMALS. Flowers unisexual, petaloid; on a branched spadix.

NARCISSALS. Flowers hermaphrodite. Ovary inferior. Perianth symmetrical.

AMOMALS. Flowers hermaphrodite. Ovary inferior. Perianth unsymmetrical. Stamens free.

ORCHIDALS. Flowers hermaphrodite. Ovary inferior. Perianth unsymmetrical. Stamens gynandrous.

JUNCALS. Flowers hermaphrodite. Ovary superior. Perianth glumaceous.

LILIALS. Flowers hermaphrodite. Ovary superior. Perianth coloured. Syncarpous.

ALISMALS. Flowers hermaphrodite. Ovary superior. Perianth coloured. Apocarpous.

THE GLUMAL ALLIANCE (V. K., p. 105.)

Natural Orders of Glumals.

Grasses (*Graminaceæ.*) Sheath of leaves slit. Embryo lateral, naked.

Sedges (*Cyperaceæ.*) Sheath of leaves not slit. Embryo basal, inclosed.

Natural Order, Grasses; *Graminaceæ* (V. K., p. 106.)

Prevailing Quality. Sweet, nutritious; very rarely narcotic.

TRITICUM. *Linnæus.*

A *spike. Spikelets* many-flowered, parallel with the zig-zag rachis. *Glumes* 2, nearly equal. *Paleæ* 2, the lower awned or not.

1. T. *repens* Linnæus.—(COUCH GRASS. QUITCH. CHIEN-DENT *Fr.*)

Spike distichous; spikelets about 5-flowered; paleæ lanceolate, 5-nerved, acuminate, not ventricose; rachis usually scabrous; leaves rough, with lines of points on the upper side. A perennial, with a creeping rhizome.

Habitat. A common weed, in neglected ground.
Quality. Rhizomes diaphoretic, aperient and refreshing.
Uses. For diet drink, and as a substitute for Sarsaparilla.

2. T. *æstivum* Linnæus.—(WHEAT.) Fig. 52.
Spike 4-cornered; spikelets about 4-flowered; paleæ ventricose, ovate, truncate, mucronate or awned, compressed under the point, rounded at the back; grain free. An annual.

Habitat. Unknown.
Quality. Grain nutritious.
Uses. The flour forms wheaten bread. Bran, which is the pericarp, is emollient and demulcent, and even purgative, owing the latter quality to its mechanical action.

53

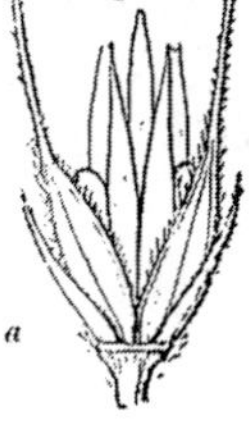

a

SECALE. *Linnæus.*

A *spike.* *Spikelets* 2-flowered, with a long-stalked rudiment of a third floret. *Glumes* subulate; otherwise like Triticum.

1. S. *cereale* Linnæus.—(RYE.) Fig. 53.
Glumes shorter than the spikelet. Rachis tough.

Habitat. Commonly cultivated.
Quality. Grain nutritious.
Uses. The flour forms an inferior kind of bread. Ergot is the ovary, diseased by the attack of a parasitical fungus. See OIDIUM, p. 14.

52

LOLIUM. *Linnæus.*

A *spike.* *Spikelets* distichous, many-flowered, placed edgewise on the rachis. *Glume* solitary, or that next the rachis rudimentary.

1. L. *temulentum* Linnæus.—(DARNEL.)
Glume as long or longer than the spikelet, which contains from 5 to 7 florets; florets when in fruit elliptical, awned; awn straight, longer than the palea. An annual.

Habitat. Corn fields and by pathways.
Quality. Grains narcotic and acrid, producing fatal consequences when mixed with flour. Darnel meal has been used for sedative poultices.

N. B.—This is the only *authentic* instance of unwholesome qualities in the order of Grasses. The cases mentioned in the "Vegetable Kingdom" are all doubtful. As to Bromus catharticus, figured under the name of Guilno by Feuillée, there can be no doubt that his statement is a blunder. The grass he has figured is not distinguishable from B. secalinus; the rhizome, in which he says that purgative qualities reside, evidently, both by the figure and description, does not belong to any grass whatever; it may possibly be that of some purgative Sisyrinchium.

Fig. 52.—Ear of Triticum æstivum; 53. Part of ear of Secale cereale; *a*, a spikelet magnified.

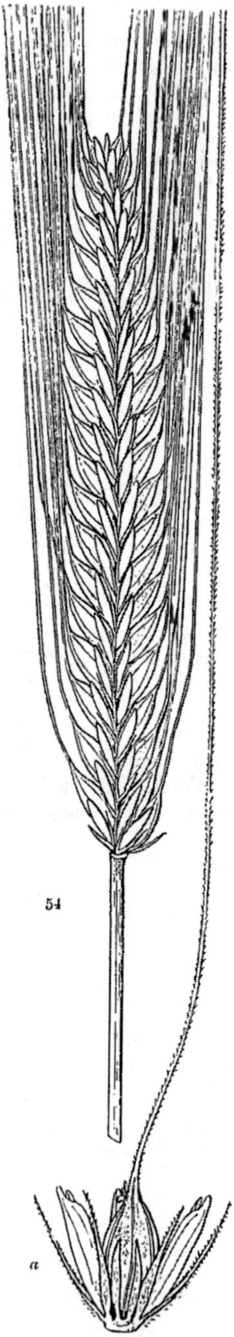

54

a

HORDEUM. *Linnæus.*

A *spike*. *Spikelets* 1-flowered, placed in threes, the lateral spikelets generally abortive or male. *Glumes* 2, placed in front of the floret. *Paleæ* 2.

1. H. *distichum* Linnæus.—(COMMON BARLEY.) Fig. 54.

Lateral spikelets male, awnless ; those in the middle hermaphrodite, awned, pressed close to the rachis ; awns stiff, erect.

Habitat. Mesopotamia ? Commonly cultivated.

Quality. The grain demulcent, emollient ; the husk slightly acrid.

Uses. Barley-water in fevers, inflammation of the lungs; the grain forms malt ; the flour forms a dark, strong-tasted unpleasant bread. Pearl Barley is the grain deprived of its skin by rubbing, and is much employed for gruel, &c.

55

Fig. 54.—Ear of Hordeum distichum ; *a*, a tierce of spikelets ; 55. Panicle of Avena sativa.

AVENA. *Linnæus.*

A *panicle.* *Glumes* 2- or more-flowered, as long as the florets. Outer *Palea* with distinct lateral nervures, 2-pointed, with a dorsal, kneed and twisted awn. *Ovary* hairy at top.

1. A. *sativa* Linnæus.—(THE COMMON OAT.) Fig. 55.

Panicle spreading, equal-sided ; glumes generally 2-flowered, longer than the florets, the uppermost 9-nerved ; florets smooth, bifid, or 2-toothed at the point.

Habitat. Mesopotamia ?
Quality. The grain nutritious.
Uses. Gruel, prepared from the skinned grains or groats, easily digested ; employed as an emollient and demulcent in cases of poisoning by acrid substances ; meal also as poultices.

SACCHARUM. *Linnæus.*

A large branched *panicle.* *Spikelets* in pairs, buried in long silky hairs, 2-flowered ; one sessile, the other stalked. *Lower floret* neuter with one palea ; *upper floret* hermaphrodite with 2 paleæ. *Glumes* 2, membranous. *Paleæ* minute, transparent, awnless.

1. S. *officinarum* Linnæus.—(The SUGAR CANE.)

Panicle very loose ; flowers triandrous ; glumes indistinctly one-nerved, with very long hairs at the back.

Habitat. Tropical countries.
Quality. Sweet ; demulcent ; emollient.
Uses. Its sugar is universally employed where sweet substances are needed. An antidote to poisoning by metallic salts.

ANDROPOGON. *Linnæus.*

Either a *spike* or *panicle.* *Spikelets* in pairs, or, if terminal, in threes ; one perfect awned, the others withering, sterile, and awnless ; perfect spikelet 2-flowered, the lower floret neuter with one palea ; the upper floret hermaphrodite with 2 paleæ. *Glumes* 2, hard.

1. A. *Schœnanthus* Linnæus.—(LEMON GRASS.) Fig. 56.

Perennial, erect ; panicle rather secund, linear, leafy ; the spikelets having a common foot-stalk furnished with a spathe ; florets all awnless ; male with but one valve.

Habitat. Bengal, Arabia.
Quality. Leaves very fragrant ; their taste acrid, aromatic, and bitter. Rhizome and flowers have similar qualities.
Uses. Roasted leaves stomachic and diaphoretic. Yields an aromatic stimulating essential oil (Grass oil), employed externally in rheumatic affections.

Fig. 56.—Portion of the inflorescence of Andropogon Schœnanthus ; *a*, a spikelet magnified.

ANATHERUM. *Palisot.*

Like *Andropogon*, but the glumes are covered with asperities bearing hairs.

1. A. *muricatum* Palisot.—(VETIVER or VETIVERT.) Fig. 57.

Erect; leaves in 2 rows, long, narrow, stiff; panicle whorled, with simple spreading branches; both hermaphrodite and male florets awnless.

Habitat. East Indies.
Quality. Rhizome has a strong penetrating agreeable odour.
Uses. Like those of Andropogon Schœnanthus.

58

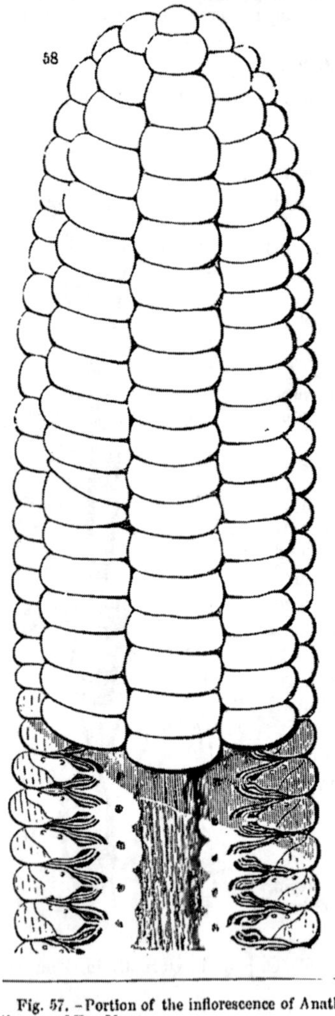

57

a

ZEA. *Linnæus.*

Inflorescence unisexual. *Male* a terminal panicle; *spikelets* 2-flowered, with a pair of nearly equal glumes. *Female* a lateral spike, enveloped in tough spathes; *spikelets* 2-flowered; the upper floret ♀, with 2 or 3 paleæ; the lower floret neuter, with only 2 paleæ. *Glumes* membranous, very broad. *Grains* roundish, compressed, naked, in perpendicular rows.

1. Z. *Mays* Linnæus.—(MAIZE. INDIAN CORN.) Fig. 58.

A coarse, upright, broad-leaved annual; paleæ shorter than the ripe grain.

Habitat. North of Mexico, or the southern districts of the Rocky Mountains, according to the botanists of the United States.
Quality. Grain very nutritious.
Uses. Employed largely as food; but is apt to cause diarrhœa.

ORYZA. *Linnæus.*

A *panicle. Spikelets* 1-flowered. *Glumes* 2, small, unequal, awnless. *Paleæ* 2, nearly equal, cartilaginous, ribbed, the lower with or without an awn. *Stamens* 6!

There are numerous varieties of this very common tropical grain, as there is in Europe of Wheat, Barley, Oats, &c.

Fig. 57.—Portion of the inflorescence of Anatherum muricatum; *a*, a spikelet magnified; 58. Part of the ear of Zea Mays.

1. O. *sativa* Linnæus.—(RICE.) Fig. 59.
Leaves linear, long, rough; panicle racemose, rather contracted; paleæ very hairy.

Habitat. Cultivated in all hot countries, in swampy places.
Quality. Grain nutritious.
Uses. Employed largely as food; but from the small proportion of protein compounds is much less nutritious than ordinary cereal grains.

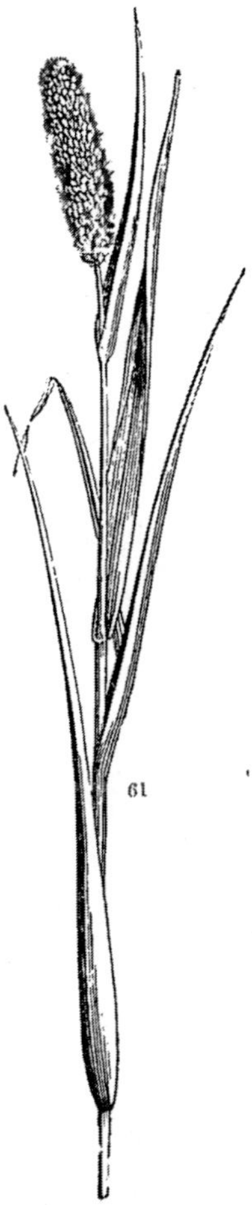

61

a 59 b c

SETARIA. *Palisot.*

A *compound* cylindrical spike. *Spikelets* 2-flowered, surrounded by an involucre of bristles. Lower *floret* rudimentary, consisting of one palea resembling the glumes.

1. S. *italica* Palisot. S. *germanica* Pal.—(GERMAN MILLET. MOHA.)

60

Fig. 59.—Inflorescence of Oryza sativa; *a*, a spikelet magnified; *b*, the awned variety; *c*, an awned spikelet; 60. Spikelets of Setaria; 61. Setaria italica, natural size.

Ear decompound, lobed; involucre rough upwards, the teeth directed forwards; paleæ of the hermaphrodite floret smoothish.

Habitat. Cultivated in Southern Europe, and India.
Quality. Grain nutritious.
Uses. Employed as food in the South and Middle of Europe, and in India, where better grain may not be had. A useful agricultural plant in such countries, because of its power of resisting drought. S. germanica is a larger variety, with a longer involucre.

Natural Order, **Sedges**; *Cyperaceæ*. (V. K., p. 117.)

Prevailing Quality. Demulcent.

Carex. *Linnæus.*

Flowers unisexual. *Males* with one glume. *Females* inclosed in a flask-shaped involucre.

1. C. *arenaria* Linnæus.—(German Sarsaparilla.) Fig. 62.

Spike decompound, oblong, or somewhat ovate; upper spikes male, lower female; stigmas 2; fruit ovate plano-convex, 7-ribbed, with a rough wing from the middle upwards; stem rough at the angles near the summit; rhizome tough, creeping.

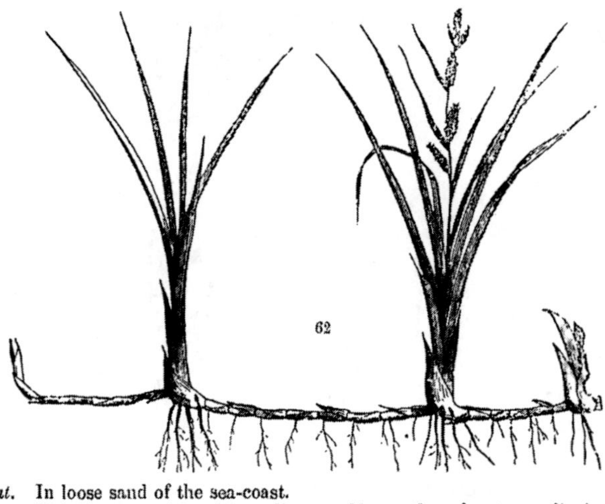

62

Habitat. In loose sand of the sea-coast.
Quality. Rhizome sweetish, with a disagreeable earthy after-taste, diaphoretic and demulcent.
Uses. A substitute for Sarsaparilla.

2. C. *hirta* Linnæus.

Covered with fine hairs; female spikes ovate or cylindrical, distant; males 2 or 3; sheaths of bracts nearly as long as the peduncles; fruit hairy, tumid, with a deeply-cloven beak.

Habitat. In wet meadows, woods, &c. Common. *Quality* and *Uses* as in the last.

Fig. 62.—Carex arenaria; diminished in size.

CYPERUS. *Linnæus.*

Flowers hermaphrodite. *Spikes* composed of numerous distichous glumes, containing no bristles or scales.

1. C. *longus* Linnæus.

Inflorescence decompound, the longer branches umbellate ; stalks of the umbels erect, of unequal length ; stigmas 3 ; glumes ovate obtuse ; rhizome creeping.

Habitat. Ditches and damp meadows on the Continent.

Quality. Rhizome bitter, astringent, aromatic ; smells slightly of violets.

Uses. Furnishes an aromatic distilled water, reckoned tonic and stomachic.

2. C. *esculentus* Linnæus.—(SOUCHET COMESTIBLE ; AMANDE DE TERRE *Fr.*) Fig. 63.

Inflorescence decompound ; spikes both stalked and sessile, clustered at the point of the longer branches, linear, compressed ; glumes obtuse; stigmas 3 ; rhizome long, creeping, bearing pendulous tubers.

Habitat. Cultivated in the South of Europe.

Quality. Nutritive, restorative, stimulant.

Uses. Eaten like nuts ; regarded as an aphrodisiac ; employed in the preparation of orgeat.

64

63

THE ARAL ALLIANCE (V. K., p. 123.)

Natural Orders of Arals.

Bulrushes (*Typhaceæ.*) Calyx = scales or hairs. Anthers on long filaments.

Arads (*Araceæ.*) Calyx 0. Anthers sessile. Spathaceous.

Natural Order, Bulrushes ; *Typhaceæ* (V. K., p. 126.)

Prevailing Character. Subastringent, nutritious.

TYPHA. *Linnæus.*

Spikes cylindrical, the uppermost male. *Stamens* 3, monadelphous, surrounded by bristles. *Ovary* stalked, bristly at the base.

Fig. 63.—Cyperus esculentus ; 64. One of the tubers of the same.

1. T. *latifolia* Linnæus.—(BULRUSH.)

Leaves linear, flat, longer than the flowering stem ; male and female spikes contiguous.

Habitat. Streams and stagnant water all over Europe.
Quality. Rhizomes astringent and diuretic ; abound in starch.
Uses. Sometimes used as food under the name of Cossack Asparagus ; employed in dysentery.

Natural Order, Arads; *Araceæ* (V. K., p. 127.)

Prevailing Quality. Acridity.

ARUM. *Linnæus.*

Spathe convolute. *Spadix* naked at the point ; *male* flowers above, *female* below, with intermediate cirrhi. *Anthers* sessile, opening by lateral slits. *Ovary* 1-celled, with 2-6 horizontal ovules.

1. A. *maculatum* Linnæus.—(CUCKOO-PINT. LORDS AND LADIES.) Fig. 65.

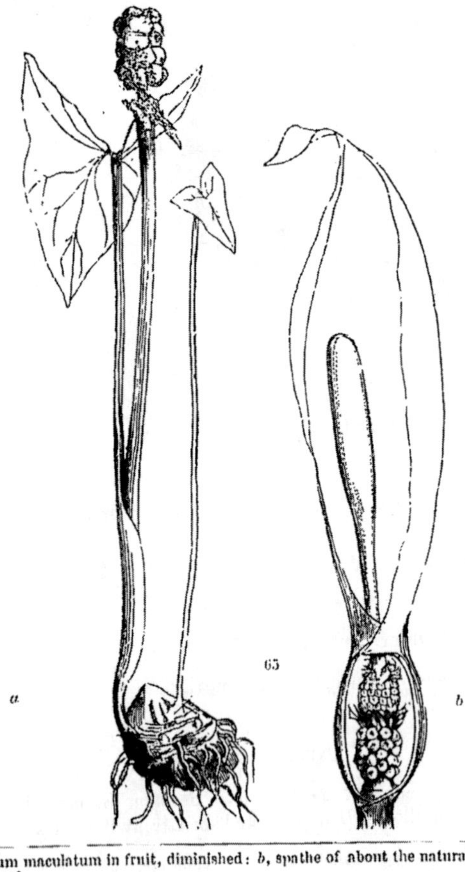

Fig. 65.—*a*, Arum maculatum in fruit, diminished : *b*, spathe of about the natural size, with the base of the spadix exposed.

Leaves hastate-sagittate ; spadix straight, elavate, shorter than the spathe.

Habitat. Hedgerows and plantations all over Europe.
Quality. Acrid ; nutritious.
Uses. Corms eatable, when the acridity is removed. Yields pure starch, known under the name of Portland Sago.

COLOCASIA. *Ray.*

Spathe erect, convolute. *Spadix* naked at the point ; *male* flowers above, *female* below, with rudimentary organs both above and below the stamens. *Anthers* opening by pores, with a very broad connective. *Ovaries* 1-celled, with 6 erect ovules rising in pairs from near the base.

1. C. *esculenta* Schott. *Caladium esculentum*, Vent.

Stemless ; leaves peltate, cordate ; spadix shorter than the ovate-lanceolate spathe.

Habitat. Tropical America.
Quality and *Uses.* As in the next.

2. C. *antiquorum* Schott. *Arum Colocasia* Linn.—(COCCO. EDDOES.)

Stemless ; leaves peltate, ovate, repand, half bifid at the base ; spathe much longer than the spadix, cylindrical, erect.

Habitat. In all tropical countries ; Egypt, Greece, &c.
Quality. Acrid.
Uses. When deprived of acridity by boiling and changing the water, the leaves are eaten as Spinach, and the corms are used for soup.

AMORPHOPHALLUS. *Blume.*

Spathe with a spreading limb. *Spadix* protruded, naked, and fungoid at the end ; *male* flowers above, *female* below, with no intermediate cirrhi. *Anthers* opening by 2 pores. *Ovary* 2-3-4-celled, with solitary erect ovules.

1. A. *campanulatus* Blume. Fig. 66.

66

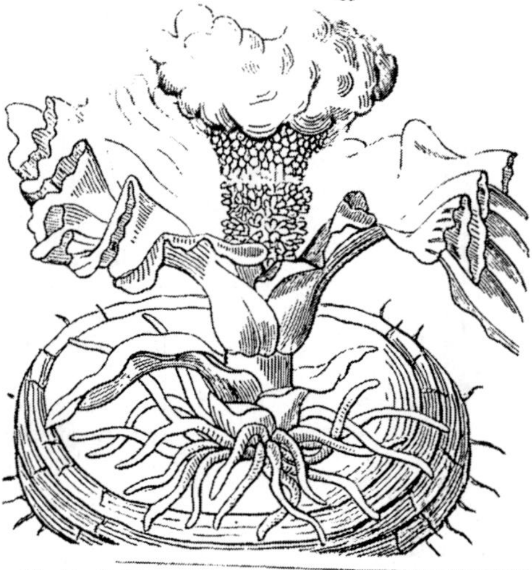

Fig. 66.—Reduced figure of Amorphophallus campanulatus.

Head of the spadix warted; style distinct; spathe sessile; petioles very rough.

Habitat. Tropical parts of Asia.
Quality. Acrid, caustic; abounds in starch.
Uses. Employed as an external stimulant; also as an emmenagogue.

DIEFFENBACHIA. *Schott.*

Spathe convolute. *Spadix* covered all over with flowers; *male* above, *female* below. *Anthers* opening by pores. *Ovaries* 1-celled, with a single erect ovule, and surrounded at the base by 3 clavate processes.

1. D. *seguina* Schott. *Arum seguinum* Linnæus.—(DUMB CANE.)

An arborescent plant; stem cylindrical, with ringed scars; leaves oblong-ovate, with a stout rib; spathe green.

Habitat. West Indies.
Quality. Juice extremely acrid and venemous.
Uses. A dangerous poison; produces dumbness when chewed.

THE PALMAL ALLIANCE (V. K., p. 133.)

Natural Order, Palms; *Palmaceæ* (V. K., p. 133.)

Prevailing Quality. Nutritious, saccharine.

PHŒNIX. *Linnæus.*

Leaves pinnate. *Flowers* diœcious. *Petals* imbricated in the females. *Ovaries* 3. *Fruit* a fleshy one-seeded false drupe.

1. P. *dactylifera* Linnæus. (The DATE PALM.) Fig. 67.

Leaves glaucous; leaflets lanceolate acuminate, very much closed up, the uppermost shorter; fruit oblong, in large bunches.

Habitat. North of Africa.
Quality. Fruit highly nutritious and sweet.
Uses. Dried fruit, a common article of food.

SAGUS. *Rumphius.*

Leaves pinnated. *Flowers* monœcious. *Spadix* branched, without any common spathe, but with numerous partial ones. *Fruit* hard, shining; its surface divided into numerous rhomboidal spaces.

1. S. *lævis* Rumphius. *Metroxylon Sagus* Rottböll.—(TRUE SAGO PALM.)

Petioles and spathes unarmed.

67

Fig. 67.—Sketch of Phœnix dactylifera

Habitat. Islands of the Indian Archipelago.
Quality. Trunk contains the fæcula called Sago.
Uses. Extremely nutritious, and easy of digestion.—Martius says that this furnishes most of the Sago sent to Europe; and thus confirms the statement of Roxburgh.

68

2. S. *Rumphii* Willdenow. — (Prickly Sago Palm.) Fig. 68.
Petioles and spathes guarded by strong prickles.

Habitat. Islands of the Indian Archipelago.
Quality and *Uses.* As in Sagus lævis.—According to Martius this sort of Sago is chiefly used in India, and is rarely exported.

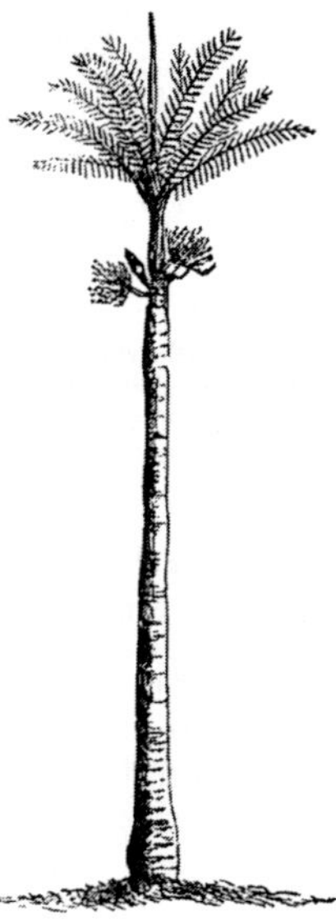

69

Areca. *Linnæus.*

Leaves pinnated. *Flowers* monœcious. *Petals* imbricated in the females, valvate in the males. *Ovary* 3-celled. *Fruit* a fibrous drupe. *Spathes* 2, membranous or fibrous.

1. A. *oleracea* Linnæus. *Oreodoxa oleracea* Martius.—(The Cabbage Palm.) Fig. 69.
Stem very tall, even; leaflets linear, very taper-pointed, bifid; spadix covered with dry ragged white deciduous downy scales.

Habitat. West Indies, especially Barbadoes.
Quality. Sweet, nutritious.
Uses. The terminal bud or "cabbage" eaten.

Fig. 68.—Sketch of Sagus Rumphii; 69. Sketch of Areca oleracea.

2. A. *Catechu* Linnæus.—(PINANG. BETEL NUT. ARECA NUT.) Fig. 70.

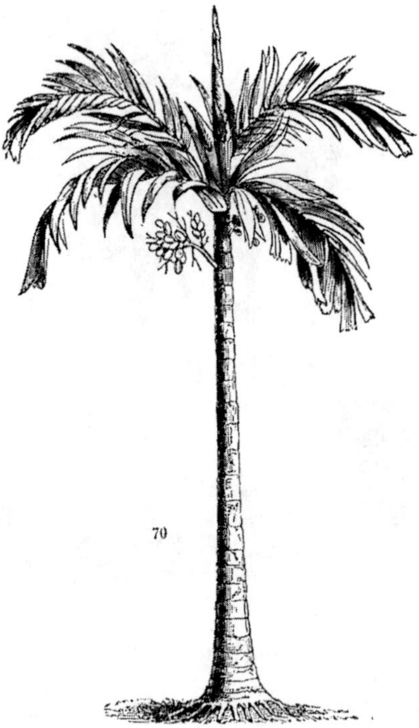

70

Unarmed ; stem tall ; leaflets broadly linear, plaited, acuminate, the upper confluent, wedge-shaped, præmorse ; fruit ovate.

Habitat. All over the East Indies.
Quality. Fruit astringent ; narcotic.
Uses. Nuts chewed, with lime and the leaves of Chavica (Piper) Betel ; yield Catechu by boiling.

SAGUERUS. *Rumphius.*

Leaves pinnate. *Sepals* of the females imbricated, convolute. *Stamens* 00. *Fruit* a globose drupe, with 2 or 3 angular stones.

1. S. *saccharifer* Blume. *Arenga saccharifera* Labill.—(The GOMUTO PALM.) Fig. 71.

Petioles not spiny ; pinnæ linear-lanceolate, acuminate, entire or emarginate, auriculate, white or silvery beneath ; branches of the spadix long, clustered, pendulous ; fruit yellowish.

Habitat. The tropical parts of Asia.
Quality. Yields a saccharine fluid abundantly ; also Sago.
Uses. Is a great source of Palm wine. The central bud or "cabbage" eatable.

Fig. 70.—Sketch of Areca Catechu.

Cocos. *Linnæus.*

Leaves pinnate. *Flowers* at the base of the branches female, the others male. *Ovary* simple, 3-celled. *Fruit* a coarse, fibrous, one-celled drupe, two cells becoming abortive.

1. C. *nucifera* Linnæus.—(The Cocoa Nut Tree.) Fig. 72.

Stem very tall, unequally ringed; leaves spreading; leaflets linear-lanceolate, acuminate; drupes very large, ovate, bluntly 3-cornered.

72

71

Habitat. Tropical islands everywhere; but only near the coast.

Quality. Root narcotic; fruit oily; stem starchy.

Uses. The nuts a universal article of food; kernel yields oil; roots chewed instead of Areca; abounds in a saccharine fluid, from which Palm wine is made.

Elæis. *Jacquin.*

Leaves pinnated. *Branches* of inflorescence unisexual. In the males the *sepals* 3, papery, dry; the *petals* membranous, lanceolate. *Ovary* 3-celled. *Fruit* an angular 1-seeded drupe, with a fibrous oily rind.

1. E. *guineensis* Linnæus.—(The Oil Palm.) Fig. 73.

Fig 71. Sketch of Saguerus saccharifer; 72. Sketch of Cocos nucifera.

Petioles spiny; leaflets linear-lanceolate, acuminate, green beneath; drupes ovate, collected in huge erect heads, deep orange-yellow.

73

Habitat. Coast of Guinea; common now in tropical America.
Quality. Drupes contain an emollient demulcent fixed oil in great abundance.
Uses. Furnishes Palm oil, employed in soap making, for frictions, &c.; also eaten as butter when quite fresh.

THE NARCISSAL ALLIANCE (V. K., p. 146.)

Natural Orders of Narcissals.

Bromeliads (*Bromeliaceæ.*) Flowers 3-petaloideous.

Amaryllids (*Amaryllidaceæ.*) Flowers hexapetaloideous. Stamens 6, introrse.

Irids (*Iridaceæ.*) Flowers hexapetaloideous. Stamens 3, extrorse.

Natural Order, **Bromeliads**; *Bromeliaceæ* (V. K., p. 147.)

Prevailing Quality. Uncertain.

Ananassa. *Lindley.*

Fruit succulent, in spikes, consolidated into a single tuberculated comose mass.

Fig. 73.—Sketch of Elæis guineensis.

1. A. *sativa* Lindley.—(The PINE-APPLE.)
Leaves glaucous, mealy ; bracts shorter than the fruits.

Habitat. Tropical America only. Introduced elsewhere.
Quality. Fruit subacrid ; sweet, and pleasantly acid.
Uses. A well-known esculent fruit.

Natural Order, **Amaryllids** ; *Amaryllidaceæ* (V. K., p. 155.)

Prevailing Quality. Emetic, narcotic, poisonous.

LEUCOIUM. *Linnæus.*

Sepals and *Petals* distinct to base, all thickened at the point. *Coronet* 0. *Stamens* equal.

1. L. *æstivum* Linnæus.—(SNOW-FLAKE.) Fig. 74.
Spathe many-flowered, style clavate ; ovary somewhat globose ; leaves long, linear, equal to the fistular scape.

Habitat. Various parts of Europe.

NARCISSUS. *Linnæus.*

Sepals and *Petals* united in a tube surmounted by a *coronet.*

1. N. *Pseudo Narcissus* Linnæus. —(DAFFODIL.)
Flowers solitary ; coronet campanulate, erect, crisp, as long as the yellow perianth.

Habitat. Thickets and grassy places all over Europe.
Quality. Acrid, poisonous.
Uses. As an emetic.

2. N. *Tazzetta* Linnæus.—(ITALIAN, or POLYANTHUS NARCISSUS.)
Flowers umbellate ; coronet campanulate, plaited, truncate, 3 times as short as the white perianth.

Habitat. North of Africa and south of Europe.
Quality and *Uses.* As the last.

AGAVE. *Linnæus.*

Caulescent. *Flowers* funnel-shaped, persistent, with erect or revolute lobes. *Capsule* coriaceous, loculicidal. *Seeds* 00, flat.

1. A. *americana* Linnæus.—(AMERICAN ALOE.)
Leaves very large, stiff, perennial, spiny at the edge ; scape lofty, branched ; stamens longer than the perianth.

Habitat. Tropical America ; introduced elsewhere.
Quality. Diuretic, antisyphilitic, detersive.
Uses. Roots a substitute for Sarsaparilla, with which they are mixed sometimes. Juice of leaves a substitute for soap. Sap of flowering branches sweet, subacid, readily ferments and forms a wine called Maguay, or Pulque.

a

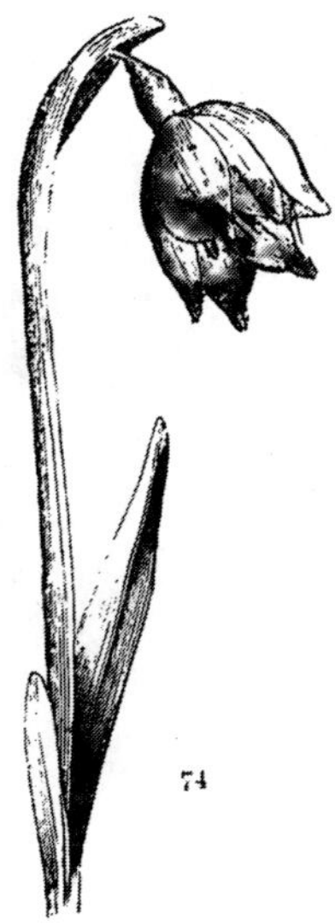
74

Fig. 74.—Leucoium æstivum ; *a*, the ripe fruit.

Natural Order, **Irids**; *Iridaceæ* (V. K., p. 159.)

Prevailing Quality. Acrid, purgative.

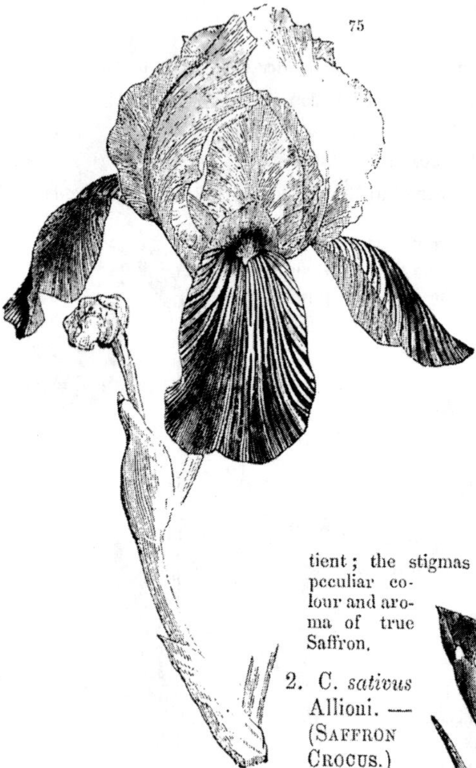

75

CROCUS. *Linnæus.*

Sepals and *petals* nearly equal, united into a long subterranean tube, with a balloon-shaped limb; funnel-shaped when expanded. *Stigma* 3-parted, plaited, convolute.

1. C. *vernus* Linnæus.—(SPRING CROCUS.) Fig. 75.

Spring flowering; stigmas erect; throat of the flower bearded; skins of corm fibrous, netted.

Habitat. Meadows of Europe.

Quality and *Uses.* The corms are acrid, and were formerly reckoned discutient; the stigmas are nearly destitute of the peculiar colour and aroma of true Saffron.

2. C. *sativus* Allioni. —(SAFFRON CROCUS.)

Autumnal flowering; stigmas as long as the limb, and hanging down on one side; skins of the corm fibrous, reticulated, with narrow meshes.

Habitat. South of Europe.

Quality. The dried stigmas are reckoned to be cordial, emmenagogue, and stimulant; they constitute Saffron.

Uses. A mere colouring or flavouring ingredient in this country. Saffron gives to water and alcohol three-fourths of its weight of an orange-red extract, largely employed in painting and dyeing. It must not be confounded with *Safflowers*, the dried florets of Carthamus tinctorius, with which it is often adulterated. The latter consisting of *corollas*, and the former of *stigmata*, a little careful observation will readily detect the difference.

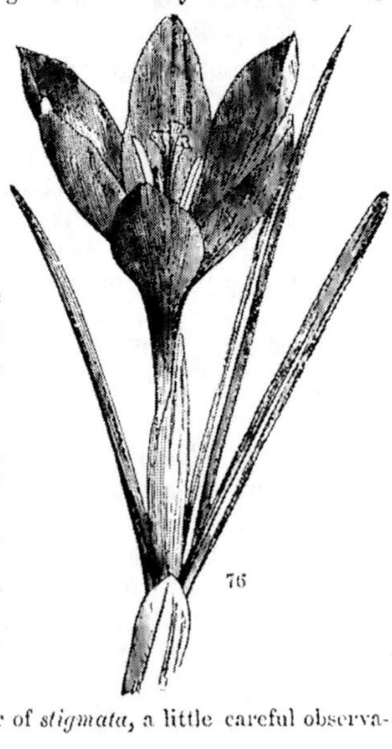

76

Fig. 75. Iris germanica; 76. Crocus vernus.

IRIS. *Linnæus.*

Sepals reflexed. *Petals* erect, arched. *Style* 3-parted, petaloid, covering the stamens.

1. I. *germanica* Linnæus. — (PURPLE ORRIS-ROOT.) Fig. 76.

Bearded; spathes membranous, herbaceous at base; tube 2 or 3 times as long as the ovary; flowers deep purple.

Habitat. Walls and dry places in Europe.
Quality. Rhizomes fragrant, bitterish, acrid.
Uses. Yields part of the fragrant orris-root.

2. I. *florentina* Linnæus. — (WHITE ORRIS-ROOT.)

Bearded; spathes herbaceous, glaucous, with a narrow scarious border; tube 2 or 3 times as long as ovary; flowers nearly white.

Habitat. Italy.
Quality and *Uses.* As in the last.

3. I. *Pseudacorus* Linnæus. — (YELLOW-FLAG.) Fig. 77.

Beardless; sepals oblong or ovate, long-stalked; petals narrower and shorter than the lobes of the style; flowers yellow.

Habitat. Ditches and rivers all over Europe.
Quality. Acrid.
Uses. Rhizome diuretic, purgative, and emetic.

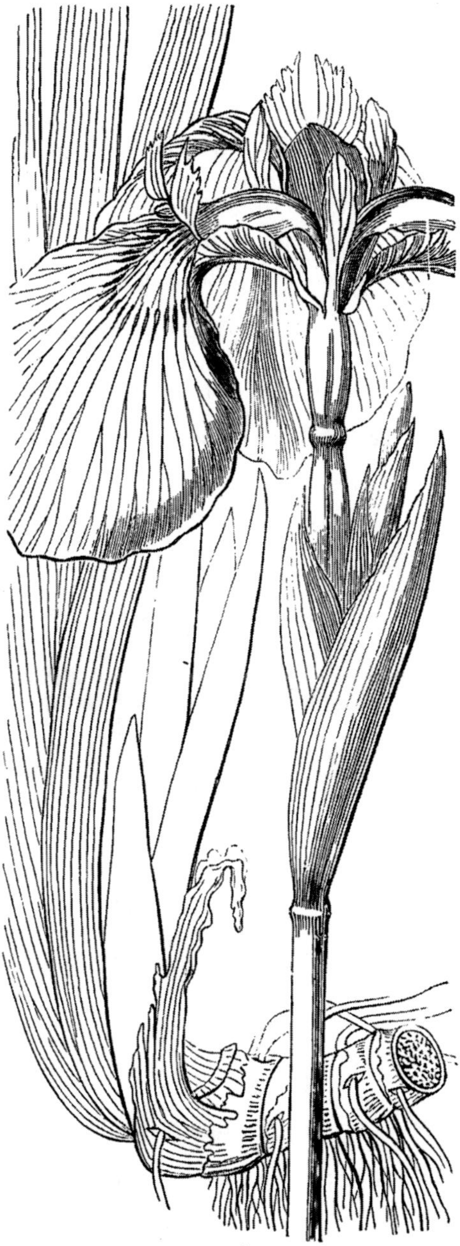

77

Fig. 77.—Iris Pseudacorus.

THE AMOMAL ALLIANCE (V. K., p. 162.)

Natural Orders of Amomals.

Musads (*Musaceæ.*) Stamens more than 1.
Gingerworts (*Zingiberaceæ.*) Stamen 1; anther complete.
Marants (*Marantaceæ.*) Stamen 1; anther halved.

Natural Order, **Musads**; *Musaceæ* (V. K., p. 163.)

Prevailing Quality. Nutritious, diaphoretic.

MUSA. *Rumphius.*

Flowers two-lipped; the lower lip tubular, 5-cleft, embracing the upper lip, which is dwarf and concave. *Ovules* 00, horizontal. *Fruit* baccate.

1. M. *sapientum* Linnæus.—(PLANTAIN. BANANA.) Fig. 78.

78

Spadix nodding; spathes deciduous, or withering.

Habitat. Tropical Asia.
Quality. Fruit sweet, soft, amylaceous.
Uses. A common fruit and nutritious esculent in all hot countries. Plantains and Bananas are mere varieties of each other.

Fig. 78.—Sketch of Musa sapientum.

Natural Order, **Gingerworts**; *Zingiberaceæ* (V.K., p. 165.)

Prevailing Quality. Aromatic, stimulating.

ZINGIBER. *Gærtner.*

Stems annual. Lateral inner lobes of *corolla* 0. *Filament* extended beyond the anther into a curved beak.

1. Z. *officinale* Roscoe.—(COMMON GINGER.) Fig. 79.

Leaves subsessile, narrowly lanceolate, smooth; spikes oblong, on a scape about 9 inches high; bracts acute; flowers whitish; lip 3-lobed, streaked with purple.

Habitat. Tropical parts of Asia.
Quality. Acrid, aromatic, stimulant, sialagogue.
Uses. A condiment; promotes digestion, relieves flatulency; chewed against tooth-ache.

79

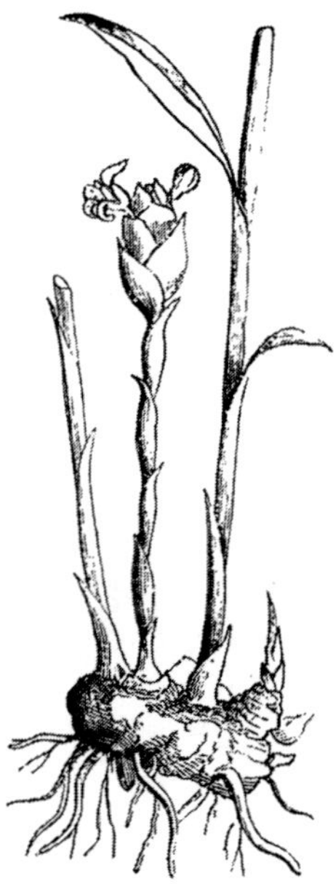

ALPINIA. *Linnæus.*

Stems perennial. *Inflorescence* terminal. Inner lateral lobes of the *corolla* small or 0. *Filament* not extended beyond the anther. *Fruit* baccate.

1. A. *racemosa* Plumier.

Stems 4-5 feet high; leaves ovate-lanceolate, recurved at the point; raceme spiked; bracts ventricose; lip trifid; flowers white.

Habitat. Tropical America.
Quality and *Uses.* As in the Galangale.

2. A. *Galanga* Linnæus.—(GALANGALE.)

Stems 6 or 7 feet high; leaves broad, sessile, with a whitish edge; panicle oblong, branched; flowers greenish-white; lip oblong, unguiculate, bifid; roots tuberous, pungent.

Habitat. Indian Archipelago.
Quality. Rhizome peppery, aromatic.
Uses. As Ginger.

AMOMUM. *Linnæus.*

Stems perennial. *Inflorescence* radical, cone-like. Inner lateral lobes of the *corolla* 0. *Lip* very large, flat. *Filament* flat, extended beyond the anther, 2-lobed, with an emarginate middle lobe.

1. A. *Cardamomum* Linnæus.—(ROUND CARDAMOM.)

Leaves short-stalked, lanceolate, acuminate, smooth; bracts villous, cinereous; lip 3-lobed, crenate, crisp, with two rosy streaks; fruit small, roundish, 3-cornered; seeds brown, angular, cuneiform.

Habitat. Indian Archipelago.
Quality. Seeds aromatic, camphoraceous.
Uses. As those of Elettaria Cardamomum.

Fig. 79.—Diminished figure of Zingiber officinale.

2. A. *Grana Paradisi* Smith. A. *Meleguetta* Roscoe.—(GRAINS OF PARADISE PLANT. MALAGUETTA PEPPER.) Fig. 80, 81.

Leaves nearly sessile, linear-lanceolate, smooth; bracts . . . ; lip entire, obovate, crenate, plaited; fruit ovate, coriaceous, (6 inches long), yellow, spotted with orange; seeds angular, light brown.

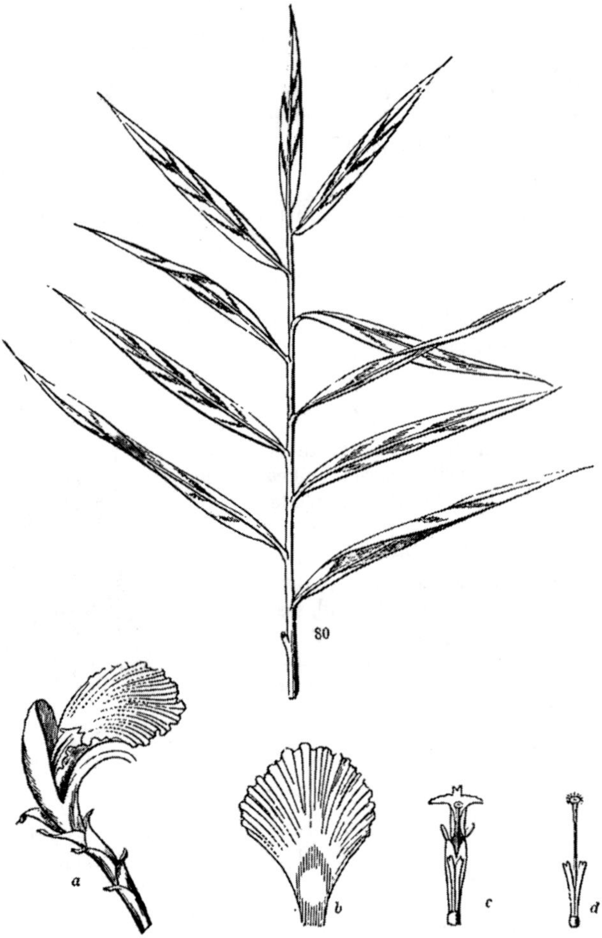

Habitat. Coast of Guinea, &c.
Quality. Seeds aromatic and excessively peppery.
Uses. A spicy condiment; used in veterinary practice; in the illegal preparation of malt liquor, &c.

Fig. 80.—Leaves and analysis of Amomum Grana Paradisi, from *Pharmaceutical Journal*; *a*, flower: *b*, lip; *c*, anther, ovary, and style; *d*, barren stamens and style.

Fig. 81.—Fruit of Amomum Grana Paradisi, from *Pharmaceutical Journal*; *a*, full-sized fruit; *b*, a transverse section.

Curcuma. *Linnæus.*

Stems annual. Lateral inner lobes of *corolla* similar to the exterior. *Filament* petaloid, 3-lobed, with a two-spurred anther on the middle lobe.

1. C. *longa* Linnæus. — (Turmeric.) Fig. 82.

Tubers long, lobed, pendulous, deep orange; leaves broad, lanceolate, whole coloured; spikes central, among the leaves, pale green, with a rose-coloured coma; flowers yellow, concealed by the bracts.

Habitat. East Indies.
Quality. A mild aromatic.
Uses. A condiment; an ingredient in curry powder; its juice a test for free alkalies.

In the East Indies Turmeric is regarded as an important bitter, aromatic, stimulant, tonic; and is employed in debilitated states of the stomach, intermittent fever, dropsy. The native practitioners consider it, in the form of powder, as an excellent application for cleansing foul ulcers. It is also used in dyeing. There seems little doubt that it was the κυπειρος ινδικος of Dioscorides, as it certainly was the Carcumaa of Avicenna; the Persians now call it *Kurkoom.* The starch of the young white tubers forms one of the East Indian arrowroots.—*Royle.* See also the pendulous tubers of several other species of Curcuma yield beautiful pure amylaceous matter, which the natives of the countries where the plants grow prepare and eat like *Arrow-root.* In Travancore this flour or starch forms a large part of the diet of the inhabitants. It is, however, to be observed, that the same tubers which yield starch when young, yield turmeric when old; the colour and aroma which give its character to the latter, appearing to be deposited in the cells at a later period of growth.

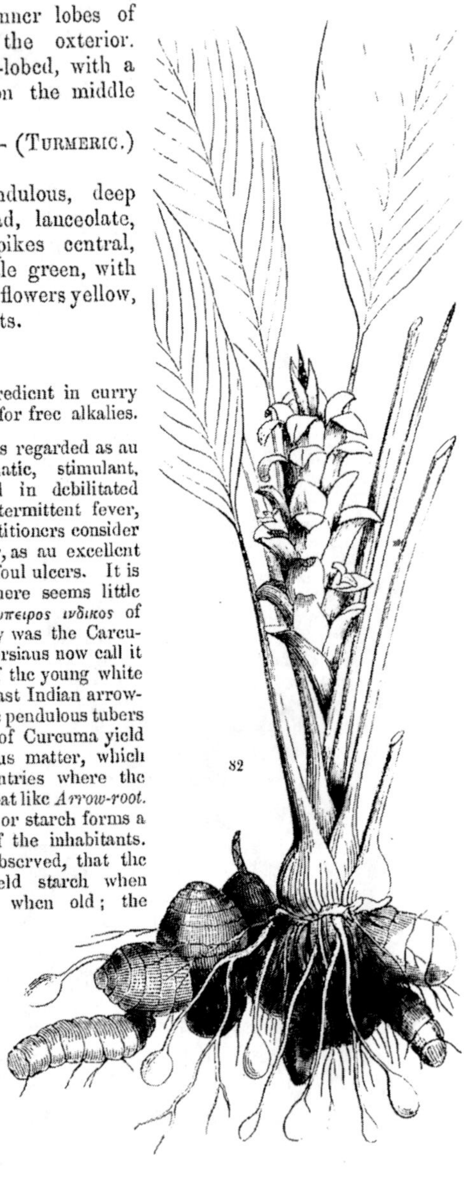

82

Fig. 82.—Diminished figure of Curcuma longa; the small oblong colourless knobs are those in which starch alone exists; the larger and darker rhizomes are the older structure containing turmeric.

2. C. *Zedoaria* Roxburgh.—(ROUND ZEDOARY.)
Tubers long, palmate, yellow inside; leaves sessile, silky beneath, broad, whole coloured; spikes radical, comose, rose-coloured; flowers pink, shorter than the bracts.

Habitat. Tropical Asia.
Quality. Rhizome warm, aromatic, bitter.
Uses. As Curcuma longa; as a tonic.

3. C. *Zerumbet* Roxburgh.—(TRUE ZEDOARY.)
Tubers palmate, straw-coloured inside; leaves green, stalked, broad, stained with purple in the middle; spike radical, comose, purple; flowers yellow, concealed by the bracts.

Habitat. East Indies.
Quality and *Uses.* As in the last.

ELETTARIA. *Rheede.*

Stems perennial. *Inflorescence* radical, loose. Lateral inner lobes of *corolla* minute. *Filament* not extended beyond the anther.

1. E. *Cardamomum* Maton.—(CARDAMOM.)
Root with fleshy fibres; leaves lanceolate, villous above, silky beneath, with villous sheaths; scapes radical, prostrate; flowers greenish-white; lip obovate, slightly 3-lobed, streaked with violet.

Habitat. Malabar.
Quality. Seeds an agreeable aromatic without acridity.
Uses. An adjunct to stimulant cordial mixtures.

Natural Order, **Marants**; *Marantaceæ* (V. K., p. 168.)

Prevailing Quality. Insipid; diaphoretic.

CANNA. *Linnæus.*

Ovules 00, horizontal. *Style* petaloid, straight. *Stems* simple. *Flowers* red, orange, or yellow.

1. C. *edulis* Ker. — (TOUS LES MOIS ?) Fig. 83.
Tuberous; stem purple; leaves broad, smooth, glaucous; corolla tripartite, erect, with oval oblong retuse segments, of which the middle one is much the shortest; lip linear, revolute, emarginate.

83

Fig. 83.—Flowers of Canna edulis.

Habitat. Peru.
Quality. Starch nutritive, emollient, demulcent.
Uses. Supposed to furnish the fæcula called Tous les Mois in the shops.

2. C. *Achiras* Gillies.—(ACHIRA.)

Tuberous; stem green, downy; leaves abruptly acuminate; corolla bipartite with linear divisions; lip revolute, lanceolate, emarginate.

Habitat. America, Mendoza to Guatemala.
Quality. See next species.
Uses. Tubers eaten as food in Chili and Peru.

MARANTA. *Plumier.*

Ovules solitary, basal. *Style* fleshy, curved downwards. *Stems* branched. *Flowers* white.

84

85

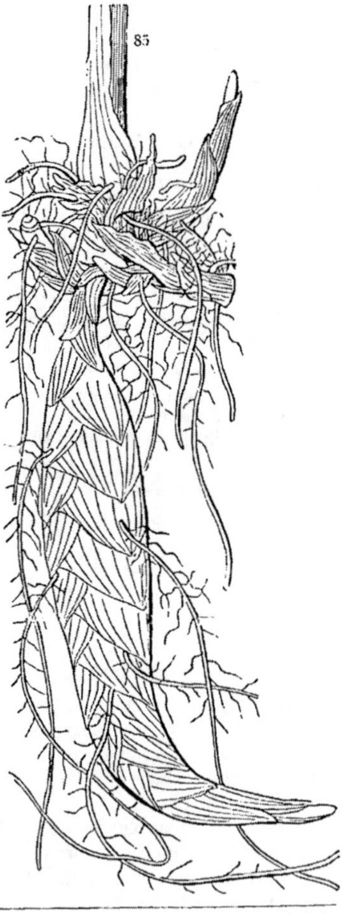

Fig. 84.—Leaf and flowers of Maranta arundinacea; *a*, the stamen and style; 85. The tubers of the same, much less than natural size.

1. M. *arundinacea* Linnæus.—(Ara-root, or Arrow-root.) Fig. 84, 85. Stem branched ; leaves ovate-lanceolate, hairy on the under side ; peduncles 2-flowered.

Habitat. Tropical America.
Quality. Starch nutritive, emollient, demulcent.
Uses. Starch an article of diet, under the name of Jamaica arrow-root.

THE ORCHIDAL ALLIANCE (V. K., p. 170.)

Natural Order, Orchids ; *Orchidaceæ* (V. K., p. 173.)

Prevailing Quality. Stimulating, fragrant.

Orchis. *Linnæus.*

Anther terminal, erect. *Lip* spurred. *Pollen* sectile. *Pollen-masses* two, with their glands inclosed in a common pouch.

1. O. *mascula* Linnæus.
Roots oblong, undivided ; leaves spotted ; spike loose, many-flowered ; sepals reflexed ; lip 3-lobed, the intermediate lobe being emarginate, with a blunt horizontal spur.

Habitat. Meadows in Europe.
Quality. Roots gummy, nutritious, emollient, demulcent.
Uses. Roots form the agreeable diet called Salep.

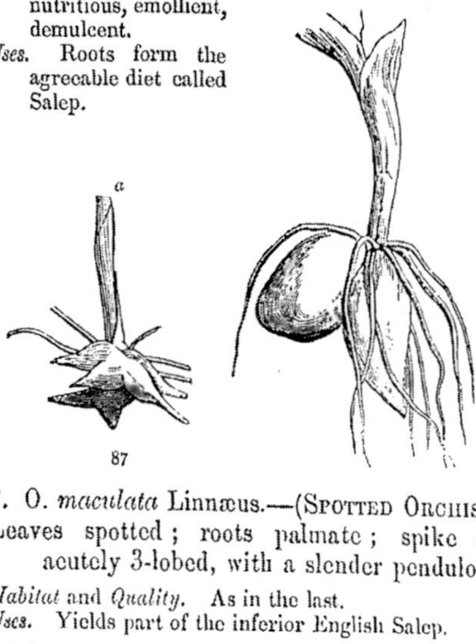

87

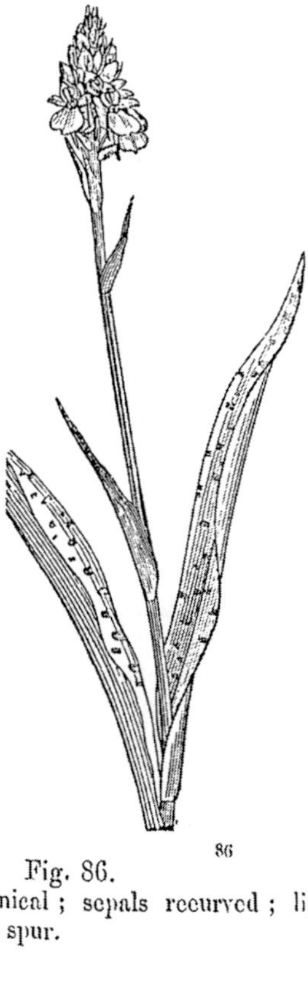

86

2. O. *maculata* Linnæus.—(Spotted Orchis.) Fig. 86.
Leaves spotted ; roots palmate ; spike conical ; sepals recurved ; lip acutely 3-lobed, with a slender pendulous spur.

Habitat and *Quality.* As in the last.
Uses. Yields part of the inferior English Salep.

Fig. 86.—Orchis maculata ; 87. *a*, root of O. maculata ; *b*, of O. mascula ; common forms of Salep.

BLETIA. *Ruiz* and *Pavon.*

Pollen-masses waxy, 8, cohering by elastic straps. *Lip* spurless, 3-lobed, sessile, articulated with the column. *Anther* 8-celled.

1. B. *verecunda* R. Brown.

Petals oblong, obtuse, arched over the column ; middle lobe of lip longer than broad, wavy, with numerous crisp veins, which are sometimes branched.

Habitat. West Indies.
Quality. Tuber somewhat fragrant, bitterish, aromatic.
Uses. In weak digestion.

VANILLA. *Plumier.*

Fruit a long pulpy pod, with round seeds not inclosed in a loose membrane.

1. V. *planifolia* Andrews.—(VANILLA.)

Leaves oblong-lanceolate, flat ; sepals and petals oblong, flat, obtuse ; lip crisp, convex, covered with numerous transverse wedge-shaped plates; fruit fragrant.

Habitat. Mexico, Guatemala, West Indies.
Quality. Aromatic, stimulant.
Uses. Asthenic fevers, rheumatism, hysteria, male impotence ; in confectionary ; in the preparation of chocolate.

THE JUNCAL ALLIANCE (V. K., p. 190.)

Natural Order, Orontiads ; *Orontiaceæ* (V. K., p. 193.)

(These are Juncals with the form of Arals.)

Prevailing Quality. Acridity.

SYMPLOCARPUS. *Salisbury.*

Spathe cucullate. *Spadix* subglobose, all covered with flowers. *Sepals* 4, becoming baccate. *Stamens* 4, opposite the sepals. *Ovary* one-celled, with one ovule.

1. S. *fœtidus* Salisbury.—(SKUNK CABBAGE.)

Stemless ; leaves ovate, cordate.

Habitat. Ditches in Canada and the United States.
Quality. Acrid, fœtid, antispasmodic, expectorant.
Uses. Seeds and rhizome palliatives in paroxysms of asthma.

CALLA. *Linnæus.*

Spathe flat. *Spadix* cylindrical, covered with a mixture of stamens and pistils. *Ovary* 1-celled, with from 6 to 8 erect ovules.

1. C. *palustris* Linnæus.

A marsh plant, with a creeping rhizome ; leaves cordate, cuspidate, stalked ; spathe white ; spadix yellow.

Habitat. Swamps of Europe, Siberia, and North America.
Quality. Acrid, caustic ; diaphoretic.
Uses. Rhizomes yield eatable starch after grinding and washing.

ACORUS. *Linnæus.*

Spathe replaced by a two-edged leaf-blade. *Scales* 6, permanent, herbaceous. *Stamens* with filiform filaments.

1. A. *Calamus* Linnæus.—(SWEET FLAG.)

Spathe a direct continuation of the scape, in no respect different from the leaves.

Habitat. Marshes all over Europe.
Quality. Rhizome an aromatic stimulant, and mild tonic ; stomachic.
Uses. An adjunct to other tonics ; asthenic fevers, ague, chronic catarrh, dyspepsia ; aromatic baths, perfumery, hair powder.

THE LILIAL ALLIANCE (V. K., p. 195.)

Natural Orders of Lilials.

Melanths (*Melanthaceæ*). Anthers extrorse. Styles separate.

Lilyworts (*Liliaceæ*). Anthers introrse. Styles united.

Natural Order, **Melanths** ; *Melanthaceæ* (V. K., p. 198.)

Prevailing Quality. Acrid-narcotic.

VERATRUM. *Tournefort.*

Flowers polygamous. *Sepals* and *petals* spreading flat, with an extremely short tube, persistent. *Stamens* perigynous; *anthers* opening transversely. *Capsule* three-horned, turgid.

1. V. *album* Linnæus.—(WHITE HELLEBORE.) Fig. 88.

Panicle much branched, downy, spreading; flowers greenish-white, longer than the lower bracts.

Habitat. Subalpine parts of Europe.
Quality. A powerful acrid poison. In small doses, emetic and purgative. Errhine.
Uses. In melancholia, mania, epilepsy, herpes, gout ; chronic affections of the brain ; against pediculi.

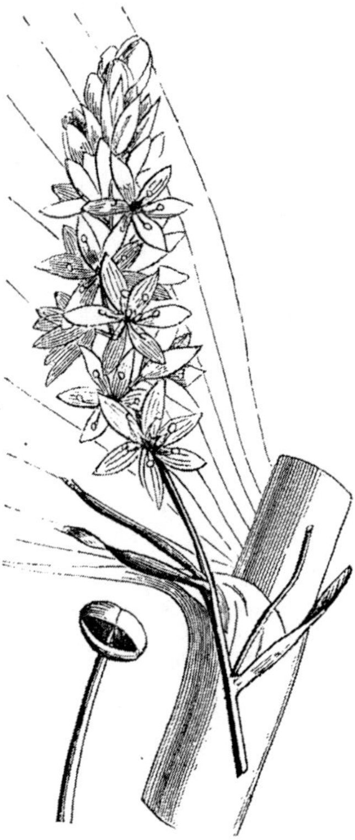

Fig. 88.—Lateral branch of the inflorescence of Veratrum album ; *a*, a stamen.

Colchicum. *Linnæus.*

Sepals and *petals* united into a subterranean tube; the limb balloon-shaped, funnel-shaped when expanded.

1. C. *autumnale* Linnæus.—(Meadow Saffron.) Fig. 89.

Leaves broad ; tube of flower five or six times as long as the limb ; styles thickened and curved at the point, as long as the stamens or longer.

89

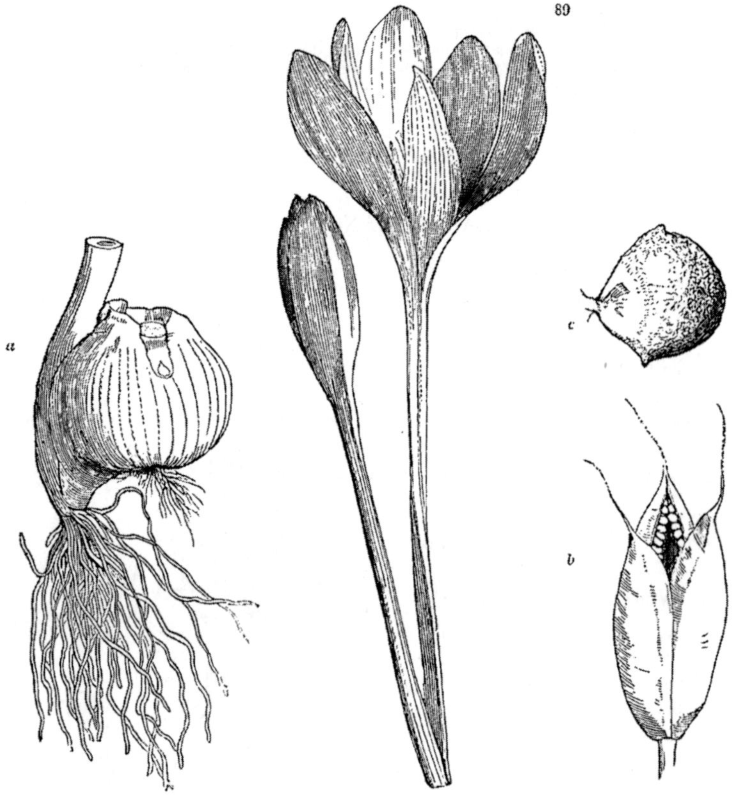

Habitat. Meadows in England, and other parts of Europe.

Quality. An acrid poison. In small doses, emetic and purgative.

Uses. Gout, lumbago, rheumatism, inflammatory diseases, humoral asthma, worm cases.

Asagræa. *Lindley.*

Flowers racemose, naked. *Sepals* and *petals* narrow, coloured, with a honey-spot at the base. *Stamens* perigynous, alternately shorter ; *anthers* bursting vertically. *Follicles* 3, acuminate, papery ; *seeds* winged.

Fig. 89.—Colchicum autumnale ; *a*, corm ; *b*, capsule ; *c*, seed much magnified.

1. A. *officinalis* Lindley.—(SABADILLA.) Fig. 90.

Fig. 90.— Part of the spike of Asagræa officinalis; *a*, anther; *b*, capsule; 91. Stenanthium frigidum *a*, ovary and stamens.

Leaves grassy, green, rough at the edge ; scape 4 or 5 feet long ; raceme very dense ; flowers white.

Habitat. Cool uplands of Mexico.
Quality. Like Veratrum, but more acrid.
Uses. Seeds anthelmintic ; against pediculi ; dangerous. A source of veratria.

STENANTHIUM. *A. Gray.*

Flowers panicled with leafy bracts. *Sepals* and *petals* adherent to the ovary at their base, campanulate, non-nectariferous. *Stamens* perigynous ; *anthers* reniform, opening at the back. *Follicles* 3, acuminate, papery ; *seeds* winged.

1. S. *frigidum* Kunth. *Helonias frigida* Schlecht.—(SEVOEJA.) Fig. 91.
Leaves linear, channelled, keeled ; panicle simple ; flowers large, dark purple, nodding, hermaphrodite.

Habitat. Cool uplands of Mexico.
Quality. Acrid, poisonous.
Uses. As Asagræa.

Natural Order, Lilyworts ; *Liliaceæ* (V. K., p. 200.)

Prevailing Quality. Subacrid, or insipid.

ALLIUM. *Linnæus.*

Bulbous. Flowers umbellate, inclosed within a spathe. *Sepals* and *petals* spreading, having the stamens inserted in their base. *Fruit* a capsule. *Seeds* angular.

1. A. *sativum* Linnæus.—(GARLICK.)
Umbel bulbiferous ; leaves obscurely keeled ; spathe 1-valved, deciduous ; bulbs compound, covered by a loose white skin.

Habitat. Sicily.
Quality. A local irritant ; tonic, stimulant, diuretic.
Uses. As a condiment in cookery ; chronic catarrh, dyspepsy ; as a liniment in hooping cough, infantile convulsions, &c.

2. A. *Scorodoprasum* Linnæus.—(ROCAMBOLE.)
Umbel bulbiferous, few-flowered ; leaf-sheaths 2-edged ; spathe 2-valved, mucronate, permanent ; lobes of the flowers ovate-lanceolate, acute, as long as the stamens ; bulbs simple.

Habitat. South of Europe.
Quality. Stimulant, diuretic.
Uses. In cookery, as a stimulating flavouring ingredient.

3. A. *Porrum* Linnæus.—(LEEK.)
Umbel not bulbiferous ; stem leafy ; spathe 1-valved, deciduous ; lobes of the flower oblong, obtuse, about as long as the stamens ; bulb simple, soft, being a mere continuation of the stem.

Habitat. South of Europe.
Quality, &c. As in the last, but much more mild.

4. A. *Ascalonicum* Linnæus.—(SHALLOT.)

Umbel not bulbiferous, globose; stem leafy at the base only; leaves subulate; spathe 2-valved; stamens 3-cuspidate, as long as the ovate lanceolate lobes of the flowers; bulbs clustered.

Habitat. Syria.
Quality, &c. As in the Rocambole.

5. A. *Cepa* Linnæus.—(COMMON ONION.)

Umbel not bulbiferous, globose; stem ventricose, leafy at the base; leaves terete; spathe reflexed; lobes of the flower obtuse, hooded, not half so long as the stamens; bulb solitary, flattened. Biennial.

Habitat. Egypt?
Quality, &c. As in the last. "Raw Onions are occasionally taken with advantage, as an expectorant, by elderly persons affected with winter cough."—*Pereira.*

6. A. *Schœnoprasum* Linnæus —(CHIVE.)

Umbel not bulbiferous, globose, compact; scape naked, as long as the subulate terete leaves; bulbs long, naked, small, clustered.

Habitat. Europe.
Quality, &c. As in the last.

7. A. *fistulosum* Linnæus.—(WELCH ONION.)

Umbel not bulbiferous, globose; scape and leaves terete, fistular; stamens twice as long as the lobes of the flower; ovary 3-cornered. Perennial.

Habitat. Siberia.
Quality, &c. As in the common Onion; but very strong.

ASPARAGUS. *Linnæus.*

Caulescent. Flowers scattered. *Sepals* and *petals* herbaceous, partially united into a tube. *Style* 1; *stigmas* 3, reflexed. *Fruit* succulent.

1. A. *officinalis* Linnæus.—(ASPARAGUS.) Fig. 92.

Stem unarmed, branched; false leaves setaceous; true leaves membranous, acute; peduncles lax, 1-flowered, drooping; roots long, thick, and unbranched; young shoots covered with scales.

Habitat. Sea coast of Europe.
Quality. Diuretic; roots aperient.
Uses. Young succulent shoots a common esculent when boiled. Roots employed on the Continent for falsifying Sarsaparilla. A spirit has been obtained from the fermented berries.

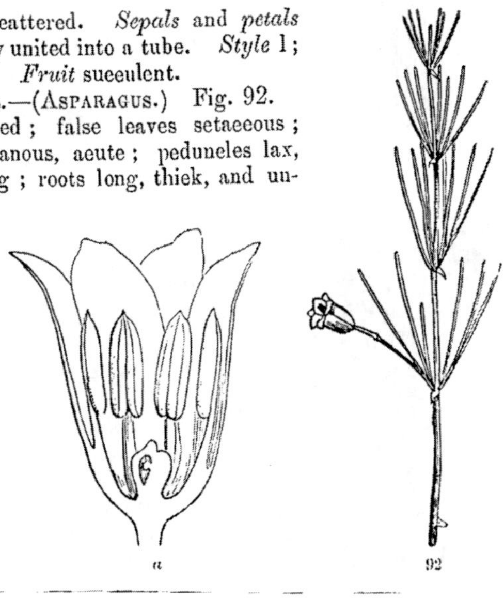

Fig. 92.—Asparagus officinalis; *a*, a section of its flower much magnified.

URGINEA. *Steinheil.*

Bulbous. *Flowers* racemose, stellate. *Stamens* distinct, perigynous. *Filaments* subulate. *Seeds* numerous, flat, ascending.

93

1. U. *Scilla* Steinheil. *Scilla maritima* Linnæus.—(OFFICINAL SQUILL.) Fig. 93.

Leaves appearing after the flowers; raceme very long, many-flowered; bracts spurred at the base; flowers rotate.

Habitat. Sea coast of the Mediterranean.
Quality. Acrid, diuretic, expectorant, emetic, purgative.
Uses. Dropsies, chronic catarrh, asthma, hooping cough.

2. U. *Pancration* Steinheil. — (PANCRATIC SQUILL.)

Leaves shorter and narrower; scape more glaucous; flowers smaller and more compact; bulbs much smaller than in the last.

Habitat. The Mediterranean coast.
Quality and *Uses.* As in the last; but said to be milder in its effects. An obscure plant, supposed to be the Παγκρατιον of Dioscorides.

CONVALLARIA. *Linnæus.*

Caulescent. *Flowers* racemose, terminal. *Sepals* and *petals* nearly separated, but forming a bell by their junction. *Stigma* obtuse, 3-cornered. *Fruit* succulent.

1. C. *majalis* Linnæus. — (LILY OF THE VALLEY.)

Scape naked; raceme one-sided; flowers campanulate, nodding.

Habitat. Europe, in thickets and woods.
Quality. Rhizomes acrid, purgative, and diuretic.
Uses. The flowers are acrid; their powder forms a kind of cephalic snuff. Their distilled water is highly esteemed under the French name of *eau d'or.*

2. C. *Polygonatum* Linnæus.—(SOLOMON'S SEAL.)

The strong-smelling rhizomes of this plant are reputed to be soporific; at least they entered into the composition of what was called Solomon's opiate.

a

Fig. 93.—Diminished figure of Urginea Scilla; *a*, a seed.

ALOE. *Linnæus.*

Caulescent. Leaves permanent, succulent. *Flowers* cylindrical. *Stamens* hypogynous. *Ovules* 00. *Fruit* a membranous capsule.

1. A. *socotrina* Haworth.—(SOCOTRINE ALOE.) Fig. 94.

Stem arborescent; leaves ensiform, green, with small white serratures; flowers yellow, or red and yellow.

94

Habitat. Island of Socotra.
Quality. Purgative, tonic, emmenagogue.
Uses. Habitual costiveness, dyspepsia, irregular menses, worms, imperfect secretion of bile.

2. A. *purpurascens* Haworth.—(CAPE ALOE ?)

Stem arborescent; leaves ensiform, glaucous, recurved at the point, with white serratures; flowers deep red.

Habitat. Cape of Good Hope.
Quality and *Uses.* As in the last.

3. A. *spicata* Linnæus.—(CAPE ALOE.)

Stem arborescent; leaves ensiform, flat, dentate, spotted with white; flowers spiked, campanulate, horizontal, whitish.

Habitat. Cape of Good Hope.
Quality and *Uses.* As in the last.

4. A. *vulgaris* Lamarck. A. *barbadensis* Miller.—(BARBADOES ALOE.)

Stem arborescent, throwing up many suckers from the base; leaves ensiform, sinuate-serrated, white-spotted; flowers yellow.

Habitat. Uncertain. Found in the East and West Indies, Italy, Sicily, and Malta.
Quality and *Uses.* As in the last.

FRITILLARIA. *Linnæus.*

Bulbous. Sepals and *petals* campanulate, distinct, with a conspicuous honey-pore at the base. *Style* trifid at apex. *Fruit* a capsule. *Seeds* 00, thin, flat. Flowers often almost unisexual.

1. F. *imperialis* Linnæus.—(CROWN IMPERIAL.) Fig. 95.

Raceme short, comose; flowers nodding, with six large white honey pores at the base inside.

Habitat. Persia? Brought to Europe from Constantinople.
Quality. Bulbs acrid, emetic, poisonous, with a peculiar heavy hircine smell.
Uses. It is said that these bulbs are equally powerful as the corms of Colchicum. Orfila killed dogs by making them swallow pieces of the bulb.

Fig. 94.—A sketch of the habit of Aloe socotrina.

96 95 96*

RUSCUS. *Linnæus.*

Caulescent. *Flowers* axillary, half unisexual. *Sepals* and *petals* distinct. *Filaments* monadelphous. *Style* 1. *Stigma* capitate. *Fruit* succulent.

1. R. *aculeatus* Linnæus.—(BUTCHER'S BROOM.) Fig. 96.

A straggling green shrub ; false-leaves ovate, acute, spiny-pointed, rigid.

Habitat. Woods in Europe.

Quality. Roots bitter, subacrid, aperient, diuretic. An imperfect substitute for Sarsaparilla.

Fig. 95.—Fritillaria imperialis; 96. Ruscus aculeatus in flower; 96*. The same in fruit

THE ALISMAL ALLIANCE (V. K., p. 207.)

Natural Orders of Alismals.

Butomads (*Butomaceæ.*) Carpels 00-seeded. Placentæ parietal.

Alismads (*Alismaceæ.*) Carpels 1-2-seeded. Placentæ basal or sutural.

Natural Order, **Butomads**; *Butomaceæ* (V. K., p. 208.)

Prevailing Quality. Acridity.

BUTOMUS. *Linnæus.*

Sepals and *petals* 6, coloured, half more petaloid than the others. *Stamens* 9. *Carpels* 6.

97

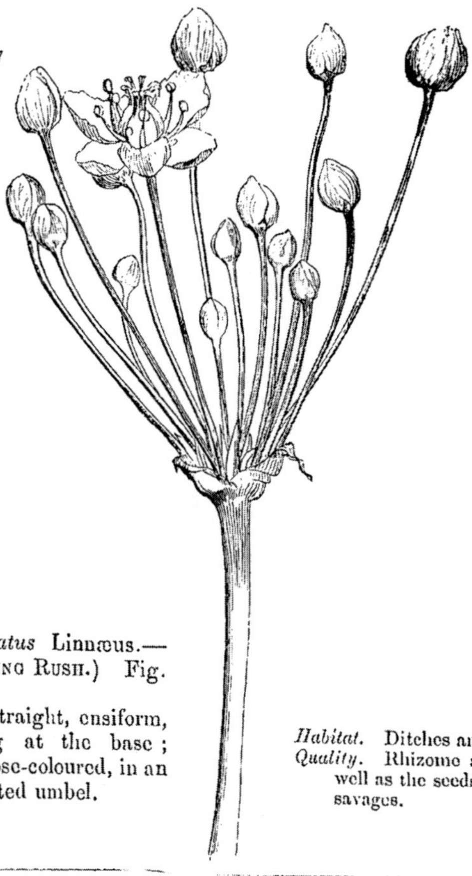

Fig. 97.—Butomus umbellatus.

1. B. *umbellatus* Linnæus.—(FLOWERING RUSH.) Fig. 97.

Leaves long, straight, ensiform, sheathing at the base; flowers rose-coloured, in an involucrated umbel.

Habitat. Ditches and ponds.

Quality. Rhizome acrid, bitter, as well as the seeds; eaten among savages.

Natural Order, 𝔄𝔩𝔦𝔰𝔪𝔞𝔡𝔰; *Alismaceæ* (V. K., p. 209.)

Prevailing Quality. Acridity.

Sagittaria. *Linnæus.*

Flowers monœcious. *Sepals* 3, herbaceous. *Petals* 3, coloured. *Stamens* 00. *Carpels* 00, compressed, one-seeded, on a globose receptacle.

1. S. *chinensis* Sims.—(Chinese Arrowhead.)

Leaves deeply sagittate, acute; the basal lobes as long as the terminal one, ovate, acute, diverging; scape branched, polygonal; male flower terminal.

Habitat. Ditches and ponds in China.
Quality. Subacrid; corms full of starch.
Uses. Sold in the markets of China and Japan as food.

2. S. *sagittifolia* Linnæus.—(Common Arrow-head.)

Uses. This has been recommended, without reason, as a cure for hydrophobia.

CLASS V. DICTYOGENS (V. K., p. 211.)

The useful species of this Class are so few that it is not worth classing them under their natural orders. The Genera alone are sufficient for the purpose of the student.

Tamus. *Linnæus.*

(*Order.* Dioscoreaceæ.)

Ovary inferior. *Fruit* succulent.

1. T. *communis* Linnæus. — (Black Bryony.) Fig. 98.

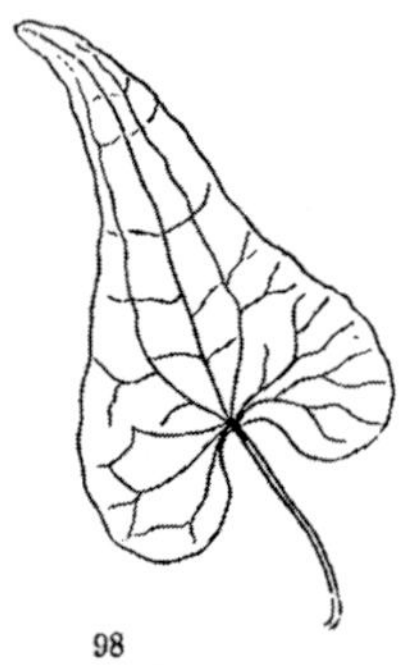

98

Leaves cordate, undivided.

Habitat. Hedges in all parts of Europe.
Quality. Acrid, purgative, emetic.
Uses. Fleshy roots used for stimulating plaisters. Nevertheless the young suckers, in which the acrid principle is not much developed, are eaten as Asparagus, after careful boiling, and changing the water.

Dioscorea. *Linnæus.*

(*Order.* Dioscoreaceæ, or Yams; V. K., p. 214.)

Ovary inferior. *Fruit* membranous, winged.

1. D. *triphylla* Linnæus.—(Three-leaved Yam.)

Somewhat prickly; leaves alternate, downy, ternate, with obovate cuspidate leaflets.

Habitat. East Indies.
Quality and *Uses.* Roots "dreadfully nauseous," even after being boiled.

Fig. 98.—Leaf of Tamus communis.

2. D. *sativa* Linnæus.—(West Indian Yam.) Fig. 99.
Not prickly; leaves alternate, smooth, cordate, with about 9 ribs, the angles of the base rounded.

Habitat. In Tropical America.

Quality and *Uses.* The great fleshy roots, filled with starch, are an important article of food in this and some other species.

99

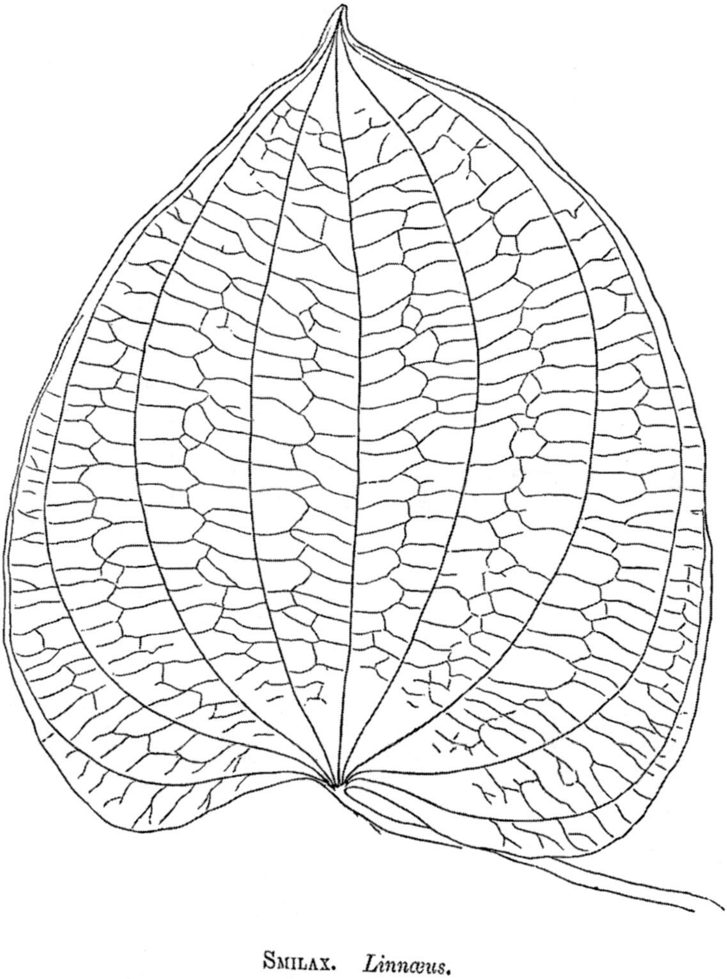

Smilax. *Linnæus.*

(*Order.* Smilaceæ; V. K., p. 215.)

Ovary superior. *Fruit* baccate.

1. S. *medica* Schlechtendahl.—(Vera Cruz Sarsaparilla.) Fig. 100.

Fig. 99.—Leaf of Dioscorea sativa.

Stem prickly, angular; leaves papery, cordate, auriculate, 5-ribbed, somewhat panduriform, occasionally tending to hastate; umbels about 12-flowered.

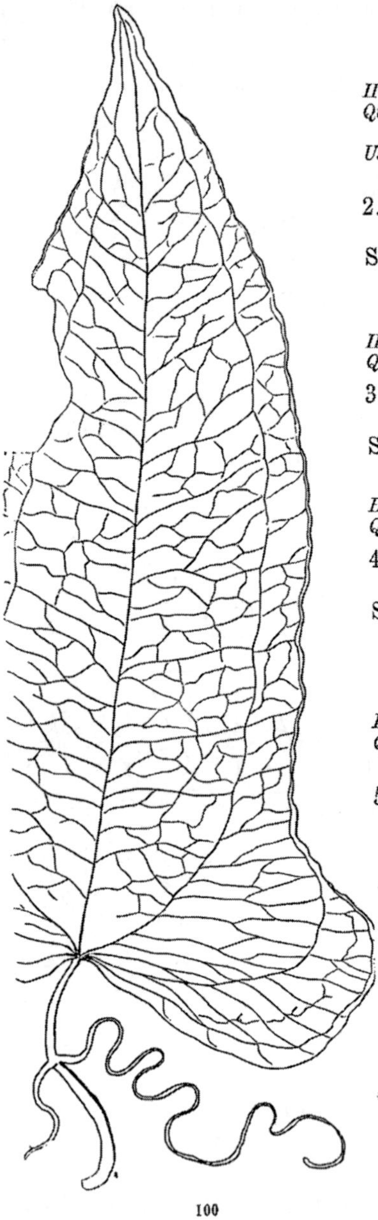

Habitat. Uplands of Mexico.
Quality. Diuretic, diaphoretic, emetic, alterative, tonic.
Uses. In venereal diseases, rheumatism, various skin diseases.

2. S. *siphilitica* Humboldt.—(Brazilian Sarsaparilla.)
Stem slightly prickly, round; leaves leathery, oblong-lanceolate, acuminate, 3-ribbed.

Habitat. Woods of Tropical America.
Quality and *Uses.* As in S. medica.

3. S. *officinalis* Kunth.—(Jamaica Sarsaparilla.)
Stem prickly, angular; leaves leathery, oblong, acute, cordate, 5-7-ribbed.

Habitat. Banks of Rio Magdalena.
Quality and *Uses.* As in S. medica.

4. S. *aspera* Linnæus.—(Italian Sarsaparilla.) Fig. 101.
Stem prickly, angular; leaves cordate, sometimes hastate; about 7-ribbed, leathery, prickly at the edge.

Habitat. South of Europe.
Quality and *Uses.* As Sarsaparilla, but of inferior quality.

5. S. *China* Linnæus.—(China Root.) Fig. 102.
Stem prickly, round; leaves thin, roundish, oblong, about 5-ribbed, acute; rhizome tuberous.

Habitat. China.
Quality. Rhizome sub-astringent, diaphoretic.
Uses. As Sarsaparilla, and also eaten as food, on account of the abundance of its starch.

⁂ According to Dr. Hancock, there is but one species of Smilax that yields genuine Sarsaparilla. This grows chiefly on the elevated lands of the Rio Imiquen, at Unturana and Caraburi; but it is constantly adulterated with inferior sorts.

100

Fig. 100 — Leaf of Smilax medica.

Dr. Hancock says that the Sarsa of the Rio Negro, which comes by way of Angostura or Para is the best, and this is certainly not Willdenow's S. siphilitica; the true species has no axillary spines.

101

It appears that of six or eight species of Smilax growing in the woods of Guayana, but one is found to manifest to the taste any of the sensible properties of the genuine medicinal Sarsa; the root being insipid and inert; *that* one Dr. Hancock describes thus:—

"The stem is round, armed with short curved spines. The leaves are oblong, pointed, distant, smooth, and glossy. The root is a tuber with numerous divergent fibres of 2 or 3 lines in thickness and several feet in length."

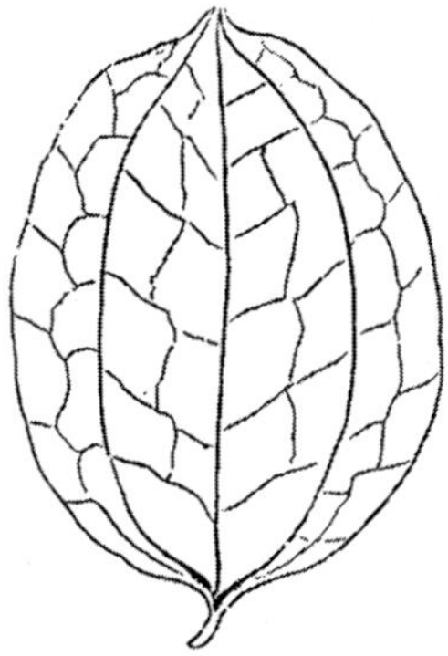

102

CLASS VI. GYMNOGENS (V. K., p. 221.)

Natural Orders of Gymnogens.

Cycads (*Cycadeaceæ.*) Stem simple.

Conifers (*Pinaceæ.*) Stem branched. Females in cones.

Taxads (*Taxaceæ.*) Stem branched. Females solitary.

Natural Order, Cycads; *Cycadeaceæ* (V. K., p. 223.)

Prevailing Quality. Bitter, nutritious.

ZAMIA. *Linnæus.*

Males and *Females* both in cones, composed of woody scales, with a truncated hexagonal apex. *Seeds* two to each scale.

1. Z. *tenuis* Willdenow.

Leaflets linear, tapering to the base, obtuse, with 1 or 2 teeth on the edge, below the end; petiole triquetrous, smooth.

Habitat. Bahamas.

Quality and *Uses.* The dwarf fleshy trunk yields with the following an abundance of pure starch, used as a fine arrow-root in the Bahamas.

2. Z. *furfuracea* Aiton.

Leaflets lanceolate, acute, pointless, serrated from the middle up to the point; petiole terete, prickly.

Habitat. The West Indies.

Quality and *Uses.* The same as in the last.

Fig 101.—Stem and leaf of Smilax aspera; 102. Leaf of Smilax China.

Dion. *Lindley.*

Female cone composed of flat lanceolate woolly scales, cordate at the base, and bearing two seeds.

1. D. *edule* Lindley.

Leaflets sword-shaped, very sharp, attached to the petiole by their whole base ; seeds as large as Chesnuts.

Habitat. Lowlands of Mexico.
Quality and *Uses.* The seeds yield a large quantity of starch, used as arrow-root in Mexico.

Cycas. *Linnæus.*

Males in cones, with an acute thickened apex. *Females* bearing woolly pinnatifid leaves, on the edge of which the seeds stand singly.

1. C. *circinalis* Linnæus.

Leaflets linear-lanceolate, flat (not revolute at the edge.)

Habitat. East Indies.
Quality. Gummy, amylaceous.
Uses. A sago-like flour extracted from the seeds ; the dry gum produces rapid suppuration in malignant ulcers.

Natural Order, Conifers ; *Pinaceæ* (V. K., p. 226.)

Prevailing Quality. Resinous, terebinthinous.

Pinus. *Linnæus.*

Flowers monœcious. *Cones* woody, with numerous 2-seeded scales, having an angular truncated apex. *Leaves* acerose, in clusters of from 2 to 5, surrounded by scarious scales at the base.

1. P. *Pinea* Linnæus.—(Stone Pine.)

Leaves in pairs ; cones obtuse, somewhat round, with unarmed scales ; seeds large, oblong.

Habitat. South of Europe.
Quality. Irritant, stimulant, diuretic.
Uses. Seeds an article of dessert, under the name of Pignons.

2. P. *sylvestris* Linnæus.—(Scotch Fir.) Fig. 103.

Leaves in pairs, rather spreading, short, glaucous ; cones ovate, conical, about as long as the leaves, recurved ; a large tree.

Habitat. North of Europe.
Quality. As in No. 1.
Uses. Yields common turpentine, tar, and pitch; also red deal.

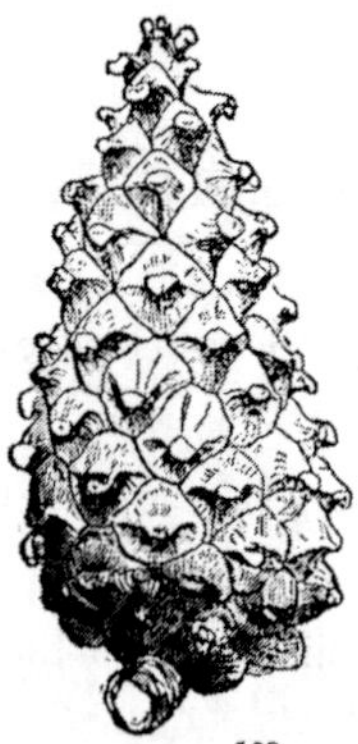
103

3. P. *Pumilio* Waldstein.—(The Mugho Pine.)

Fig. 103.—Cone of Pinus sylvestris.

Leaves in pairs, adpressed, short, dark green; cones short, ovate, obtuse, erect; a trailing bush.

Habitat. Alps of Europe.
Quality. As in No. 1.
Uses. Yields Hungarian Balsam.

4. P. *Pinaster* Aiton.—(CLUSTER PINE.)

Leaves in pairs, dark green, long, stiff, and straight; cones oblong, conical, whorled, pendulous, with prickly scales.

Habitat. South of Europe.
Quality. As in No. 1.
Uses. Yields Bourdeaux turpentine. Wood of bad quality.

ABIES. *Tournefort.*

Flowers monœcious. *Cones* woody, with numerous 2-seeded scales, having a flat rounded apex. *Leaves* single, or, if clustered, indefinite in number.

1. A. *excelsa* Lindley.—(COMMON SPRUCE.)

Leaves 4-cornered, distichous; cones cylindrical, long, pendulous, with the end rounded and uneven.

Habitat. North of Europe.
Quality. As in Pinus No. 1.
Uses. Yields frankincense and Burgundy pitch; also white Norway deal. Leaf buds in scurvy, rheumatism, gout.

2. A. *Larix* Lamarck.—(The LARCH.) Fig. 104.

Leaves deciduous, clustered; cones short, ovate-oblong, loose, erect, with uneven-edged scales.

Habitat. Alps of Europe.
Quality. As in Pinus No. 1.
Uses. Yields Venice turpentine, Larch manna, Oxenburgh gum. Bark abounds in tannin.

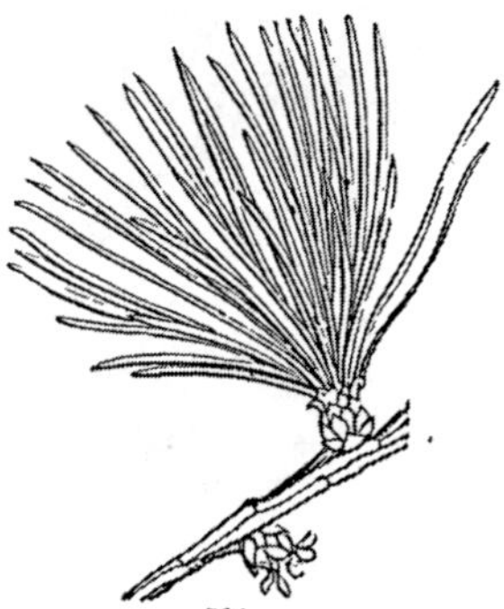
104

3. A. *nigra* Lindley.—(BLACK SPRUCE.)

Leaves 4-cornered, straight, erect, dark glaucous green; cones short, ovate-oblong, pendulous, with scales uneven at the edge.

Habitat. North America.
Quality. Diuretic, antiscorbutic.
Uses. Yields essence of spruce; the basis of spruce-beer.

4. A. *Balsamea* Marshall.—(BALM OF GILEAD FIR.)

Leaves flat, somewhat pectinate in arrangement, emarginate, whitish beneath; cones cylindrical, erect, with acuminate reflexed scales.

Habitat. North America.
Quality. As in Pinus No. 1.
Uses. Yields Canada Balsam.

5. A. *Picea* Lindley. A. *pectinata* DC.—(SILVER FIR.)

Fig. 104.—Cluster of leaves of Abies Larix.

Leaves flat, emarginate, pectinately arranged, very white beneath; cones erect, cylindrical, with very blunt close-pressed scales.

Habitat. Alps of Europe.
Quality. As in Pinus No. 1.
Uses. Yields Strasburgh turpentine.

JUNIPERUS. *Linnæus.*

Flowers diœcious or monœcious. *Cones* ripening into fleshy galbules.

1. J. *communis* Linnæus.—(COMMON JUNIPER.) Fig. 105.

Leaves spreading, acerose spiny-pointed, glaucous above, green below, in whorls of 3; galbule glaucous.

Habitat. Woods of the North of Europe.
Quality. Fruit irritant, diuretic, diaphoretic.
Uses. Dropsy, gonorrhœa, leucorrhœa. Galbules employed in flavouring gin.

2. J. *Sabina* Linnæus.—(SAVIN.) Fig. 106.

Leaves close-pressed, ovate, short, scale-like, dull green; galbule nearly black.

Habitat. Europe, and temperate parts of Asia.
Quality. Acrid poison, local irritant, anthelmintic, emetic, purgative, stimulant, emmenagogue.

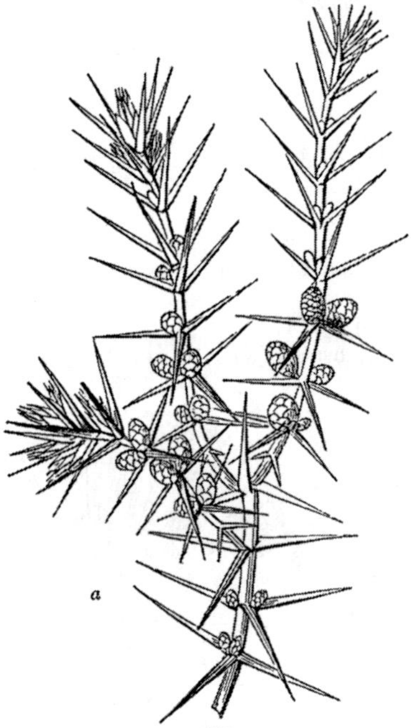

a

105

b

106

Uses. Amenorrhœa, chlorosis, chronic rheumatism; perpetual blisters.

Fig. 105.—Juniperus communis; *a*, in male flower; *b*, fruit; 106. Juniperus Sabina in fruit.

3. J. *Oxycedrus* Linnæus.

Leaves spreading, broad, 3-veined, pungent, in whorls of 3 and 4; galbule rufous, the size of a hazel-nut.

Habitat. South of Europe.
Quality. Diuretic, emmenagogue, powerfully stimulant.
Uses. Its fœtid oil employed in veterinary practice. Wood durable and fragrant.

CALLITRIS. *Ventenat.*

Flowers monœcious. *Cones* woody, of 4-6 scales, with from 3 to 6 seeds to each scale.

1. C. *quadrivalvis* Ventenat.—(SANDARACH TREE.)

A vast tree; branches straggling, jointed, brittle, naked, furrowed, with whorls of small ovate scales at the joints; cones small, purple, glaucous, 4-lobed.

Habitat. Mountains of Morocco.
Quality. Dry juice a brittle resin.
Uses. Resin forms pounce when bruised; varnishes; timber fragrant, hard, durable, mahogany-coloured.

Natural Order, Taxads; *Taxaceæ* (V. K., p. 230.)

Prevailing Quality. Narcotic.

TAXUS. *Linnæus.*

Seed solitary, terminal, surrounded by a succulent cup.

1. T. *baccata* Linnæus.—(YEW TREE.) Fig. 107.

Leaves distichous, flat, linear, with a stout midrib.

Habitat. All the north of Europe.
Quality. Narcotic; pulp of fruit harmless; seeds and leaves a dangerous poison.
Uses. Leaves a substitute for Digitalis. Wood very durable, tough, and elastic.

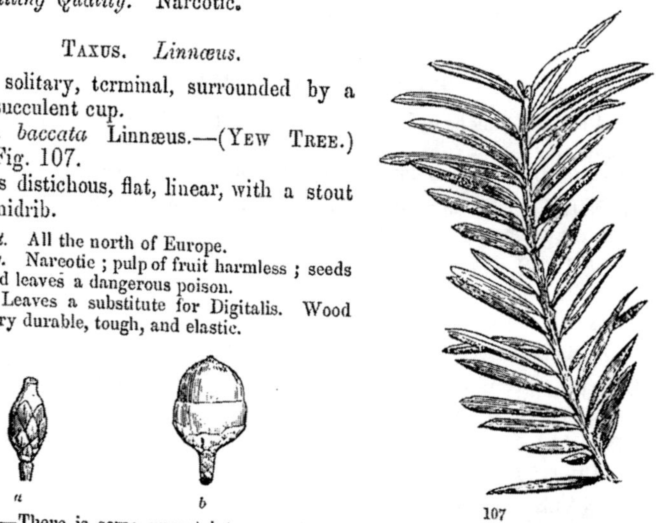

N.B.—There is some uncertainty concerning the circumstances under which the Yew is deleterious. The pulpy covering of the seeds is certainly harmless; and it is asserted that the seeds themselves are innoxious; but although their hard bony shell may guard the kernel so as to prevent its action on the stomach, it is unquestionable that they are a dangerous poison when crushed. Sheep and other animals browse on the leaves in winter with impunity; but if the leaves become partially dried they acquire even then their noxious qualities. The whole question demands renewed and careful experiment.

107.—Taxus baccata; *a*, young fruit; *b*, older fruit.

CLASS VII. EXOGENS (V. K., p. 235.)

AMONG the numerous Alliances in this Class, the following only demand the early attention of the student. They are here distinguished by their *usual* characters, which suffice for the identification of common plants :—

* Sub-class 1. Diclinous.

AMENTALS. Males in catkins. Carpels more than 1, superior, consolidated. Albumen 0.

URTICALS. Males scattered. Carpels 1 only, superior.

EUPHORBIALS. Males scattered. Carpels more than 1 (3), superior, consolidated. Albumen much.

MENISPERMALS. Males scattered. Carpels more than 1, superior, disunited.

QUERNALS. Males in catkins. Carpels more than 1, inferior, consolidated. Placentæ axile.

CUCURBITALS. Males scattered. Carpels more than 1, inferior, consolidated. Placentæ parietal.

Sub-class 2. Hypogynous.

VIOLALS. Stamens definite, equal. Flowers dichlamydeous. Placentæ parietal.

(*Brassicaceæ.* Stamens definite, tetradynamous ; *see* Cistals.)

SAPINDALS. Stamens definite. Flowers dichlamydeous, unsymmetrical in the stamens. Placentæ axile.

BERBERALS. Stamens definite. Flowers dichlamydeous, unsymmetrical in the ovary.

ERICALS. Stamens definite. Flowers dichlamydeous, symmetrical. Anthers porous.

RUTALS. Stamens definite. Flowers dichlamydeous, symmetrical. Anthers slit. Petals flat.

GERANIALS. Stamens definite. Flowers dichlamydeous, symmetrical. Anthers slit. Petals twisted.

SILENALS. Stamens definite. Flowers monodichlamydeous. Carpels more than 1. Placenta free central.

CHENOPODALS. Stamens definite. Flowers monochlamydeous. Carpel only 1. Placenta free central.

PIPERALS. Stamens definite. Flowers achlamydeous.

MALVALS. Stamens 00. Flowers dichlamydeous. Calyx valvate.

---

* There are 4 Sub-classes of Exogens. 1. *Diclinous*, in which the flowers are always unisexual: 2, 3, 4, in which they are usually hermaphrodite, 2 having the stamens *Hypogynous*, 3 *Perigynous*, and 4 *Epigynous*.

CISTALS. Stamens 00. Flowers dichlamydeous. Calyx imbricated. Embryo curved or spiral, exalbuminous. Carpels consolidated.

RANALS. Stamens 00. Flowers monodichlamydeous. Calyx imbricated. Embryo minute, straight. Carpels usually disunited.

GUTTIFERALS. Stamens 00. Flowers dichlamydeous. Calyx imbricated. Placentæ axile.

Sub-class 3. Perigynous.

FICOIDALS. Polypetalous or apetalous, with an external annular embryo.

DAPHNALS. Apetalous. Carpel solitary.

RHAMNALS. Polypetalous or apetalous. Seeds definite. Carpels consolidated.

ROSALS. Polypetalous or apetalous. Seeds definite. Carpels disunited.

SAXIFRAGALS. Polypetalous or apetalous. Seeds 00.

GENTIANALS. Monopetalous. Placentæ parietal.

CORTUSALS. Monopetalous. Placentæ free central.

SOLANALS. Monopetalous. Capsular or baccate. Symmetrical. Placentæ axile.

BIGNONIALS. Monopetalous. Capsular or baccate. Unsymmetrical. Placentæ axile.

ECHIALS. Monopetalous. Nucamentaceous.

Sub-class 4. Epigynous.

CAMPANALS. Monopetalous. Styles with collecting hairs. Albumen little.

CINCHONALS. Monopetalous. Styles without collecting hairs. Albumen much.

MYRTALS. Polypetalous. Stamens indefinite. Placentæ axile.

CACTALS. Polypetalous. Stamens indefinite. Placentæ parietal.

GROSSALS. Polypetalous. Stamens definite. Seeds 00.

UMBELLALS. Polypetalous. Stamens definite. Seeds solitary or nearly so.

ASARALS. Apetalous.

THE AMENTAL ALLIANCE (V. K., p. 248.)

Natural Orders of Amentals.

Birchworts (*Betulaceæ.*) Ovary 2-celled. Ovule 1, pendulous.
Liquidambars (*Altingiaceæ.*) Ovary 2-celled. Ovules 00, winged.
Willow-worts (*Salicaceæ.*) Ovary 1-celled. Ovules 00, cottony.

Natural Order, Birchworts; *Betulaceæ* (V. K., p. 251.)

Prevailing Quality. Astringent, subacrid.

ALNUS. *Tournefort.*

Stamens 4. *Nuts* angular, wingless.

1. A. *glutinosa* Gærtner.—(The ALDER TREE.) Fig. 108.
Leaves roundish, very obtuse, wedge-shaped at the base, with the axils of the veins of the underside bearded.

Habitat. Marshy places in Europe.
Quality. Bark astringent, tonic.
Uses. Gargles, ague, &c.

BETULA. *Linnæus.*

Stamens, 10-12. *Nuts* winged.

1. B. *nigra* Linnæus. — (The BLACK BIRCH.)
Leaves ovate-rhomboid, doubly serrated, downy beneath, entire at the base; scales of female eatkins downy.

Habitat. North America.
Quality. Bark acrid, balsamic.
Uses. Yields birch camphor; timber very hard and valuable.

2. B. *alba* Linnæus. — (The COMMON BIRCH.)
Leaves somewhat rhomboid, doubly serrated, smooth; scales of female catkins hairless.

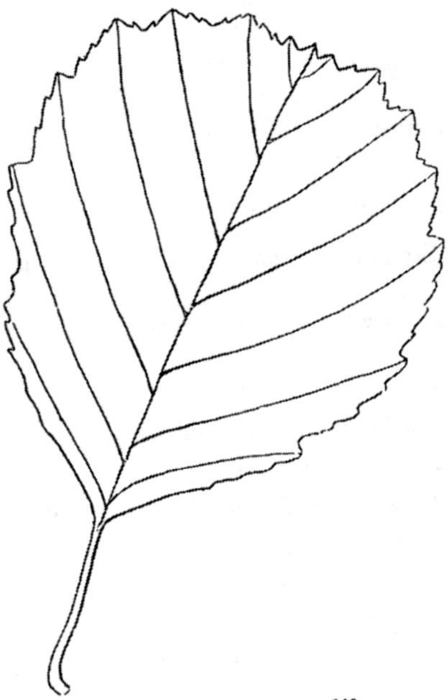

108

Habitat. North of Europe.
Quality. Bark yields the empyreumatic birch oil.
Uses. Employed in dressing Russia leather; timber of inferior quality.

Fig. 108.—Alnus glutinosa.

Natural Order, **Liquidambars**; *Altingiaceæ* (V. K., p. 253.)

Prevailing Quality. Acrid aromatic.

LIQUIDAMBAR. *Linnæus.*

Flowers monœcious. *Stamens* 00. *Female* catkins globose. *Capsules* 2-celled, many-seeded.

1. L. *styraciflua* Linnæus.—(AMERICAN LIQUIDAMBAR TREE.) Fig. 109.

Leaves palmate, lobed, with the axils of the veins of the underside bearded.

Habitat. United States and Mexico.
Quality. Balsamic, aromatic, acrid, bitter.
Uses. Yields a balsamic fluid called oil of liquidambar; and American storax.

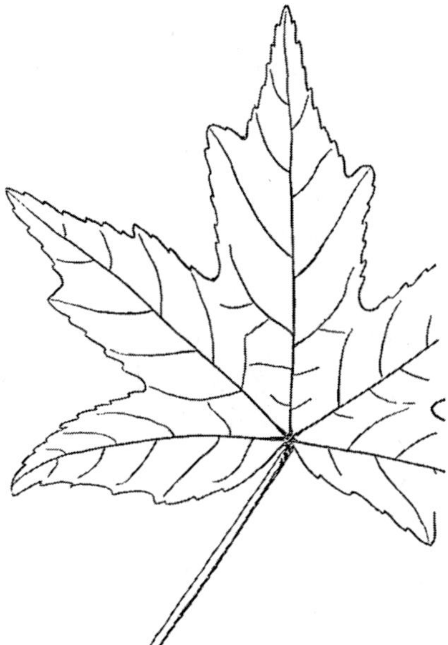

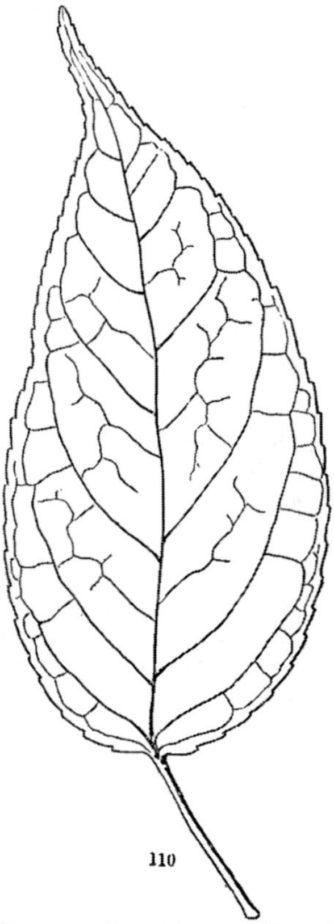

110

2. L. *orientale* Linnæus.—(ORIENTAL LIQUIDAMBAR TREE.)

Leaves palmate, lobed, with the axils of the veins of the underside hairless.

Habitat. Levant.
109 *Quality.* Bark hot, bitter, stomachic.
Uses. Yields common liquid storax, a stimulant expectorant; used in gonorrhœa, leucorrhœa, amenorrhœa, phthisis, asthma, &c.

3. L. *Altingia*, Blume. Fig. 110.

Leaves ovate, lanceolate, acuminate, serrated.

Habitat. Woods of Java.
Quality and *Uses.* Yields the fragrant stimulating liquid storax, or Rasamala of the Malay Archipelago.

Fig. 109.—Liquidambar styraciflua; 110. Liquidambar Altingia.

Natural Order, Willow-worts ; *Salicaceæ* (V. K., p. 254.)

Prevailing Quality. Tonic, astringent, aromatic.

POPULUS. *Linnæus.*

Flowers furnished with an oblique, cup-shaped calyx. *Stamens* 8 or more.

1. P. *nigra* Linnæus.—(BLACK POPLAR TREE.) Fig. 111.

Leaves smooth on each side, serrated, somewhat rhomboid, acuminate, longer than broad.

Habitat. North of Europe.
Quality. Buds aromatic, bitter, resinous ; diuretic, antispasmodic.
Uses. In ointment against tumours, wounds, and burns ; balsam and tincture against colic.

2. P. *balsamifera* Linnæus.—(BALSAM POPLAR.)

Leaves ovate, acuminate, with close serratures, white and netted beneath ; buds very resinous.

Habitat. United States.
Quality and *Uses.* As in the last ; Tacamahac obtained from the buds.

SALIX. *Linnæus.*

Flowers absolutely naked. *Stamens* 1-5.

1. S. *pentandra* Linnæus.—(SWEET WILLOW.) Fig. 112.

Stamens 5-10 ; scales of catkins whole-coloured, deciduous ; leaves ovate-oblong, closely serrated, very smooth, with ovate-oblong equilateral straight stipules.

111

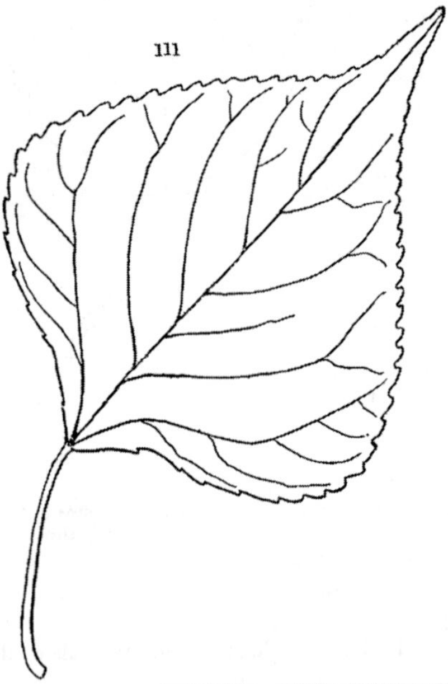

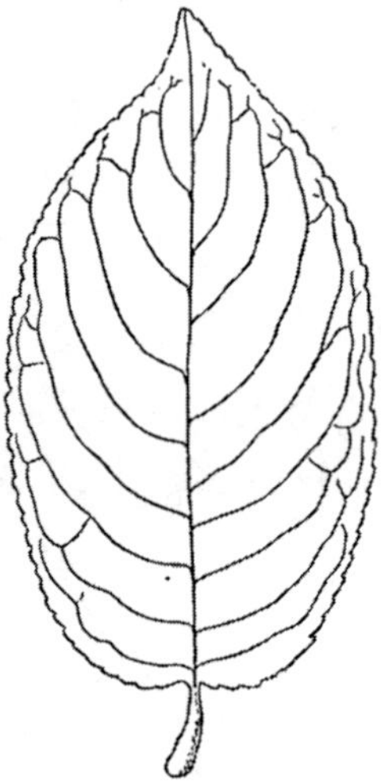

112

Habitat. Woods of Europe.
Quality. Bark bitter, astringent, tonic, febrifugal.
Uses. Dyspepsia, intermittents, &c. The most aromatic of the Willows.

Fig. 111.—Populus nigra ; 112. Salix pentandra.

2. S. *Russelliana* Smith.—(Bedford Willow.) Fig. 113.
Stamens 2; scales of catkins whole coloured, deciduous; the hypogynous gland much shorter than its stalk; leaves lanceolate, acuminate, smooth, silky only when young, serrated.

Habitat. Woods and meadows of England.
Quality and *Uses.* As in No. 1. By some regarded as the best medicinal Willow.

113

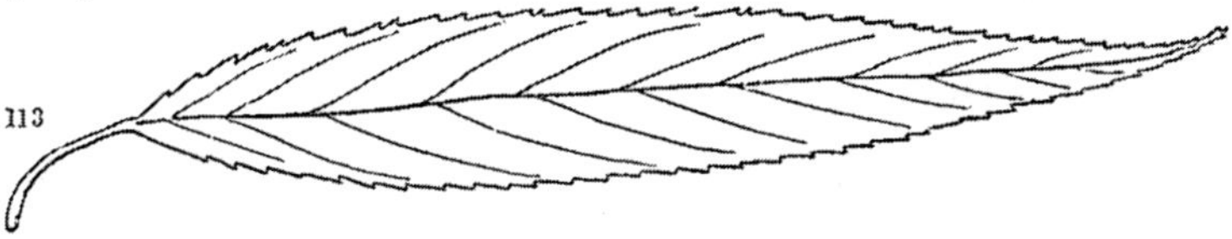

a

115

114

3. S. *vitellina* Linnæus.—(Golden Willow). Fig. 114.
Stamens 2; scales of catkin whole coloured, deciduous; hypogynous gland very short, yet as long as its stalk; leaves lanceolate, acuminate, serrulate, silky on both sides; branches bright yellow.

Habitat. Meadows of Europe.
Quality and *Uses.* As in No. 1, but weaker.

4. S. *purpurea* Linnæus. S. *Helix* Linn. a variety. Fig. 115.
Stamen 1; gland longer than the base of the ovary; leaves lanceolate, finely serrulate, smooth, flat.

Habitat. Meadows of Europe.
Quality and *Uses.* As in No. 1; its bark intensely bitter.

Fig. 113.—Salix Russelliana; 114. Diminished figure of Salix vitellina; *a*, the under side of a leaf natural size; 115. Salix purpurea.

THE URTICAL ALLIANCE (V. K., p. 258.)

Natural Orders of Urticals.

Nettleworts (*Urticaceæ.*) Ovule erect. Embryo straight. Juice watery.

Hempworts (*Cannabinaceæ.*) Ovule suspended. Juice watery.

Morads (*Moraceæ.*) Ovule suspended. Juice milky.

Artocarpads (*Artocarpaceæ.*) Ovule suspended. Embryo straight. Juice milky.

Natural Order, **Nettleworts**; *Urticaceæ* (V. K., p. 260.)

Prevailing Quality. Acrid; narcotic.

URTICA. *Linnæus.*

Male: *Calyx* 4-parted. *Stamens* 4, elastic. Female: *Calyx* 2-parted. *Stigma* sessile, capitate, pencilled.

1. U. *dioica* Linnæus.—(LARGER STINGING NETTLE.)

Leaves coarsely serrated; panicles axillary, longer than the petioles.

Habitat. Waste places.
Quality. Poisonous, acrid; astringent, diuretic.
Uses. Young shoots in broth; flogging with nettles in arthritis and paralysis.

PARIETARIA. *Linnæus.*

Calyx of both sexes 4-parted. *Style* filiform. *Stigma* capitate, pencilled.

1. P. *officinalis* Linnæus.—(WALL PELLITORY.)

Leaves ovate; stems prostrate, spreading, branched.

Habitat. Old walls all over Europe.
Quality. Diuretic, lithontriptic.
Uses. In calculous and urinary affections; in dropsies.

Natural Order, **Hempworts**; *Cannabinaceæ* (V. K., p. 265.)

Prevailing Quality. Narcotic.

CANNABIS. *Linnæus.*

Flowers diœcious. Male: *Calyx* 5-parted. *Stamens* 5. Female: *Calyx* 1-leaved, rolled up. *Styles* 2.

1. C. *sativa* Linnæus.—(HEMP.) Fig. 116.

Leaves digitate, serrated; flowers axillary.

Habitat. India and Persia.
Quality. Stimulant, narcotic; allays pain; excites appetite; a certain aphrodisiac; produces catalepsy; causes delirium.
Uses. Rheumatism, tetanus, hydrophobia, cholera.

Humulus. *Linnæus.*

Flowers diœcious. Male: *Calyx* 5-parted. *Stamens* 5. Female: *Flowers* in cones. *Calyx* scale-like, partially rolled up.

1. H. *Lupulus* Linnæus.—(The Hop.) Fig. 117.

A twiner; leaves undivided, coarsely serrated, with harsh hairs.

Habitat. Hedges of Europe.
Quality. Narcotic; lupuline, aromatic, tonic; sedative.
Uses. Pillows of hops in mania and restlessness; dyspepsia; in the preparation of malt liquor.

116 ♂

a

117

♀

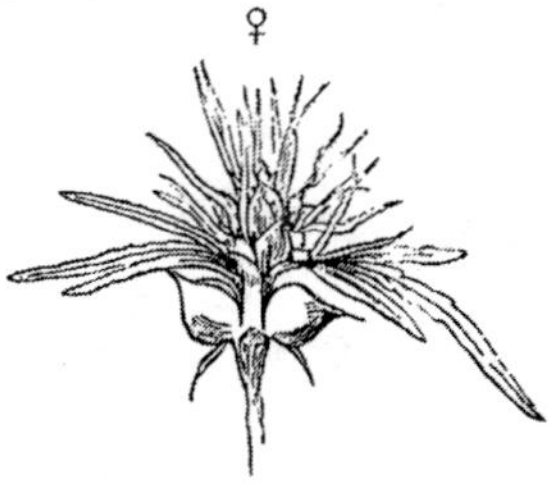

b

Natural Order, Morads; *Moraceæ* (V. K., p. 266.)

Prevailing Quality. Acrid; narcotic; with elastic gum.

Ficus. *Linnæus.*

Flowers within a closed turbinate fleshy receptacle.

1. F. *elastica* Roxburgh.—(Bengal India-rubber Tree.)

Leaves stalked, oblong, acute, glossy, with numerous fine diverging veins and a stout midrib; fruit not eatable.

Habitat. Forests of Sylhet.
Quality. Milky juice acrid, tenacious.
Uses. Forms a large part of the caoutchouc exported from Bengal.

Fig. 116.—*a*, Male inflorescence of Cannabis sativa; *b*, female ditto; 117. Humulus Lupulus in fruit.

2. F. *Carica* Linnæus.—(COMMON FIG.)

Leaves cordate, palmate, scabrous on the upper side, downy on the lower.

Habitat. Levant.

Quality. Fruit nutritive, emollient, demulcent, laxative; apt to disorder the stomach; when young, acrid.

Uses. Food; heated and split open for gumboil, &c.; added to barley water in pulmonary and nephritic affections; forms part of the confection of senna.

MORUS. *Linnæus.*

Calyx 4-parted. *Stamens* 4. *Styles* 2. *Fruit*, a spike, composed of the succulent conglomerated axis, calyxes and carpels.

1. M. *nigra* Linnæus.—(The MULBERRY TREE.)

Leaves cordate, ovate, undivided or lobed, serrated, rough; fruit sessile, purple.

Habitat. Persia.

Quality. Fruits alimentary; allay thirst; diminish febrile heat; laxative.

Uses. A common dessert fruit; as a colouring substance.

DORSTENIA. *Linnæus.*

Flowers monœcious, naked, plunged in sockets of a plane receptacle. *Stamens* 2. *Carpels* becoming dry loose achænia.

1. D. *Contrayerva* Linnæus.—(CONTRAYERVA.) Fig. 118.

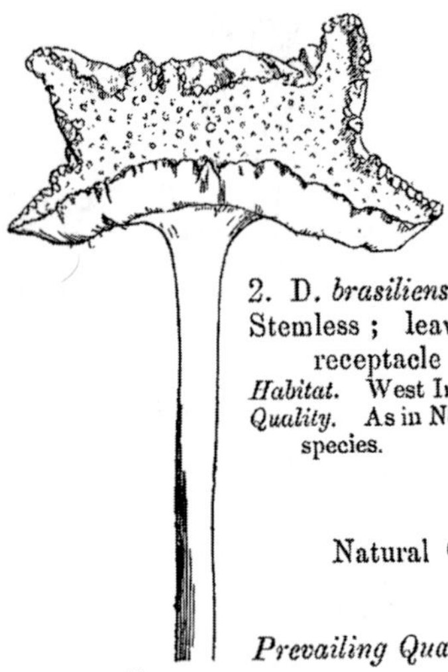

118

Caulescent; leaves palmate, with deeply serrated or almost pinnatifid acuminate lobes; receptacle somewhat quadrangular.

Habitat. Tropical America.

Quality. Rhizome stimulant, tonic, diaphoretic; emetic; keeps badly.

Uses. Low fevers, and where mild stimulants are required.

2. D. *brasiliensis* Lamarck.

Stemless; leaves cordate, oblong, obtuse, serrated; receptacle circular, crenated.

Habitat. West Indies and Brazil.

Quality. As in No. 1; also emetic. Said to be the most energetic species.

Natural Order, Artocarpads; *Artocarpaceæ*
(V. K., p. 269.)

Prevailing Quality. Acrid; narcotic.

ARTOCARPUS. *Linnæus.*

Male Flowers in catkins, with 1 stamen and 2 sepals. *Females* naked, becoming a roundish fleshy tuberculated fruit.

Fig. 118.—Receptacle of Dorstenia Contrayerva.

1. A. *incisa* Linnæus.—(The BREAD-FRUIT TREE.)

Leaves pinnatifid, sinuated, scabrous, downy on the under side; male catkins nodding.

Habitat. Islands of the Pacific, and Indian Archipelago.

Uses. The large fruit nutritious when sliced and dried; filled with a tenacious white milk before becoming ripe.

ANTIARIS. *Leschenault.*

Males on a mushroom-like receptacle, with 3 or 4 sepals, and as many sessile anthers. *Females* solitary, becoming fleshy drupes.

1. A. *toxicaria* Leschenault.—(The UPAS TREE.) Fig. 119.

Leaves oval-oblong, acute, hairy on both sides, especially on the main veins, slightly serrated; male receptacles stalked.

Habitat. East Indies.

Quality. Juice an acrid poison; emetic; causes convulsions.

Uses. Poisons weapons; its fibre woven into coarse linen.

N.B.—The fables current concerning the action of the plant are a mixture of truth connected with distinct natural phenomena in Java, and the real properties of the plant. That its emanations are occasionally noxious is an undoubted fact, though excessively exaggerated.

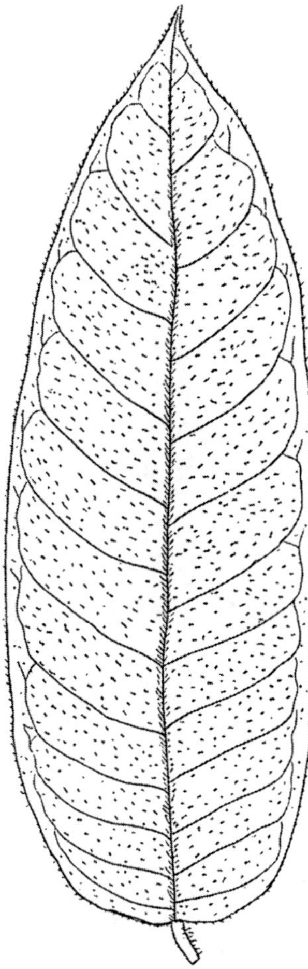

119

THE EUPHORBIAL ALLIANCE
(V. K., p. 272.)

Natural Order, Spurgeworts; *Euphorbiaceæ* (V. K., p. 274.)

Prevailing Quality. Acrid; emetic.

EUPHORBIA. *Linnæus.*

Monœcious. Flowers naked; *males* monandrous, surrounding a 3-coccous stalked female; the whole placed within a cup-shaped involucre.

Fig. 119.—Leaf of Antiaris toxicaria.

1. E. *officinarum* Linnæus.—(EUPHORBIUM BUSH.) Fig 120.

Leafless; stems succulent, tufted, prickly, with many angles; prickles in pairs, thick and strong.

Habitat. West of Barbary.
Quality. Resin violently acrid, narcotic, emetic, drastic.
Uses. Dropsy; as an errhine in chronic affections of the ears, eyes, or brain. Dangerous; mixed with cantharides, it forms "gout plaister."

120

2. E. *antiquorum* Linnæus.—(EUPHORBIUM BUSH.)

Leafless; branches succulent, spreading, triangular, or quadrangular, with sinuated angles; prickles in pairs.

Habitat. West of Barbary.
Quality. Resin violently acrid, narcotic, emetic, drastic.
Uses. Dropsy; as an errhine in chronic affections of the ears, eyes, or brain. Dangerous; mixed with cantharides, it forms "gout plaister."

3. E. *hiberna* Linnæus.—(WINTER SPURGE.)

Bracts and leaves ovate or elliptical, entire, obtuse; glands of involucre reniform; capsule muricated; seeds smooth, somewhat shining.

Habitat. Ireland, and south-west of England.
Uses. Root in venereal diseases. A fish-poison.

4. E. *Peplus* Linnæus.—(PETTY SPURGE.)

Leaves broadly ovate, somewhat emarginate, stalked; glands of involucre with long horns; capsule smooth, with a double keel at each angle; seeds pitted.

Habitat. A common weed everywhere.
Uses. Dried herb once used as a powerful purgative.

5. E. *Cyparissias* Linnæus.—(CYPRESS SPURGE.)

Leaves linear, entire, smooth; glands of involucre with 2 horns; capsule dotted and rough at the angles; seeds smooth.

Habitat. Woods of Europe.
Quality. A virulent poison.
Uses. An unsafe purgative.

6. E. *Gerardiana* Jacquin.

Leaves glaucous, linear, mucronate, entire, smooth; glands of involucre undivided; capsule smooth or nearly so; seeds smooth.

Habitat. Europe, in fields and by roadsides.
Quality. Root cathartic and emetic.
Uses. Said to be the best of the European Euphorbias.

7. E. *Ipecacuanha* Linnæus.—(AMERICAN IPECACUANHA.)

Roots deep, large and fleshy; leaves opposite, sessile, oblong, smooth; involucres solitary, axillary, on long stalks, with 5 blunt segments and 5 intermediate glands.

Habitat. Sandy places in North America.
Quality. Emetic, purgative, diaphoretic; apt to produce hypercatharsis.

Fig. 120.—A diminished fig e of Euphorbia officinarum.

S. E. *Lathyris* Linnæus.—(Caper Spurge.) Fig. 121.

Leaves opposite, decussate, oblong, sessile, the uppermost cordate; glands of involucre with 2 horns; capsules wrinkled, spongy; seeds articulated, wrinkled.

Habitat. Europe, in woods; common in cottagers' gardens.
Quality. Violently acrid, narcotic.
Uses. Oil of seeds a substitute for Croton oil. Bark of root also employed.

121

Buxus. *Linnæus.*

Flowers monœcious. *Males:* with about 4 unequal membranous sepals. *Stamens* 4. *Females:* central, of several imbricated scales. *Capsule* 3-horned.

1. B. *sempervirens* Linnæus. —(The Box Tree.) Fig. 122.

Leaves evergreen, roundish, smooth, with a separable inferior epidermis; petioles and young branches slightly downy.

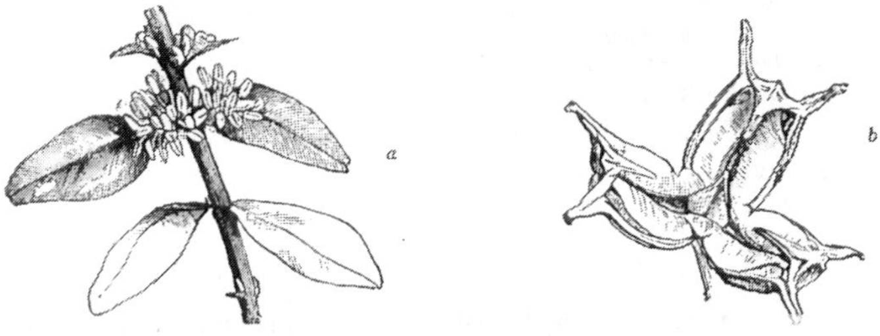

122

Habitat. Chalk-hills of Europe.
Quality. Bitter, nauseous, sudorific, purgative, acrid.
Uses. The empyreumatic oil and chips of wood in syphilis and chronic rheumatism; the first against toothache.

Mercurialis. *Linnæus.*

Sepals 3-4. *Stamens* 8 or more, with distinct filaments, and separate roundish anther-lobes. *Styles* 2. *Fruit* 2-celled, with solitary seeds.

Fig. 121.—Euphorbia Lathyris in fruit; 122. *a*, Buxus sempervirens; a branch in flower; *b*, the ripe fruit split open.

1. M. *perennis* Linnæus.—(Dog Mercury.) Fig. 123.

Leaves ovate-oblong or lanceolate; female flowers on long stalks.

123

Habitat. Waste places and plantations.
Quality. Very poisonous.
Uses. A dangerous emetic.

2. M. *annua* Linnæus.

Leaves ovate-lanceolate or ovate; female flowers nearly sessile.

Habitat. Waste places among rubbish.
Quality and *Uses.* As in the last, but more mild.

Janipha. *Kunth.*

Calyx campanulate, 5-parted. *Stamens* 10, distinct, alternately shorter. *Stigmas* 3, many-lobed. *Fruit* 3-celled, with solitary seeds.

1. J. *Manihot* Kunth. *Jatropha Manihot* Linnæus. *Manihot utilissima* Pohl.—(Mandioc Plant.) Fig. 124.

Root large, tuberous; leaves stalked, palmate, with lanceolate acuminate entire segments, glaucous beneath.

Habitat. Brazil.
Quality. Recent juice narcotic, acrid; fæcula nutritious, emollient, demulcent.
Uses. The fæcula forms cassava, tapioca, light digestible substances.

Croton. *Linnæus.*

Calyx 5-parted. *Petals* 5 in the males. *Stamens* 10 or more, distinct.

Fig. 123.—Mercurialis perennis; *a*, ♀; *b*, ♂; *c*, male flower magnified; *d*, female flower ditto; *e*, perpendicular section of ditto; *f*, ripe fruit.

Styles 3, forked or many-parted. *Capsules* 3-coccous, with one seed in each cell.

1. C. *Eleuteria* Swartz.—(CASCARILLA BUSH.) Fig. 125.

A small tree; leaves ovate, obtuse, entire, beneath silvery and densely downy; racemes axillary and terminal, compound; flowers subsessile, monœcious.

Habitat. Bahamas.
Quality. Bark aromatic, bitter, tonic; narcotic?
Uses. Dyspepsia, low fevers, intermittents, diarrhœa, dysentery.

2. C. *pseudo-China* Schlechtendahl.—(COPALCHE BUSH.)

Leaves cordate-ovate, acuminate, 3-5-nerved, nearly entire, silvery on the under side; racemes axillary, simple, monœcious.

Habitat. Mexico.
Quality and *Uses.* As in No. 1.

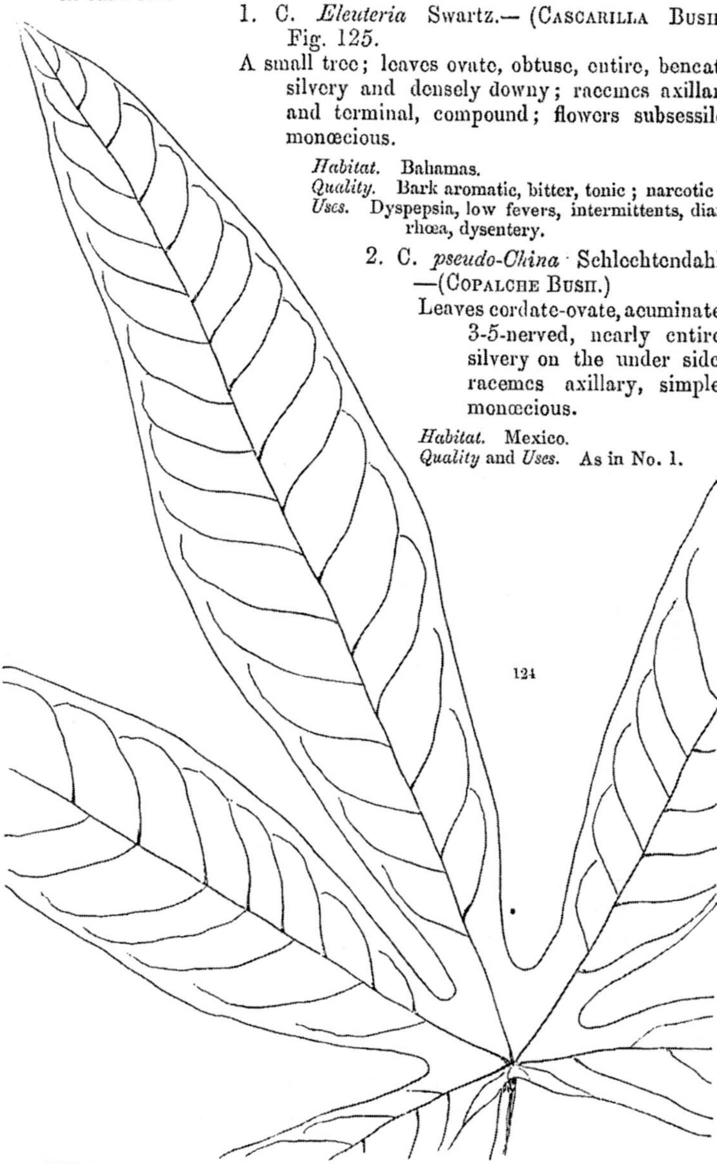

124

Fig. 124.—Leaf of Janipha Manihot.

3. C. *Tiglium* Lamarck. Fig. 126.

A small tree; leaves oval-oblong, acute, 3-5-nerved, acuminate, with shallow glandular serratures, thin and membranous, with 2 glands at the base.

Habitat. East Indies.
Quality. Oil of seeds a powerful irritant, drastic, cathartic; poisonous.
Uses. Mania, obstinate constipation, stercoraceous vomiting, paralysis, hydrocephalus.

125

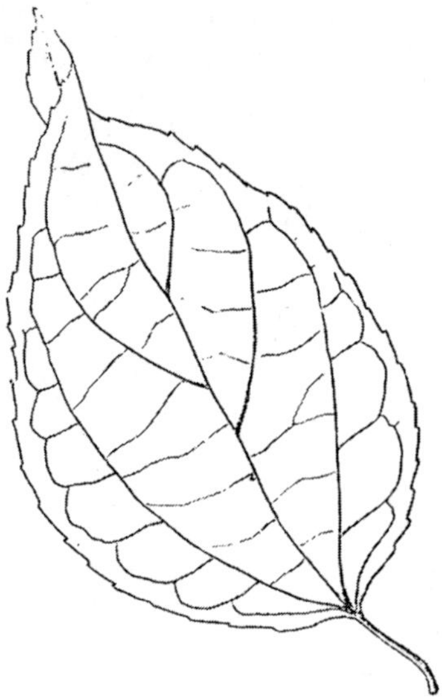

126

4. C. *Pavana* Hamilton.—(TILLY SEED.)

Leaves ovate, obtuse at the base, smooth, with all the veins alternate.

Habitat. Indian Archipelago, Ava, &c.
Quality and *Uses.* As in C. Tiglium.

5. C. *Draco* Schlechtendahl.—(MEXICAN DRAGON'S BLOOD.)

All covered with starry coarse hairs; leaves cordate, acuminate, minutely toothed, with 4 glands at the apex of the petiole; racemes very long, interrupted.

Habitat. Mexico.
Quality. Juice hardens into a kind of Dragon's blood, called Sangre del drago.
Uses. A vulnerary and astringent in Mexico.

6. C. *balsamiferum* Linnæus.

A rusty, downy, resinous, balsamic shrub; leaves ovate-lanceolate, obtuse, with two cup-shaped glands at the base; spikes terminal compact.

Habitat. West Indies.
Uses. A spirituous liquor, called Eau de Mantes, used in irregular menstruation, is distilled from it.

Fig. 125.—Croton Eleuteria; 126. Leaf of Croton Tiglium.

Crozophora. *Necker.*

Calyx 5-parted, and *Petals* 5 in the males. *Stamens* 5-10, unequal, connate. *Calyx* 10-parted, and *Petals* 0 in the females. *Styles* 3, forked. *Capsule* tricoccous, with 1 seed in each cell.

1. C. *tinctoria* Necker. *Croton tinctorium* Linnæus.—(Turnsole.) Fig. 127.

A hoary annual; leaves ovate-rhomboidal, repand, toothed, and crisp at the edge, about the same length as their stalk.

Habitat. Coast of the Mediterranean.
Quality. Juice rendered blue by ammonia and air.
Uses. Linen dipped in it a test for acids.

127

Ricinus. *Linnæus.*

Calyx 3-5-parted. *Petals* 0. *Stamens* 00, with the filaments irregularly united into branches. *Style* 1. *Stigma* forked, feathery. *Capsule* 3-coccous, with 1 seed in each cell.

1. R. *communis* Linnæus.—(Castor Oil Plant.) Fig. 128.

Stem glaucous; leaves peltate, deeply divided into 7 ovate, serrated, acuminate segments; flowers in long glaucous racemes.

Habitat. East Indies.
Quality. Oil of seeds purgative.
Uses. Inflammation of the bowels, obstructions, worms, &c., wherever a brisk purgative is demanded.

a *b*

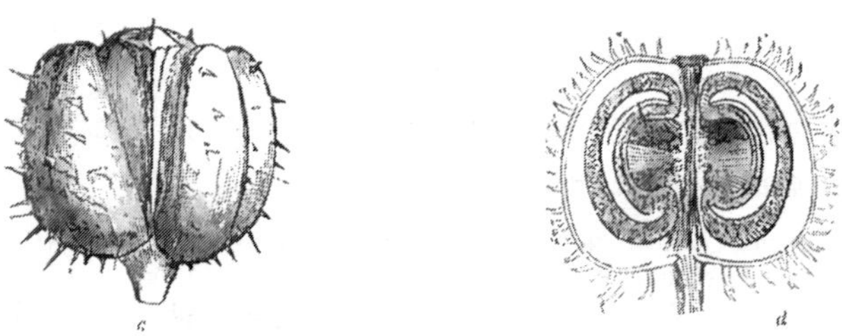

c *d*

128

Fig. 127.—Leaf of Crozophora tinctoria ; 128. *a*, Male flower of Ricinus communis ; *b*, female ditto ; *c*, ripe fruit ; *d*, the same, cut perpendicularly.

CURCAS. *Adanson.*

129

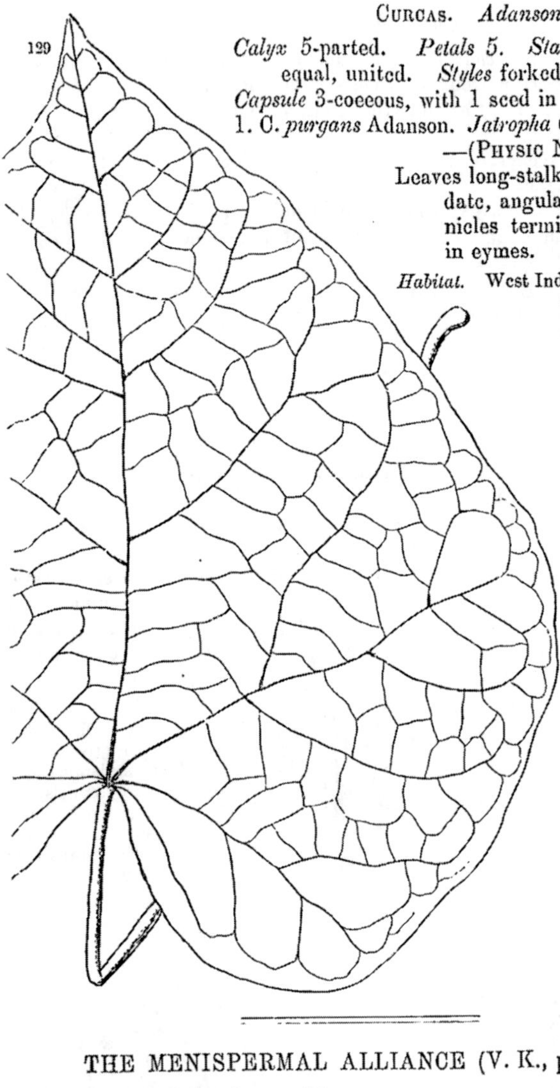

Calyx 5-parted. *Petals* 5. *Stamens* 8-10, unequal, united. *Styles* forked.

Capsule 3-coccous, with 1 seed in each cell.

1. C. *purgans* Adanson. *Jatropha Curcas* Linnæus. —(PHYSIC NUT.) Fig. 129.

Leaves long-stalked, broadly cordate, angular, roundish; panicles terminal or axillary, in cymes.

Habitat. West Indies.

Quality and *Uses.* Like C. Tiglium; in large doses a dangerous poison. Acrid oil in chronic rheumatism. Milky juice dyes linen black.

THE MENISPERMAL ALLIANCE (V. K., p. 297.)

Natural Orders of Menispermals.

Nutmegs (*Myristicaceæ.*) Albumen ruminated. Sepals united into a valvate cup.

Menispermads (*Menispermaceæ.*) Albumen solid. Sepals distinct, imbricated.

Fig. 129.—Leaf of Curcas purgans.

Natural Order, **Nutmegs**; *Myristicaceæ* (V. K., p. 301.)

Prevailing Quality. Aroma, acridity.

MYRISTICA. *Linnæus.*

Calyx 3-toothed. *Anthers* 6-10, connate. *Stigma* sessile. *Seed* surrounded by an aril, within a fleshy 2-valved pericarp.

1. M. *moschata* Thunberg.—(NUTMEG TREE.)

Leaves oblong, acuminate, smooth, with simple veins; fruit solitary, smooth.

Habitat. Moluccas.
Quality. Seed aromatic, acrid, narcotic, oily.
Uses. As spice; the seed is the Nutmeg; its arillus is Mace.

2. M. *fatua* Swartz.

Leaves oblong-lanceolate, covered with stellate hairs on the under side; veins simple; fruit racemose, downy.

Habitat. West Indies and Surinam.
Quality and *Uses.* Seeds acrid, purgative, oily; produce nausea, fulness, and flatus.

Natural Order, **Menispermads**; *Menispermaceæ* (V. K., p. 307.)

Prevailing Quality. Bitter; narcotic.

CISSAMPELOS. *Linnæus.*

Males: Sepals 8, the inner formed into a cup. *Stamens* 2, monadelphous; *anthers* dehiscing horizontally. *Females: Sepals* 2. *Ovary* 1.

1. C. *Pareira* Linnæus.—(PAREIRA BRAVA. VELVET LEAF.) Fig. 130.

Leaves orbicular, peltate, aristate, when full grown covered with silky down; fruit scarlet, round, hairy.

130

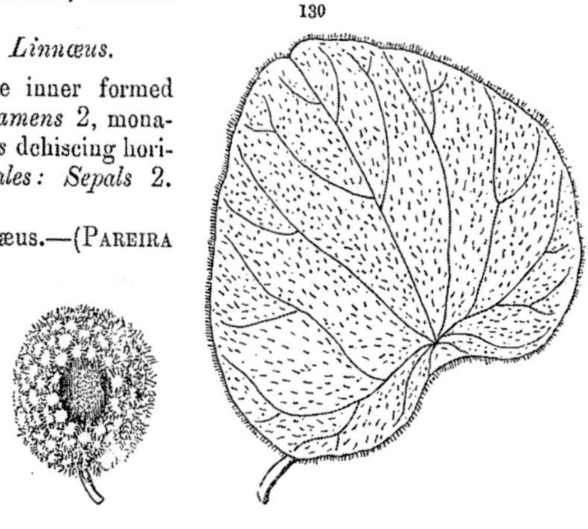

Habitat. Brazil.
Quality. Root tonic, diuretic.
Uses. Gonorrhœa, leucorrhœa, chronic inflammation of the bladder.

COCCULUS. *De Candolle.*

Sepals 12, in 4 rows. *Stamens* 6; *anthers* dehiscing vertically. *Ovaries* 3 or more.

1. C. *palmatus* De Candolle.—(CALUMBA.)

Root large, fleshy; leaves circular, palmate, hairy, with 5 to 7 entire lobes; their stalk covered with glandular hairs.

Fig. 130.—Cissampelos Pareira; *a*, leaf; *b*, ripe fruit

Habitat. Mozambique.
Quality. Tonic, not stimulant, demulcent, stomachic.
Uses. Dyspepsia, diarrhœa, dysentery, to allay vomiting.

ANAMIRTA. *Colebrooke.*

Sepals 6. *Stamens* monadelphous ; *anthers* 00, forming a globose head. *Drupes* 1-3.

1. A. *Cocculus* Colebrooke.—(COCCULUS INDICUS.)

Leaves roundish, acute, hard, leathery, shining, smooth, with 5 radiating ribs ; fruit globose.

Habitat. East Indies.
Quality. Poisonous, acrid, intoxicating.
Uses. Powdered seeds against pediculi ; porrigo ; in the adulteration of malt liquor.

131

THE QUERNAL ALLIANCE (V. K., p. 289.)

Natural Orders of Quernals.

Mastworts (*Corylaceæ.*) Ovary 2 or more celled. Ovules pendulous.

Juglands (*Juglandaceæ.*) Ovary 1-celled. Ovule erect.

Natural Order, **Mastworts**; *Corylaceæ* (V. K., p. 290.)

Prevailing Quality. Astringent, tonic.

QUERCUS. *Linnæus.*

Ovary 3-celled. *Fruit* in a scaly truncated cup : acorn round.

1. Q. *Suber* Linnæus.—(CORK OAK.) Fig. 131.

Bark corky; leaves evergreen, hard, oblong, hoary beneath.

Habitat. South-west of Europe.
Uses. The bark forms the Cork of commerce.

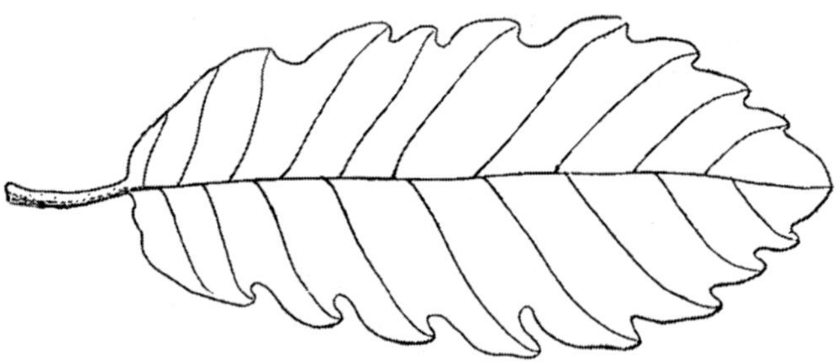

132

2. Q. *infectoria* Linnæus.—(NUTGALL OAK.) Fig. 132.

Leaves evergreen, oblong, coarsely and bluntly serrated, mucronate, smooth on both sides.

Fig. 131.—Under side of leaf of Quercus Suber; 132. Leaf of Q. infectoria.

Habitat. Levant.
Quality. Galls excessively astringent.
Uses. Hemorrhages, old diarrhœas, antidote to poisons, gargles, gleet, leucorrhœa; in the preparation of ink.

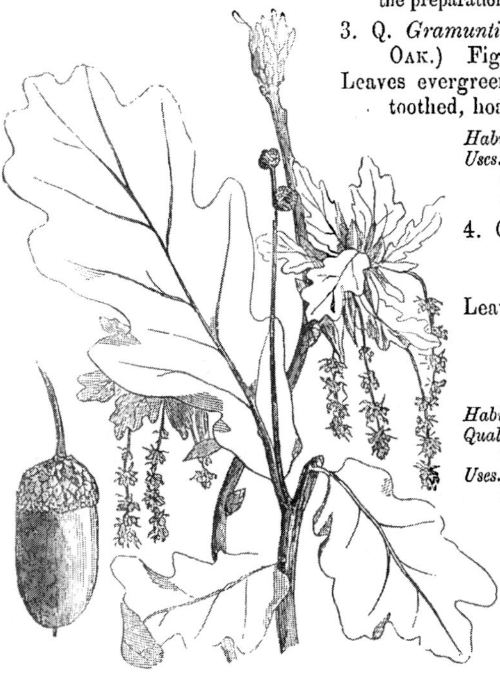

133

3. Q. *Gramuntia* Linnæus.—(Belote Oak.) Fig. 134.

Leaves evergreen, hard, round, spiny-toothed, hoary beneath.

Habitat. Spain.
Uses. Acorns sweet, eatable, used as food in Spain under the name of Belotes.

4. Q. *pedunculata* Willdenow.—(Long-stalked Oak.) Fig. 133.

Leaves deciduous, obovate-oblong, sinuated, short-stalked; acorns long-stalked.

Habitat. North of Europe.
Quality. Bark very astringent, tonic; abounds in tannin.
Uses. Decoction in gargles; a wash in ulcers; an injection in leucorrhœa, &c.; an astringent in diarrhœas; a poultice in mortification. Timber invaluable.

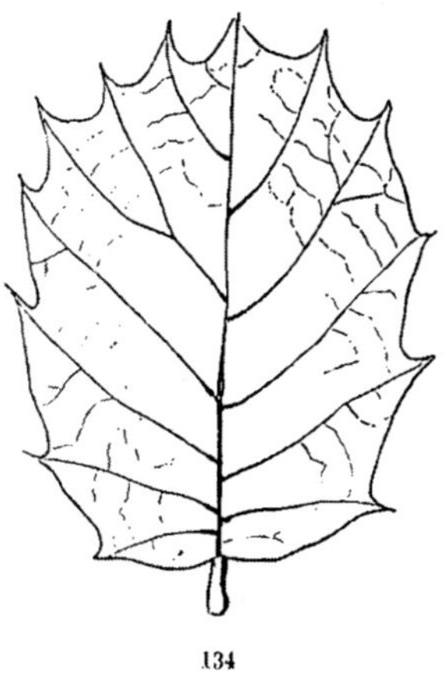

134

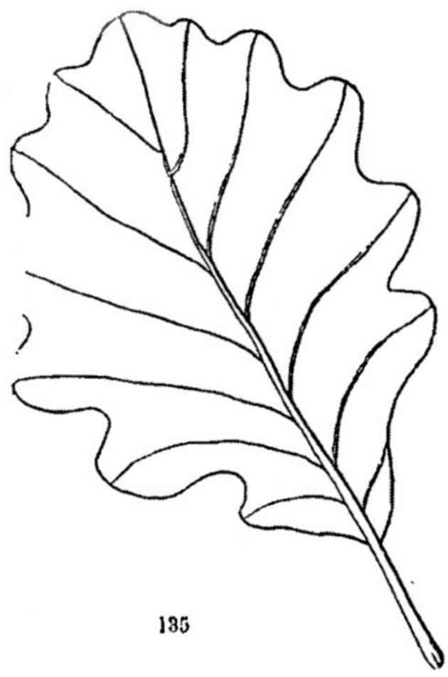

135

Fig. 133.—Q. pedunculata in flower and fruit; 134. Leaf of Q. Gramuntia; 135. Q. sessiliflora.

5. Q. *sessiliflora* Smith.—(SHORT-STALKED OAK.) Fig. 135.
Leaves deciduous, obovate, sinuated, long-stalked ; acorns nearly sessile.

Habitat. Europe, especially the central and southern parts.
Quality and *Uses.* As in the last. Acorns sometimes sweet and eatable, like Chesnuts.

FAGUS. *Linnæus.*

Ovary 3-celled. *Fruit* in a prickly husk ; *mast* triangular.
1. F. *sylvatica* Linnæus.—(The BEECH TREE.) Fig. 136.
Leaves ovate, smooth, toothed, slightly ciliated.

136

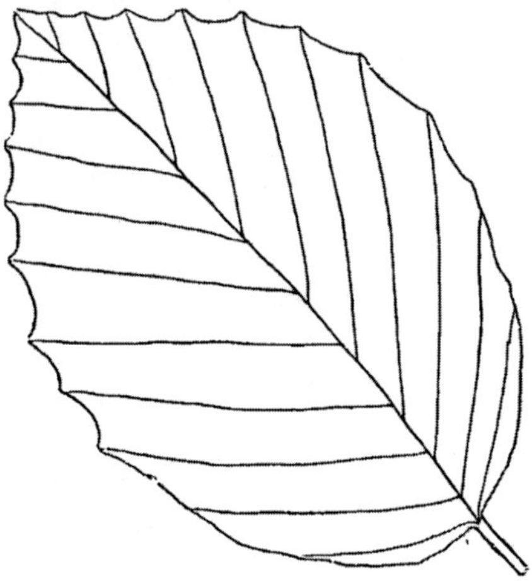

137

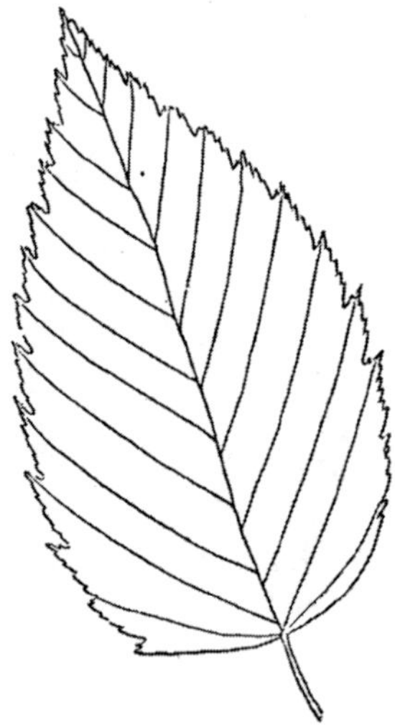

Habitat. Europe.
Quality. Husks narcotic ? Nuts eatable.
Uses. Oil in abundance in the Nuts. Timber hard, but not durable.

CARPINUS. *Linnæus.*

Ovary 2-celled. *Fruit* within flat leafy bracts ; *nut* small, striated.
1. C. *Betulus* Linnæus.—(The HORNBEAM.) Fig. 137.
Leaves ovate, doubly serrated ; bracts 3-lobed.

Habitat. Woods of Europe.
Quality. Tonic.
Uses. Bark a bad febrifuge. Timber coarse and of little value ?

CORYLUS. *Linnæus.*

Ovary 2-celled. *Fruit* rolled in a leafy lobed unarmed husk ; *nut* round.
1. C. *Avellana* Linnæus.—(HAZEL NUT.)
Leaves roundish, cordate, acuminate ; husks campanulate, spreading and lacerated at the end.

Habitat. Woods of Europe.
Uses. Nuts sweet, eatable.

Fig. 136.—Leaf of Fagus sylvatica ; 137. Ditto of Carpinus Betulus.

CASTANEA. *Tournefort.*

Ovary 5-8-celled. *Fruit* within a spiny husk ; *nuts* compressed, rounded.

1. C. *vesca* Tournefort.—(SWEET CHESNUT.) Fig. 138.

Leaves oblong-lanceolate, acuminate, mucronate-serrated, shining above, downy or smooth beneath.

Habitat. South of Europe.
Uses. Nuts sweet, eatable. Timber useful.

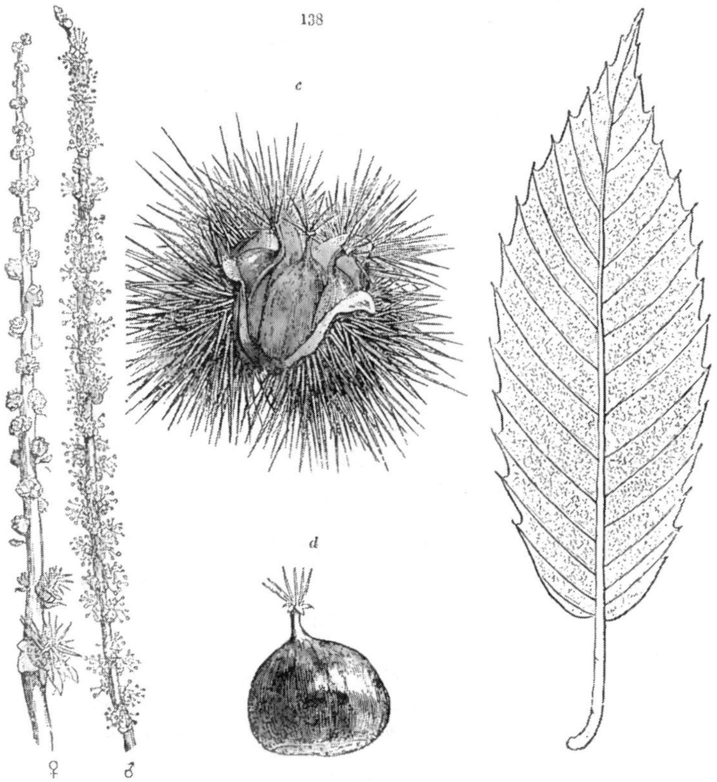

Natural Order, 𝔍𝔲𝔤𝔩𝔞𝔫𝔡𝔰 ; *Juglandaceæ* (V. K., p. 292.)

Prevailing Quality. Acridity, aroma.

JUGLANS. *Linnæus.*

Flowers monœcious. *Stamens* 18-24. *Drupe* with a 2-valved deciduous sarcocarp, or rind ; and a deeply wrinkled putamen or shell.

Fig. 138.—*a*, ♂ and ♀, inflorescence of Castanea vesca ; *b*, under side of leaf ; *c*, cupule or husk opening ; *d*, nut.

1. J. *regia* Linnæus.—(Walnut Tree.) Fig. 139.

A large tree; leaves pinnated, smooth, with about 9 oval nearly entire aromatic leaflets.

Habitat. Persia.
Quality. Sub-acrid, laxative.
Uses. Nuts abound in sweet drying oil; bark of root purgative, as is young fruit when preserved with sugar. Timber strong and durable; rising sap employed in pulmonary affections and in general debility.

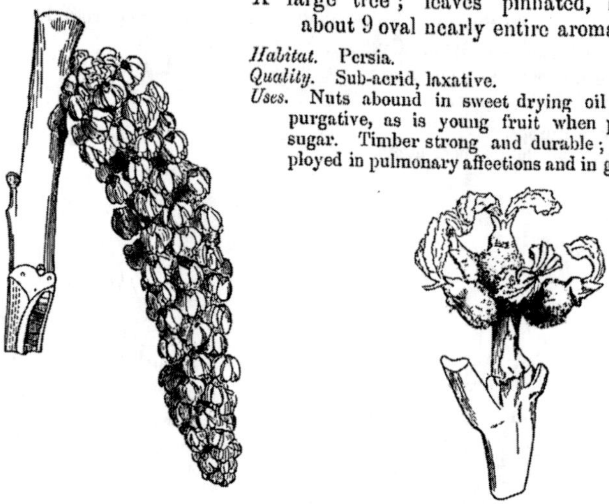

a 139 b

THE CUCURBITAL ALLIANCE (V. K., p. 310.)

Natural Order, **Cucurbits**; *Cucurbitaceæ* (V. K., p. 311.)

Prevailing Quality. Narcotic, purgative.

Lagenaria. *Seringe.*

Tendrils lateral. *Stamens* triadelphous. *Anthers* distinct. *Style* undivided. *Stigmas* 3. *Seeds* thick-edged, 2-lobed.

1. L. *vulgaris* Seringe. *Cucurbita lagenaria* Linnæus.—(Trumpet Gourd. Bottle Gourd.)

Covered with soft down; stem climbing; leaves cordate, nearly entire, somewhat glaucous, with 2 glands at the base; flowers monœcious; fruit downy, smooth when ripe, long, and swollen at one end.

Habitat. East Indies.
Quality. A dangerous purgative; poisonous.
Uses. The ripe fruit hollowed out, cleared of the pulp, and deprived of all soluble matter by washing, is used as bottles.

Cucumis. *Linnæus.*

Tendrils lateral. *Stamens* triadelphous. *Anthers* distinct. *Style* undivided. *Stigmas* 3. *Seeds* thin-edged.

1. C. *Melo* Linnæus.—(The Melon.)

Leaves rounded, angular; anthers shorter than the connective; fruit ovate or roundish, downy when young, furrowed, with a sweet pulp.

Habitat. Persia.
Quality and *Uses.* Laxative, refrigerant; a grateful and delicious fruit.

Fig. 139.—*a*, Male flowers of Juglans regia; *b*, female flowers.

2. C. *Colocynthis* Linnæus.—(COLOCYNTH GOURD.) Fig. 140.
Leaves cordate-ovate, multifid, hoary with hairs on the under side, the lobes somewhat acute ; fruit globose, whole-coloured, with an intensely bitter pulp.

140

141

Habitat. Syria and India.
Quality. Bitter, acrid, emetic, purgative, drastic cathartic, hydragogue.
Uses. Constipation, dropsy, alvine obstructions, amenorrhœa.

3. C. *pseudo-colocynthis* Royle.—(HIMALAYAN COLOCYNTH.) Fig. 141.
Leaves scabrous, 5-lobed, the lobes and angles both rounded, the middle lobe usually 3-lobed ; fruit oblong, striped, with an intensely bitter pulp.
Habitat. Plains of Northern India.
Quality and *Uses.* As the last.

MOMORDICA. *Linnæus.*

Tendrils lateral. *Stamens* triadelphous. *Anthers* connate. *Calyx* of the males short. *Fruit* finally splitting.

Fig. 140.—Cucumis Colocynthis ; 141. Fruit of C. pseudo-colocynthis.

1. M. *Balsamina* Linnæus.—(BALSAM APPLE.) Fig. 142.

Leaves 5-lobed, palmate, toothed, smooth, shining; fruit oblong, acuminate, with rows of oblong tubercles separated by crowds of roundish ones.

Habitat. South of Europe.
Quality. Fruit a dangerous poison when ripe.
Uses. Where hydragogues are required.

CUCURBITA. *Linnæus.*

Tendrils lateral. *Stamens* triadelphous. *Anthers* connate. *Calyx* of the males campanulate.

1. C. *Pepo* Linnæus.—(The COMMON GOURD.)

Stem prostrate; leaves cordate, obtuse, somewhat 5-lobed, toothletted; calyx with a neck below the limb; fruit roundish or oblong, smooth.

Habitat. The Levant.
Quality. Slightly laxative; nutritious.
Uses. An agreeable kitchen fruit; the young leaves and shoots the best of Spinach. The Vegetable Marrow is a variety of this.

2. C. *Melopepo* Linnæus.—(The SQUASH GOURD.)

Stem erect; leaves cordate, obtuse, somewhat 5-lobed, toothletted; calyx short, hemispherical, campanulate, with a very wide mouth; fruit erect, depressed, circular, crenated.

Habitat. Unknown.
Quality and *Uses.* As the last.

3. C. *maxima* De Candolle.—(The SPANISH GOURD.)

Stem prostrate; leaves cordate, large, very rugose; calyx with a short neck below the limb; fruit very large, rather rough.

Habitat. Unknown. Common in gardens under the French name of Potiron.
Quality and *Uses.* As in No. 1. Fruit sometimes weighs, even in England, 200 lbs.

4. C. *aurantia* Willdenow.—(The ORANGE GOURD. FALSE COLOCYNTH.)

Stem prostrate, extremely scabrous; leaves somewhat cordate, 3-lobed, cuspidate, finely toothletted; fruit globose, smooth, deep orange, with a bitter pulp.

Habitat. Unknown.
Quality and *Uses.* Like those of Colocynth, but milder

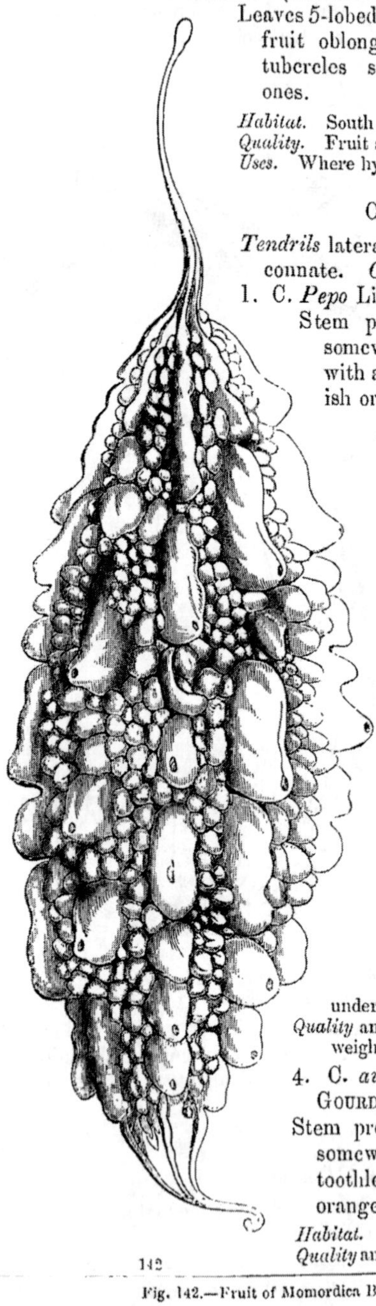

142

Fig. 142.—Fruit of Momordica Balsamina just before it splits.

BRYONIA. *Linnæus.*

Tendrils lateral. *Stamens* triadelphous. *Anthers* distinct. *Style* trifid. *Fruit* not corticated, few-seeded.

1. B. *dioica* Jacquin.—(COMMON BRYONY.) Fig. 143.

Stem climbing; leaves cordate, palmate, 5-lobed, toothed, with callous points, the terminal lobe longer; tendrils simple; flowers racemose, diœcious.

143

Habitat. Hedgerows.
Quality. Root emetic, purgative, acrid, poisonous.
Uses. Root applied topically to bruises.

ECBALIUM. *Richard.*

Tendrils lateral. *Stamens* monadelphous. *Anthers* connate. *Calyx* of males campanulate. *Fruit* expelling the seeds with elasticity, indehiscent.

1. E. *agreste* Richard. *Momordica Elaterium* Linnæus.—(SPIRTING CUCUMBER.)

Hispid, scabrous; stem dwarf, without tendrils; leaves cordate, somewhat lobed, crenate-toothed, very rugose, on long stalks.

Habitat. Rubbish and old walls in the south of Europe.
Quality. Juice an acrid irritant, drastic, hydragogue.
Uses. Dropsy, apoplexy, obstinate constipation, gout.

Fig. 143.—Bryonia dioica in flower and fruit.

THE VIOLAL ALLIANCE (V. K., p. 326.)

Natural Orders of Violals.

Bixads (*Flacourtiaceæ.*) Stamens 00.

Passionworts (*Passifloraceæ.*) Stamens definite. Flowers coronetted.

Moringads (*Moringaceæ.*) Stamens definite. Anthers 1-celled. Fruit siliquose.

Violetworts (*Violaceæ.*) Stamens definite. Anthers crested.

Natural Order, Bixads; *Flacourtiaceæ* (V. K., p. 327).

Prevailing Quality. Uncertain.

FLACOURTIA. *Commerson.*

Sexes imperfect. *Petals* 0. *Calyx* deciduous. *Berry* globose.

1. F. *Ramontchi* L'Heritier.

Leaves roundish-ovate, acute, crenated.

Habitat. Madagascar.
Quality and *Uses.* Fruits resembling black Plums, eatable and wholesome.

BIXA. *Linnæus.*

Flowers complete. *Stigma* simple. *Calyx* hispid. *Sepals* 5. *Capsule* 2-valved.

1. B. *orellana* Linnæus.—(ARNOTTO TREE.)

Leaves smooth on each side, &c.

Habitat. Tropical America.
Quality. Seeds cordial, astringent, febrifugal.
Uses. Seeds covered with a red pulp, called Arnotto, used in dyeing cheeses, and in chocolate making ; a supposed antidote to the poison of Janipha.

Natural Order, Passionworts; *Passifloraceæ* (V. K., p. 328.)

Prevailing Quality. Narcotic.

PASSIFLORA. *Linnæus.*

Ovary stipitate. *Fruit* pulpy. *Coronet* in several rows, of which the interior are very short.

1. P. *quadrangularis* Linnæus.—(GRANADILLA.)

Leaves smooth, cordate, ovate, acuminate; petioles with from 4 to 6 glands, stipules ovate entire as well as the bracts ; branches with 4 winged angles.

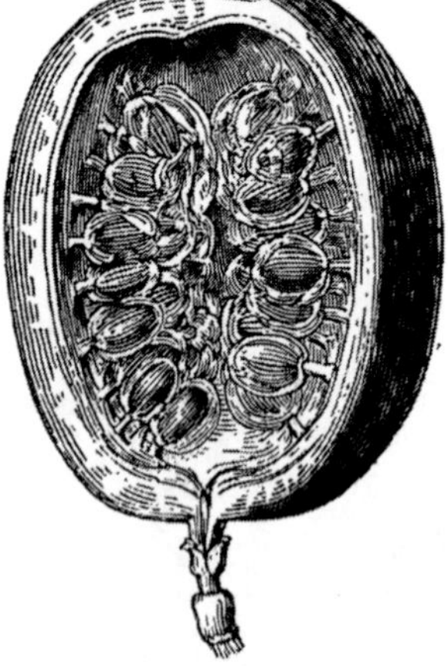

144

Fig. 144.—Longitudinal section of the fruit of a Passiflora.

Habitat. Tropical America.
Quality. Root emetic, narcotic.
Uses. Fruit common in tropical desserts: subacid pulp of the seeds the part used.

2. P. *rubra* Linnæus.—(DUTCHMAN'S LAUDANUM.)

Leaves velvety, cordate and 2-lobed at the base, awned at the sinus, beneath without glands as well as the petiole; pedicels solitary; ovary hairy, roundish.

Habitat. Jamaica.
Quality. Narcotic.
Uses. Tincture of the flowers a substitute for opium.

3. P. *fœtida* Linnæus. Fig. 145.

Stem and petioles hispid; leaves villous on both sides, 5-nerved, cordate, 3-lobed; lobes acute, the lateral very short; involucre divided into fine glandular entangled arms.

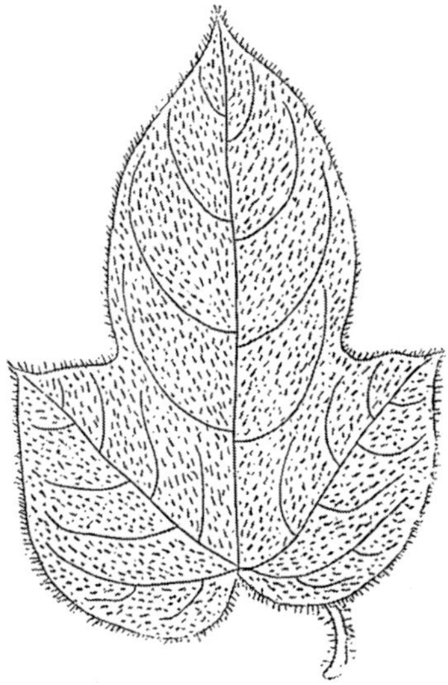

145

Habitat. West Indies.
Quality. Flowers pectoral; leaves emollient; narcotic? emmenagogue?
Uses. Leaves in poultices; flowers in hysterics; root in amenorrhœa.

Natural Order, Moringads; *Moringaceæ* (V. K., p. 336.)

Prevailing Quality. Pungency.

MORINGA. *Burmann.*

Fruit long, siliquose, 3-angular, 3-valved, bearing the amygdaloid *seeds* on the face of the valves.

1. M. *pterygosperma* Gærtner.—(HORSERADISH TREE.)

Leaves twice or thrice pinnate, with an odd leaflet; leaflets roundish oblong; flowers white, in naked terminal panicles; seed roundish, with 3 membranous wings.

Habitat. India and Arabia.
Quality. Acrid, pungent, aromatic, stimulant.
Uses. Leaves in curries, as sinapisms; paralysis; intermittents; seeds yield oil of Ben; the bark a gum like Tragacanth.

Natural Order, Violetworts; *Violaceæ* (V. K., p. 338.)

Prevailing Quality. Emetic, depurative.

VIOLA. *Linnæus.*

Sepals nearly equal, extended backwards at the base. *Petals* very unequal. the lowermost spurred. *Stamens* distinct.

Fig. 145.—Leaf of Passiflora fœtida.

1. V. *odorata* Linnæus.—(SWEET VIOLET.)

Stemless, throwing off runners ; leaves broadly cordate, pubescent ; sepals blunt ; flowers very sweet-scented.

Habitat. Hedgerows and woods.
Quality. Seeds and roots emetic and purgative ; flowers anodyne, produce faintness.
Uses. As a substitute for ipecacuanha ; flowers form an official syrup used as a test ; acids redden it ; alkalies render it green.

2. V. *canina* Linnæus.—(DOG VIOLET.)

Stems procumbent ; leaves cordate, oblong-ovate ; stipules on the middle of the stem fringed, much shorter than the petiole ; flowers scentless.

Habitat. Hedgerows and woods.
Quality and *Uses.* Roots and seeds as in No. 1.

3. V. *tricolor* Linnæus.—(The PANSY or HEARTSEASE.)

Stems prostrate and ascending ; leaves crenate ; stipules leafy, lyrate, pinnatifid, with the middle lobe crenated ; flowers 3-coloured, scentless.

Habitat. Corn-fields and gardens.
Quality. Acrid.
Uses. Leaves bruised, against cutaneous affections (*tinea capitis.*)

IONIDIUM. *Ventenat.*

Sepals not extended at the base. *Petals* very unequal ; 1 large and spreading, 4 small. *Stamens* separate.

146

1. I. *Itubu* Aublet. I. *Ipecacuanha* Auguste de St. Hilaire.—(WHITE IPECACUANHA.) Fig. 146.

Extremely hairy ; leaves oblong, serrated ; sepals fringed.

Habitat. Brazil and Guiana.
Quality. Roots emetic.
Uses. Substitute for Ipecacuanha.

2. I. *microphyllum* Humboldt. — (CUICHUNCHULLI.)

Half shrubby, smooth ; leaves minute, oblong, acute, serrated ; peduncles 3 times as long as the leaves.

Habitat. Peru.
Quality. Emetic, purgative ; very active.
Uses. Said to be a certain cure for the elephantiasis tuberculata.

Fig. 146.—Ionidium Itubu.

THE SAPINDAL ALLIANCE (V. K., p. 373.)

Natural Orders of Sapindals.

Milkworts (*Polygalaceæ.*) Anthers 1-celled. Calyx 2-winged.

Soapworts (*Sapindaceæ.*) Anthers 2-celled. Petals with an appendage. Ovules erect.

Erythroxyls (*Erythroxylaceæ.*) Anthers 2-celled. Petals with an appendage. Ovules pendulous.

Natural Order, **Milkworts**; *Polygalaceæ* (V. K., p. 375).

Prevailing Quality. Bitter, emetic.

POLYGALA. *Linnæus.*

Calyx persistent. *Stamens* 8. *Capsule* mucronate. *Seeds* hairy or carunculate.

1. P. *vulgaris* Linnæus.—(MILKWORT.) Fig. 147.

Leaves linear-lanceolate, more or less obtuse; stems ascending; wings ovate, shorter than the corolla, longer than the capsule; ovary nearly sessile.

Habitat. Heaths and open downs.
Quality. Bitter.
Uses. Pulmonary affections; spitting of blood.

2. P. *rubella* Pursh.

Leaves lanceolate-linear, mucronate; stem erect, furrowed, branched at the top; racemes somewhat spiked, lax, elongate: wings oval.

Habitat. United States.
Quality. Bitter, tonic, stimulant, diaphoretic.
Uses. Dyspepsia.

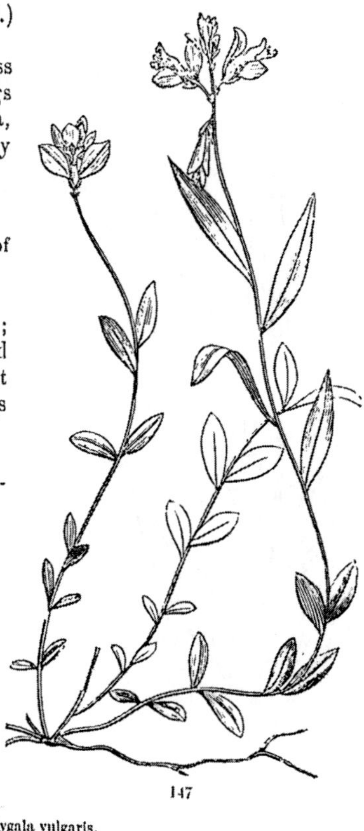

147

Fig. 147.—Polygala vulgaris.

3. P. *Senega* Linnæus.—(The SENEKA.) Fig. 148.

Leaves ovate-lanceolate, the upper acuminate ; stems tufted, erect, simple ; racemes spicate ; wings circular ; capsule elliptical, emarginate.

Habitat. United States.
Quality. Root diaphoretic, diuretic, expectorant ; emetic, purgative, emmenagogue.
Uses. Bronchial or pulmonary inflammation, chronic catarrh, humoral asthma, low fever, rheumatism, dropsy, amenorrhœa.

4. P. *Chamæbuxus* Linnæus.

Stems dwarf, branched, procumbent, shrubby ; leaves evergreen, oblong-lanceolate, mucronate ; flowers single or in pairs ; keel crested.

Habitat. Mountains of Europe.
Quality and *Uses.* Like the last.

5. P. *Poaya* Martius.

Stem shrubby ; leaves coriaceous, 5-nerved ; racemes spicate ; wings oblong or obovate, obtuse ; corolla crested ; seeds clavate, shaggy.

Habitat. Brazil.
Quality. An active emetic.
Uses. Root, when fresh, in bilious fevers.

SOULAMEA. *Lamarck.*

Stamens 6. *Fruit* samaroid.

1. S. *amara* Lamarck.—(The BITTER KING.) Fig. 149.

A shrub or small tree ; leaves large, oval oblong ; racemes axillary, little longer than the petioles; fruit obcordate.

148

149

Fig. 148.—Polygala Senega ; 149. Soulamea amara.

Habitat. Indian Archipelago.
Quality. Intensely bitter.
Uses. Cholera, pleurisy, intermittents.

150

KRAMERIA. *Loefling.*

Calyx without wings. *Stamens* 4, or fewer. *Fruit* a 1-celled drupe covered with hooked spines.

1. K. *triandra* Ruiz and Pavon.—(RATTANY-ROOT.) Fig. 150.

Leaves oblong, softly hairy, rather acute; flowers in short racemes.

Habitat. Peru and Chili.
Quality. Root a powerful astringent.
Uses. Mucous discharges, passive hemorrhage; tooth-powder, mouth-washes.

Natural Order, Soapworts; *Sapindaceæ* (V. K., p. 382.)

Prevailing Quality. Acridity.

NEPHELIUM. *Linnæus.*

Leaves alternate. *Flowers* regular. *Calyx* 5-6-toothed. *Ovules* solitary. *Fruit* indehiscent, brittle; seed covered with pulp.

1. N. *Longan.*—(The LONGAN.) Fig. 151.

Leaves pinnated; leaflets with the midrib very prominent below; fruit globose, areolate.

Habitat. China.
Quality and *Uses.* A very agreeable fruit, often imported from China, for the sake of the sweet subacid vinous pulp which covers the seed.

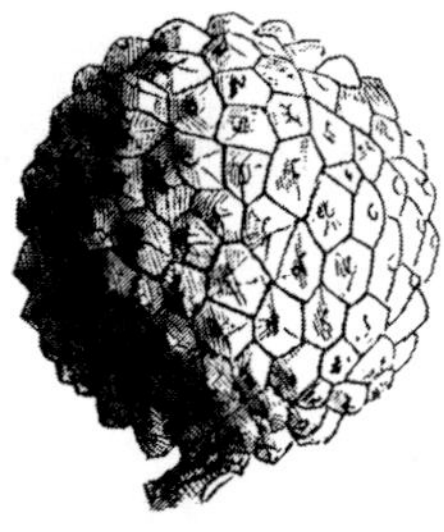

151

PAULLINIA. *Linnæus.*

Leaves alternate. *Flowers* irregular. *Stamens* 8. *Fruit* capsular, wingless, 3-cornered. *Seeds* solitary.

1. P. *pinnata* Linnæus.

Leaves pinnate, in 2 pairs with an odd one; leaflets ovate-lanceolate, sessile, crenated; petiole winged; fruit pyriform.

Habitat. Brazil.
Quality. Extremely acrid and poisonous.

2. P. *sorbilis* Martius.

An uncertain plant, of which no description has been hitherto published.

Habitat. Brazil.
Quality. Astringent, febrifugal, stomachic, a certain aphrodisiac.
Uses. Pounded seeds form Guarana bread, cakes of which are consumed largely in Brazil; said to contain theine. (Specimens of this bread will be found in the Museum of the Royal Botanic Garden, Kew.)

Fig. 150.—Krameria triandra in fruit; 151. Fruit of Nephelium Longan.

ÆSCULUS. *Linnæus.*

Leaves opposite, digitate. *Petals* 5, spreading, with short claws, unequal. *Stamens* 7, declinate. *Fruit* leathery, 3-valved.

1. Æs. *Hippocastanum* Linnæus.—(HORSE-CHESNUT.) Fig. 152. Leaflets 7, obovate, cuneate, acute, toothed ; fruit prickly.

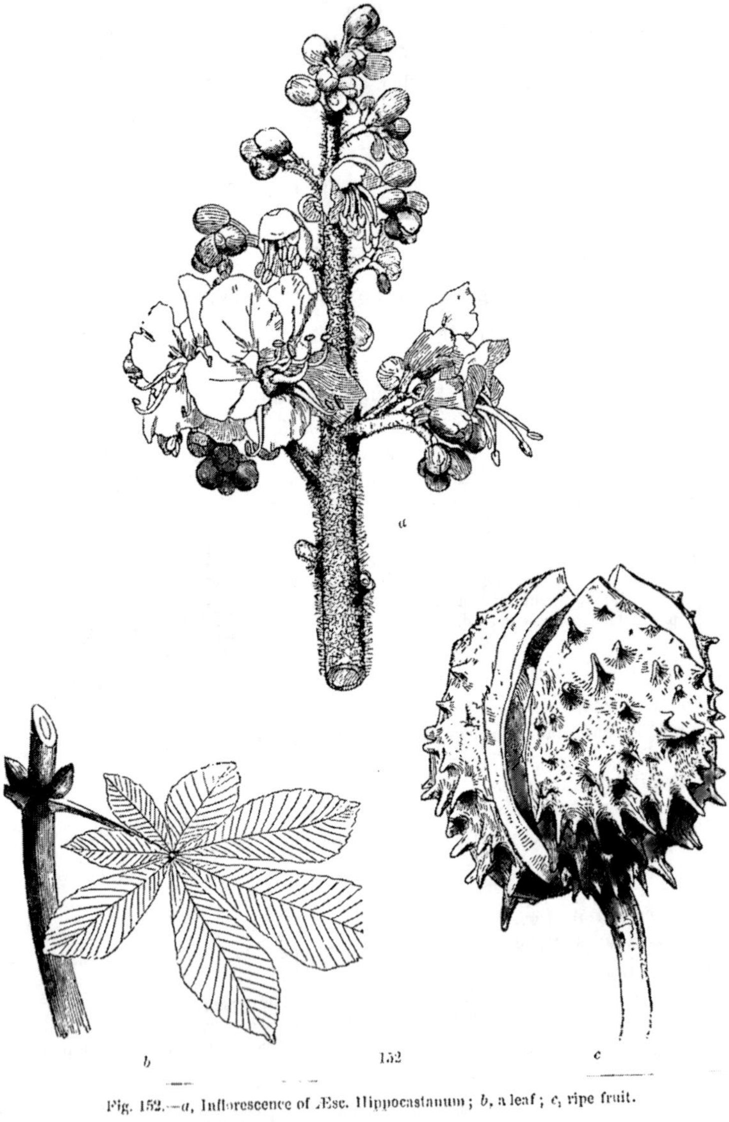

Fig. 152.—*a*, Inflorescence of Æsc. Hippocastanum ; *b*, a leaf ; *c*, ripe fruit.

Habitat. Persia ?
Quality. Seeds acrid ; bark febrifugal.
Uses. Seeds a good sheep-food ; yield abundance of nutritious starch when washed.

SAPINDUS. *Linnæus.*

Leaves alternate. *Flowers* regular. *Disk* complete. *Stamens* 8-10. *Fruit* wingless, indehiscent, fleshy.

1. S. *saponaria* Linnæus.—(SOAP-BERRY.)

Leaves abruptly pinnate ; leaflets oblique, entire, lanceolate, in 3 or 4 pairs, with a broad-winged petiole ; flowers in large panicles.

Habitat. West Indies.
Quality. Fruit detersive, acrid, narcotic.
Uses. Used instead of soap; intoxicates fish.

Natural Order, **Erythroxyls** ; *Erythroxylaceæ* (V. K., p. 391.)

Prevailing Quality. Uncertain.

ERYTHROXYLON. *Linnæus.*

Calyx 5-parted, pentangular at the base. *Styles* 3, distinct.

1. E. *Coca* Lamarck.—(The COCA.) Fig. 153.

153

Leaves oval, acute, thin, with 3 slight line-like ribs near the midrib.

Habitat. Peru.
Quality. Stimulating, narcotic.
Uses. Leaves largely chewed instead of Opium.

THE BERBERAL ALLIANCE (V. K., p. 432.)

Natural Orders of Berberals.

Berberids (*Berberidaceæ.*) Anther valves recurved.

Vineworts (*Vitaceæ.*) Anther valves straight.

Natural Order, **Berberids** ; *Berberidaceæ* (V. K., p. 437.)

Prevailing Quality. Astringency, slight acridity.

BERBERIS. *Linnæus.*

Flowers complete, ∛. *Stamens* 6. *Petals* with 2 glands at the base. *Stigma* sessile, peltate. *Fruit* fleshy.

Fig. 153.—Erythroxylon Coca.

1. B. *vulgaris* Linnæus.—(BERBERRY BUSH.) Fig. 154.

154

Leaves obovate with ciliated serratures; racemes pendulous, many-flowered.

Habitat. Europe.
Quality. Berries acid, astringent; bark astringent.
Uses. Wood used by dyers, for its yellow colour.

2. B. *Lycium* Royle.

Spines 3-parted, conical; leaves coriaceous, oblong-lanceolate, tapering to the base, mucronate, with spiny teeth; flowering racemes erect; flowers small; racemes of fruit pendulous.

Habitat. North of India.
Quality. Very astringent.
Uses. Extract valuable in ophthalmia.

Natural Order, Vineworts; *Vitaceæ* (V. K., p. 439.)

Prevailing Quality. Acridity.

VITIS. *Linnæus.*

Flowers ⚥/5. *Calyx* 5-toothed. *Petals* calyptrate.

1. V. *vinifera* Linnæus.—(The VINE.)

Leaves angular, lobed, sinuated, toothed, more or less hairy; fruit sweet.

Habitat. North of India?
Quality. Ripe fruit nutritious, refrigerant, diuretic, laxative; skin astringent, indigestible; colouring matter deranges the stomach.
Uses. Preparation of grape wine and brandy; fevers, inflammatory complaints, dysentery, phthisis.

CISSUS. *Linnæus.*

Flowers ⚥/4. *Calyx* 4-toothed. *Petals* horned.

1. C. *cordata* Roxburgh. C. *repens* Lamarck.

Stem creeping; leaves cordate-ovate, somewhat toothed, smooth; flowers in umbels, dark purple.

Habitat. Malabar.
Quality. Acrid.
Uses. Leaves as poultices for indolent tumours.

GENUS OF UNCERTAIN AFFINITY.

Somewhere in the neighbourhood of the Berberal Alliance must stand the following genus, whose structure is imperfectly ascertained.

CANELLA. *P. Browne.*

Sepals 3, blunt, imbricated. *Petals* 5. *Stamens* 20, on a truncated hypogynous cone; *anthers* linear. *Ovary* one-celled, with 2 or 3 pendulous anatropal ovules. *Stigma* emarginate.

Fig. 154. Leaves of Berberis vulgaris.

1. C. *alba* Swartz.—(WHITE WOOD. WILD CINNAMON.) Fig. 155.
Leaves alternate, dotted, obovate, obtuse, glaucous beneath.

Habitat. West Indies.
Quality. Hot, aromatic, with a flavour between that of fennel and cinnamon, stimulant.
Uses. Scurvy; in addition to tonics or purgatives.

155

THE ERICAL ALLIANCE
(V. K., p. 446.)

Natural Orders of Ericals.

Winter Greens (*Pyrolaceæ.*) Flowers nearly polypetalous. Embryo minute. Herbs.

Heathworts (*Ericaceæ.*) Flowers (usually) quite monopetalous. Embryo axile. Shrubs.

Natural Order, Winter Greens; *Pyrolaceæ* (V. K., p. 450.)

Prevailing Quality. Diuretic.

CHIMAPHILA. *Nuttall.*

Stamens 10, with the filaments triangular at the base. *Anthers* hornless. *Stigma* nearly sessile, peltate. *Capsule* bursting at the upper end.

1. C. *umbellata* Nuttall.—(UMBELLED WINTER GREEN.) Fig. 156.
Leaves cuneate-lanceolate, whole coloured; filaments smooth.

Habitat. North America.
Quality. Acrid, tonic, diuretic, narcotic.
Uses. Dropsies, cystirrhœa, calculus, dysury, strangury, gonorrhœa, &c., scrofula.

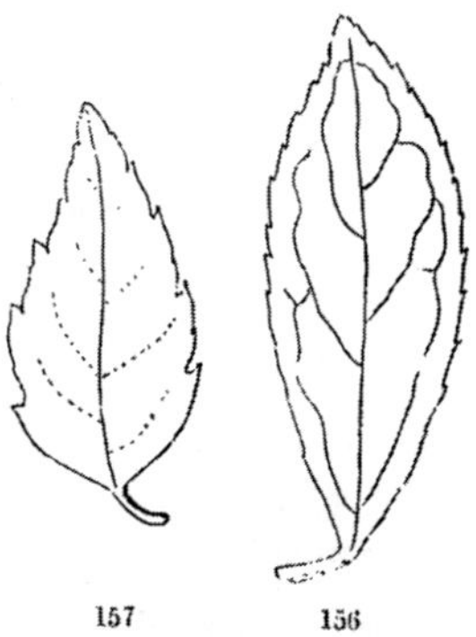

157 156

2. C. *maculata* Pursh.—(VARIEGATED WINTER GREEN.) Fig. 157.
Leaves ovate-lanceolate, blotched with white; filaments shaggy below the dilated part.

Habitat. United States.
Quality and *Uses.* Like the last.

Fig. 155. Canella alba; 156. Chimaphila umbellata; 157. Chimaphila maculata.

Natural Order, Heathworts; *Ericaceæ* (V. K., p. 453.)
Prevailing Quality. Narcotic.

ARCTOSTAPHYLOS. *Adanson.*

Drupe with 5 distinct 1-seeded stones. *Corolla* urceolate, with a revolute limb. *Stamens* included. *Anthers* with 2 spurs at the back.

1. A. *Uva Ursi* Sprengel.—(BEAR-BERRY.) Fig. 158.
Procumbent; leaves obovate, obtuse, entire, shining; flowers in short terminal racemes.

Habitat. North of Europe, Asia, and America.
Quality. Astringent, diuretic, emetic.
Uses. Chronic affections of the bladder, calculus, bronchial affections.

158

GAULTHERIA. *Kalm.*

Capsule 5-valved, loculicidal, covered over by the fleshy tube of the calyx.

1. G. *procumbens* Linnæus.—(PARTRIDGE-BERRY.)
Smooth, procumbent, rooting; leaves obovate, with setaceous serratures, acute at either end; pedicels bearing 1-2 nodding flowers; anthers with 4 bristles.

Habitat. North America.
Quality. Aromatic, astringent.
Uses. Emmenagogue; unsafe. Oil is a flavouring substance.

ANDROMEDA. *Linnæus.*

Capsule 5-valved, loculicidal, naked. *Anthers* 2-awned. *Corolla* globose, with a contracted orifice.

1. A. *polifolia* Linnæus.
Leaves alternate, oval or linear-lanceolate, entire, revolute, glaucous beneath; flowers terminal, almost umbellate.

Habitat. Bogs of North of Europe and North America.
Quality. An acrid dangerous narcotic; kills sheep.

LEDUM. *Linnæus.*

Petals 5, nearly, or quite, distinct. *Stamens* 5 to 10. *Style* short, straight, thick. *Capsule* septicidal.

1. L. *latifolium* Aiton.—(LABRADOR TEA.)
Leaves oblong; stamens 5, the length of the corolla.

Habitat. North of Europe and America.
Quality. Narcotic.
Uses. Agues, dysentery, diarrhœa; renders beer heady.

KALMIA. *Linnæus.*

Corolla campanulate, with 10 honey-pores sunk in its sides, each holding back an anther, which finally rises by the elasticity of its arched filament.

Fig. 158.—Arctostaphylos Uva Ursi.

1. K. *latifolia* Linnæus.—(Calico Bush.)

Leaves on long stalks, scattered, elliptical, acute at either end, shining and smooth ; corymbs terminal, viscid.

Habitat. United States.
Uses. Narcotic, poisonous, dangerous.

Azalea. *Linnæus.*

Stamens 5, loose. *Calyx* leafy. *Corolla* funnel-shaped. *Capsule* septicidal.

1. A. *pontica* Linnæus.

Leaves obovate-oblong, lanceolate, ciliated, very much wrinkled ; flowers bright yellow, glutinous externally.

Habitat. Caucasus, Asia Minor.
Quality. Dangerous, narcotic, honey highly poisonous.

Rhododendron. *Linnæus.*

Stamens 10. *Calyx* obsolete. *Corolla* campanulate, or between campanulate and funnel-shaped. *Capsule* septicidal.

1. R. *chrysanthum* Pallas.—(Golden Rhododendron.) Fig. 159.

Leaves oblong-lanceolate, revolute, rugose, tapering to the base, pallid beneath; flowers campanulate, corymbose, bright yellow.

Habitat. Siberia, Dauria.
Quality. A powerful narcotic.
Uses. Chronic rheumatism; venereal affections. The leaves largely employed by the Russians. It is said, that a long use of this narcotic does not affect the general health.

2. R. *ferrugineum* Linnæus.

Leaves small, oblong, tapering to each end, above shining and smooth, beneath covered with ferruginous scales ; corolla funnel-shaped, with resinous dots externally.

Habitat. Switzerland.
Quality. Narcotic.
Uses. Oil of the buds, called Olio di Marmotta, in pains of the joints.

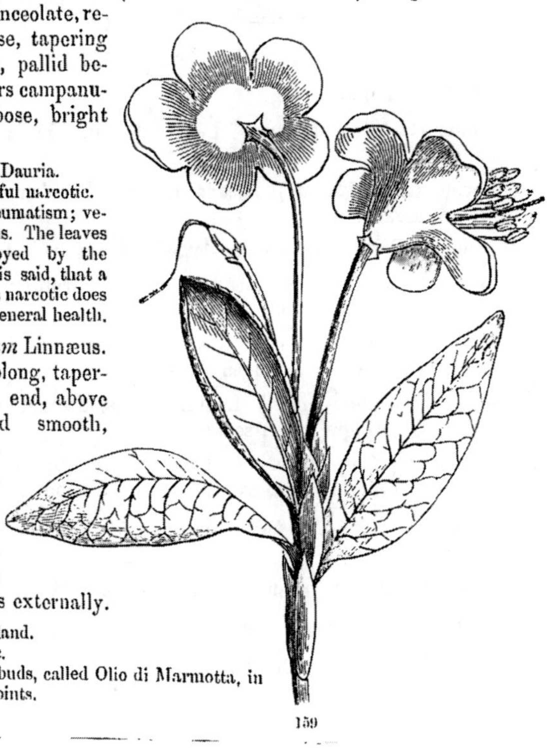

159

Fig. 159. Rhododendron chrysanthum.

ARBUTUS. *Linnæus.*

Fruit fleshy, 5-celled, many-seeded, usually papillose. *Corolla* urceolate, 5-toothed. *Hypogynous Scales* 0.

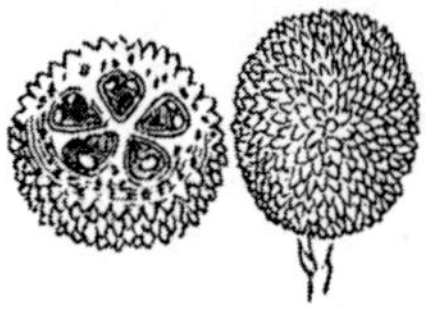

160

1. A. *Unedo* Linnæus.—(STRAWBERRY TREE.) Fig. 160, 161.

A tree ; bark rugged ; leaves oblong-lanceolate, bluntly serrated, smooth, shining; fruit globose, muricated.

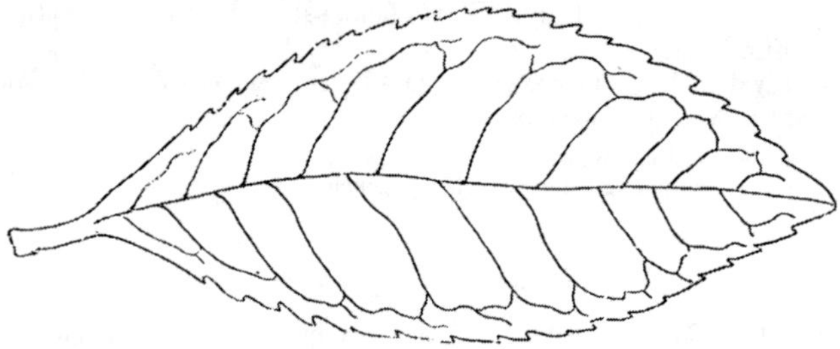

161

Habitat. South and west of Europe.
Quality. Fruit narcotic ; bark and leaves astringent.

THE RUTAL ALLIANCE (V. K., p. 456.)

Natural Order of Rutals.

Citronworts (*Aurantiaceæ.*) Leaves dotted. Fruit succulent, many-celled.

Amyrids (*Amyridaceæ.*) Leaves dotted. Fruit dry. Petals valvate. Ovules sessile.

Rueworts (*Rutaceæ.*) Leaves dotted. Fruit dry. Petals imbricated.

Cedrelads (*Cedrelaceæ.*) Leaves dotless. Fruit dry. Petals imbricated. Ovules sessile.

Anacards (*Anacardiaceæ.*) Leaves dotless. Fruit somewhat dry. Ovule supported by a long umbilical cord.

Quassiads (*Simarubaceæ.*) Leaves dotless, alternate, without stipules. Stamens springing from scales.

Beancapers (*Zygophyllaceæ.*) Leaves dotless, opposite, with stipules. Stamens springing from scales. Branches jointed.

Meliads (*Meliaceæ.*) Leaves dotless. Stamens forming a tube.

Natural Order, **Citronworts** ; *Aurantiaceæ* (V. K., p. 457.)

Prevailing Quality. Aromatic, tonic.

CITRUS. *Linnæus.*

Stamens 20 or more, irregularly polyadelphous. *Fruit* pulpy with a spongy rind. *Seeds* smooth.

Fig. 160.—Fruit of Arbutus Unedo ; 161. Leaf of ditto.

1. C. *Aurantium* Risso.—(COMMON SWEET ORANGE.) Fig. 162.
Leaves ovate-oblong, acute, sometimes serrulate; petiole more or less winged; flowers white; fruit roundish, occasionally mammose, with the cortical vesicles convex.

Habitat. Forests of the Himalayas and China, whence all the genus has also been derived.
Quality. Pulp of fruit sweet, refrigerant; rind aromatic, tonic; leaves bitter and aromatic.
Uses. A grateful fruit; rind and flowers inferior to those of C. Bigaradia, though similar.

2. C. *Bigaradia* Risso.—(The SEVILLE ORANGE.)
Spiny; leaves elliptical acute; petiole winged; flowers snow-white; fruit middle-sized, roundish, smooth or wrinkled, deep yellow; with concave oil-cysts.

Habitat. Cultivated in the south of Europe.
Quality. Pulp of fruit acid and bitter. Rind very bitter; aromatic, tonic.
Uses. Flowers yield oil of Neroli; crushed fruit, boiled in sugar, forms marmalade. Essential oil stimulant, and antispasmodic; Orange-flower water chiefly obtained from this.

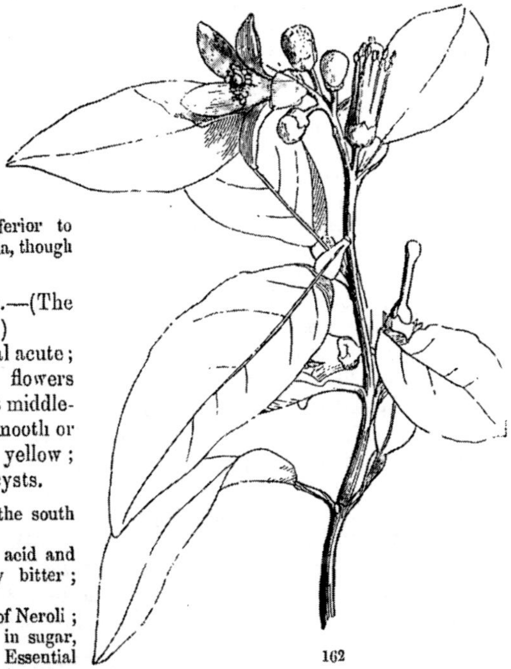

162

3. C. *Bergamia* Risso.—(The BERGAMOT ORANGE. MELLAROSA.)
Branches brittle; leaves oblong, with a winged petiole, dark green above, pale beneath; flowers small, white; fruit pyriform, smooth, pale yellow, with a green subacid firm fragrant pulp.

Habitat. Cultivated in the south of Europe.
Quality. Rind extremely fragrant.
Uses. The essential oil (oil of Bergamot) of both the flowers and fruit largely employed by perfumers; rind formed, by violent pressure in proper moulds, into small boxes.

4. C. *Limonum* Risso.—(The COMMON LEMON.)
Leaves ovate or oblong, usually serrulate, pale green, with a winged petiole; flowers middle-sized, red outside; fruit oblong, wrinkled or furrowed, pale yellow, with the oil-cysts concave; rind tolerably thin; pulp very acid.

Habitat. Forests of north of India.
Quality. Rind having a peculiar fragrance, bitter, stomachic, and aromatic; juice agreeably acid, refrigerant, antiscorbutic.

Fig. 162.—Citrus Aurantium.

Uses. In febrile complaints, preparation of effervescing draughts, lemonade ; against scurvy in the navy.

5. C. *Lumia* Risso.—(The Sweet Lemon.)

Branches and leaves those of the lemon ; flowers red outside ; fruit like the lemon, but with a sweet pulp.

Habitat. Cultivated in the south of Europe.
Quality and *Uses.* As those of the lemon, of which, however, its juice wants the peculiar sharpness.

6. C. *acida* Roxburgh.—(The Lime.)

Spiny; leaves oval, oblong, or ovate-oblong, crenate, obtuse, petioles winged ; petals generally 4 ; fruit small, blunt, oval, or oblong, with a thin rind, and an extremely acid juice.

Habitat. North of India and China.
Quality. Rind aromatic; pulp antiscorbutic.
Uses. As the lemon, but the acidity sharper, and rather more agreeable.

7. C. *Limetta* Risso.—(The Sweet Lime.)

Leaves ovate, obovate, and oblong ; petiole almost wingless ; flowers small, white ; fruit ovate or roundish, pale yellow, with a raised point, and concave cysts of oil ; pulp subacid.

Habitat. Cultivated in the south of Europe. *Quality.* An inferior description of lemon.

8. C. *decumanus* Risso.—(The Shaddock.)

Leaves large, with a broad-winged petiole; flowers very large, white ; fruit very large, roundish, pale yellow ; rind with flat or convex oil-cysts, white and spongy ; pulp greenish, subacid, watery.

Habitat. Cultivated in the south of Europe.
Quality. Sub-aromatic, subacid. *Uses.* In preserves ; as a pleasant cooling fruit.

9. C. *Medica* Risso.—(The Citron and Cedrate.)

Leaves oblong, toothed ; flowers violet outside ; fruit large, warted and furrowed ; rind very thick, tender ; pulp subacid.

Habitat. Forests of north of India.
Quality. Pulp refrigerant ; rind and leaves aromatic, tonic.
Uses. Rind only used in preserving. Furnishes the fragrant perfumer's *huile de Cedrat ;* juice employed in flavouring punch, &c.

Natural Order, Amyrids ; *Amyridaceæ* (V. K., p. 459.)

Prevailing Quality. Stimulating.

Boswellia. *Roxburgh.*

Cal. 5-toothed. *Pet.* 5. *Stam.* 10. *Disk* fleshy, longer than the calyx. *Fruit* triangular, 3-celled, 3-valved, septicidal. *Seed* winged.

1. B. *thurifera* Colebrooke. B. *serrata.*—(Olibanum Tree.)

Leaves pinnated ; leaflets ovate, acuminate, serrated, downy ; racemes axillary, simple.

Habitat. Mountains of India. *Quality.* Yields a stimulating oleo-resin.
Uses. Chronic diarrhœa, old catarrhs, leucorrhœa, gleet, hæmoptysis ; stimulating plasters ; fumigation.

Balsamodendron. *Kunth.*

Sexes sometimes imperfect. *Calyx* 4-toothed, cup-shaped. *Petals* 4. induplicate-valvate. *Stamens* 8, arising from without a fleshy disk. *Ovary* 2-celled. *Drupe* bony, 1-2 celled.

1. B. *Myrrha* Nees.—(MYRRH TREE.)*

"Stem shrubby, arborescent; branches squarrose, spinescent; leaves ternate; leaflets obovate, obtuse, bluntly toothletted at the apex, the lateral smooth; fruit acuminate."—*Nees.*

Habitat. Abyssinia.
Quality. Stimulant, irritant, astringent, tonic (tonico-balsamic.)
Uses. Dyspepsia, flatulence, amenorrhœa, chlorosis, secretions from the mucous membranes; dentifrice, gargles, foul ulcers.

Natural Order, Rutworts; *Rutaceæ* (V. K., p. 469.)

Prevailing Quality. Acrid, stimulating.

RUTA. *Linnæus.*

Petals 4-5, spoon-shaped. *Stamens* 8-10, all perfect. *Anthers* glandless, spreading equally. *Ovary* 4-lobed, with several ovules in each cell.

1. R. *graveolens* Linnæus.—(COMMON RUE.) Fig. 163.

Leaves supra-decompound; lobes oblong, that at the end obovate; petals entire, or somewhat toothed.

Habitat. South of Europe.
Quality. Acrid, antispasmodic, emmenagogue, anthelmintic, stimulant, narcotic, heavy-smelling.
Uses. Flatulent colic, infantile convulsions, hysteria, amenorrhœa.

2. R. *montana* Clusius.

Leaves supra-decompound; lobes all linear; petals entire.

Habitat. Spain.
Quality. Excessively acrid, dangerous to handle.

164

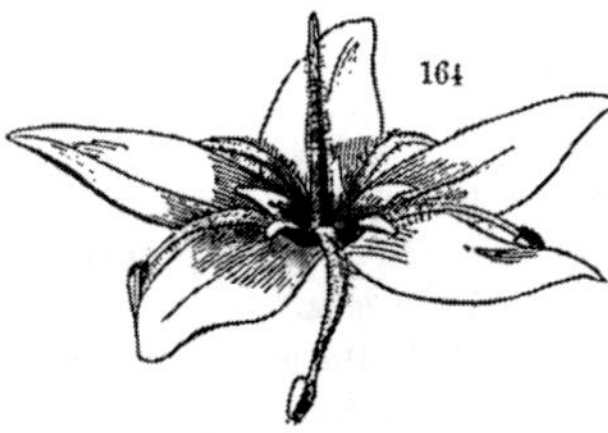

a

163

BAROSMA. *Willdenow.*

Flowers regular. *Petals* 5. Stamens 10, of which half are sterile and squamiform; *anthers* with a minute terminal gland. *Ovary* 5-lobed. Fig. 164.

Fig. 163.—Ruta graveolens; *a*, its capsule; 164. Flower of Barosma magnified.

* Elemi, Bdellium, and similar drugs are obtained from plants related to this; but there is so little certainty upon the botanical part of the subject, that the reader is referred to Pereira, ed. 2, II. 1628, and Royle, p. 342, for such information as at present exists about them. The history of Myrrh itself is still far from clear. Balm of Gilead is said to be another Balsamodendron.

1. B. *crenata* Willdenow. *Diosma crenata* Linnæus.—(BUCKU.) Fig. 165 *b*.
Leaves ovate, acute, dotted, with glandular serratures; pedicels solitary, rather leafy.

165

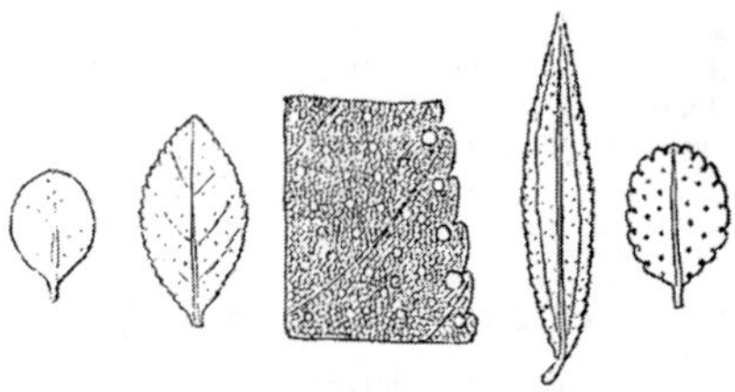

a *b* *b** *c* *d*

Habitat. Cape of Good Hope.
Quality. Aromatic, stimulant, tonic; diuretic, diaphoretic.
Uses. Chronic inflammation of the bladder, stricture, gleet, prostatic affections, rheumatism, dyspepsia.

GALIPEA. *Aublet.*

Flowers regular. *Petals* 5, partially adhering to each other and the stamens. *Stamens* 4-8, of which a part are sterile, in no regular order; the filaments bearded.

1. G. *Cusparia* A. de St. Hilaire.—(ANGOSTURA BARK TREE.)
Leaflets 3; racemes stalked, terminal or nearly so, calyx 5-toothed; sterile stamens 3.

Habitat. South America.
Quality. Powerfully aromatic and stimulant; stomachic, diaphoretic.
Uses. Intermittents, remittents, dyspepsia, diarrhœa, &c.

2. G. *officinalis* Hancock.—(ANGOSTURA BARK TREE.)
Leaflets 3; racemes stalked, axillary or terminal; sterile stamens 5.

Habitat. Banks of the Oronoco.
Quality and *Uses.* As in the last.

DICTAMNUS. *Linnæus.*

Flowers irregular. *Stamens* 10 declinate, with the filaments covered with glands. *Follicles* 5, united at the base, each with 2-3 seeds.

166

1. D. *albus* Linnæus.—(FRAXINELLA. BASTARD DITTANY.) Fig. 166.
Leaves alternate, unequally pinnated; stems glandular at the point; racemes terminal; flowers white or purple.

Habitat. South of Europe.
Quality. Aromatic, tonic, antispasmodic, diuretic, emmenagogue.
Uses. Root in intermittents, epilepsy, hysteria, amenorrhœa, chlorosis, worms.

Fig. 165.—Leaves of various species of Bucku: *a*, Barosma graveolens; *b*, Diosma crenata; *b**, a magnified view of a portion of it; *c*, D. serratifolia; *d*, B. venusta. 166. Fruit of Dictamnus.

Natural Order, **Cedrelads**; *Cedrelaceæ* (V. K., p. 461.)

Prevailing Quality. Tonic, astringent.

CEDRELA. *Linnæus.*

Stamens distinct. *Capsule* 5-celled, 5-valved; *seeds* suspended, winged at the back.

1. C. *Toona* Roxburgh.

Leaves abruptly pinnated; leaflets ovate-lanceolate, acuminate, somewhat serrated, pallid beneath, glaucous; racemes axillary, panicled.

Habitat. Bengal, Indian Islands.
Quality. Tonic, astringent.
Uses. Bark in epidemic fevers, diarrhœa, dysentery.

SOYMIDA. *Adrien de Jussieu.*

Stamens united in a short, cup-shaped tube. *Seeds* winged all round.

1. S. *febrifuga* Adrien de Jussieu.

Leaves alternate, abruptly pinnated; leaflets in about 4 pairs, oval, obtuse or emarginate, rather oblique at the base; racemes forming a terminal panicle.

Habitat. Mountains of India.
Quality. Tonic, febrifugal.
Uses. Bark in jungle fevers, typhus, gangrene.

Natural Order, **Anacards**; *Anacardiaceæ* (V. K., p. 465.)

Prevailing Quality. Dangerous acridity.

ANACARDIUM. *Linnæus.*

Fruit a kidney-shaped nut, seated on the end of a pyriform fleshy peduncle.

1. A. *occidentale* Linnæus.—(CASHEW NUT.) Fig. 167.

A large tree; leaves oval, very blunt or emarginate, little narrowed to the base, rather longer than broad.

Habitat. Tropics of both hemispheres.
Quality. Acrid, venomous; gum astringent.
Uses. Seeds oily, when roasted eatable and wholesome; but dangerous to roast on account of the acrid vapour.

167

MANGIFERA. *Linnæus.*

Fruit a naked fleshy drupe. *Petals* 4-5. *Stamens* 5, mostly sterile. *Ovary* 1, seated in a fleshy disk.

1. M. *indica* Linnæus.—(MANGO TREE.)

A tree; leaves oblong-lanceolate, stalked; panicles erect; petals spreading at the point; stamen 1 perfect; fruit smooth.

Habitat. East Indies.
Quality. Terebinthinous; gum-resin bitter, sub-acrid.
Uses. Fruit of much value for the dessert, sweet, luscious.

Fig. 167.—Fruit of Anacardium occidentale.

PISTACIA. *Linnæus.*

Flowers apetalous, imperfectly diœcious. *Fruit* a dry drupe, with a bony stone.

1. P. *atlantica* Desfontaines.—(BARBARY MASTICH.) Fig. 168.

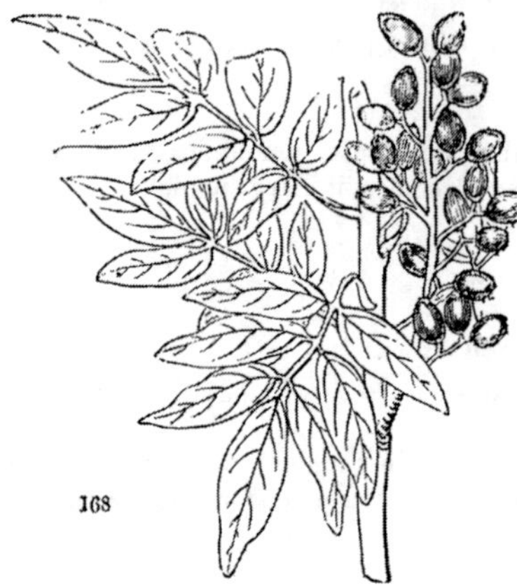

168

Leaves unequally pinnated, deciduous; leaflets about 9, lanceolate, rather tapering to the base; the petiole winged between the terminal pairs.

Habitat. North of Africa; Levant.
Quality and *Uses.* As in the next.

2. P. *Lentiscus* Linnæus.—(THE MASTICH TREE.)

Leaves abruptly pinnated, evergreen; leaflets 8, lanceolate; petiole winged.

Habitat. South of Europe, North of Africa, Levant.
Quality. Resin fragrant, astringent.
Uses. Occasionally in gleet, leucorrhœa, &c.; chiefly by dentists, and as a varnish.

3. P. *Terebinthus* Linnæus.—(TURPENTINE PISTACIA.)

Leaves unequally pinnate, deciduous; leaflets about 7, ovate-lanceolate, rounded at the base, acute, mucronate.

Habitat. South of Europe, North of Africa, Levant.
Quality. Like that of other turpentines.
Uses. Yields Scio or Cyprus turpentine.

4. P. *vera* Linnæus.—(PISTACIA NUT.)

Leaves unequally pinnate, deciduous; leaflets ovate, somewhat narrowed to the base, rather mucronate, in fives, threes, or single.

Habitat. Syria.
Quality. Fruit oily.
Uses. At dessert: and for astringent emulsions.

RHUS. *Linnæus.*

Flowers ☿/. *Petals* from beneath a large orbicular disk. *Ovary* sessile; *styles* 3. *Drupe* nearly dry, with a bony stone.

1. R. *Toxicodendron* Linnæus.—(POISON OAK.)

Leaves trifoliolate; leaflets ovate, oblong, thin, cut and angular, pubescent; sometimes entire.

Habitat. United States.
Quality. Excessively acrid, and narcotic: even its gaseous emanations; a dangerous poison.
Uses. Paralysis, chronic rheumatism, amaurosis.

2. R. *Metopium* Linnæus.—(HOG GUM.)

Leaves unequally pinnate, in 2 pairs, very smooth; leaflets with a short stalk, ovate, entire.

Habitat. West Indies.
Quality. Gum astringent, vulnerary, diuretic.
Uses. Fresh wounds, sores, colic, gonorrhœa; as a plaster in gout and rheumatism.

Fig. 168.—Pistacia atlantica in fruit.

3. R. *Cotinus* Linnæus.—(Venetian Sumac. Wig Tree.) Fig. 169.
Leaves simple, obovate; branches of the panicle sterile, becoming covered with long hairs.

Habitat. South of Europe.
Quality. Wood astringent, dyes yellow.
Uses. As a dye under the name of Young Fustic.

169

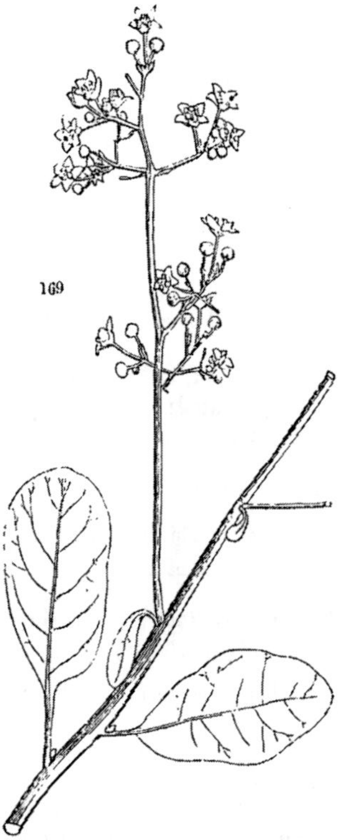

170

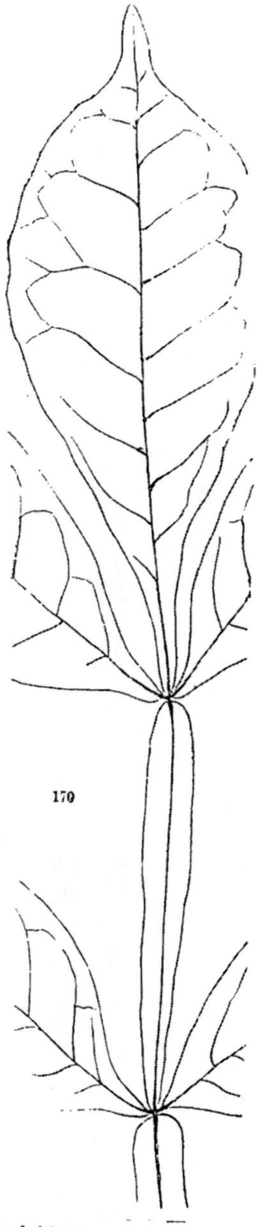

Natural Order, **Quassiads**; *Simarubaceæ*
(V. K., p. 477.)

Prevailing Quality. Bitter, narcotic.

Quassia. *Linnæus.*

Flowers hermaphrodite. *Stamens* 10, longer than the petals. *Drupes* 5.

1. Q. *amara* Linnæus.—(Surinam Quassia.) Fig. 170.
Leaves unequally pinnate, with a broad-winged jointed petiole; racemes terminal; flowers large, red.

Fig. 169.—Branch in flower of Rhus Cotinus; 170. Leaf of Quassia amara.

Habitat. Surinam.
Quality. Intensely bitter.
Uses. Like those of Simaruba ; its infusion a fly-poison.

SIMARUBA. *Aublet.*

Flowers polygamous. *Stamens* 10, as long as the petals. *Carpels* 5, surrounded by 10 short hairy scales.

1. S. *amara* Aublet.—(MOUNTAIN DAMSON.) Fig. 171.
Leaves abruptly pinnate ; leaflets alternate, on short stalks, downy on the under side.

Habitat. West Indies.
Quality. Bitter, tonic ; emetic, purgative.
Uses. Dysentery, diarrhœa, anorexia, intermittents.

171

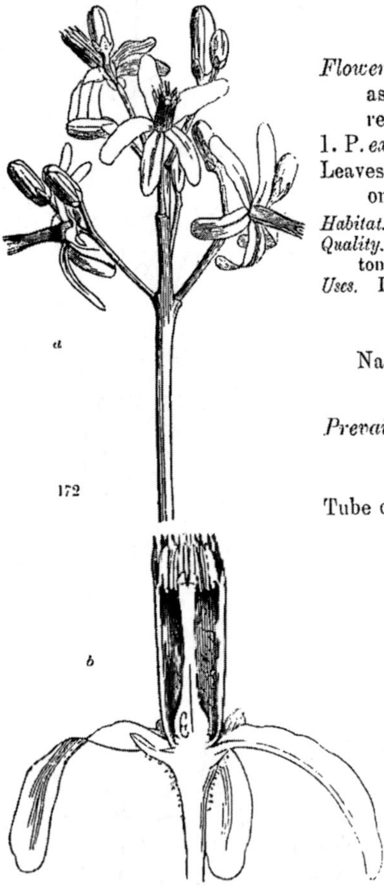

172

PICRÆNA. *Lindley.*

Flowers polygamous. *Stamens* 5, as long as the petals. *Carpels* 3, on a tumid receptacle.

1. P. *excelsa* Lindley.—(JAMAICA QUASSIA.)
Leaves unequally pinnated; leaflets opposite, on short stalks.

Habitat. Jamaica.
Quality. Narcotic poison ; bitter, stomachic, tonic ; antiseptic.
Uses. Dyspepsia, anorexia, intermittents.

Natural Order, Meliads; *Meliaceæ*
(V. K., p. 463.)

Prevailing Quality. Bitterness.

MELIA. *Linnæus.*

Tube of *stamens* very long, ten-cleft, bearing 10 anthers below the end. *Fruit* a drupe, with a 1-celled bony nut.

1. M. *Azedarach* Linnæus. Fig. 172.
Leaves unequally bipinnate with opposite ovate acute serrated leaflets ; fruit the size of an Olive.

Habitat. Syria ; commonly cultivated in the south of Europe.
Quality. Root bitter, nauseous. Pulp of fruit suspicious.
Uses. Root as an anthelmintic in the United States.

Fig. 171.—Portion of inflorescence of Simaruba amara ; 172. *a*, do. of Melia Azedarach : *b*, a perpendicular section of a flower magnified.

Natural Order, **Beancapers**; *Zygophyllaceæ* (V. K., p. 478.)

Prevailing Quality. Acridity.

ZYGOPHYLLUM. *Linnæus.*

Style tapering. *Flowers* complete, $\frac{5}{}$/. *Seeds* in 2 rows in each cell.

1. Z. *Fabago* Linnæus.—(BEAN CAPER.)

Leaflets in pairs, obovate ; pedicels erect ; calyxes smooth ; petals undivided.

Habitat. Syria.
Quality. Anthelmintic.
Uses. As a vermifuge ; flower-buds a substitute for capers.

GUAIACUM. *Plumier.*

Style tapering. *Flowers* complete, $\frac{5}{}$/. *Seeds* (by abortion) solitary. *Capsules* fleshy, stipitate.

1. G. *officinale* Linnæus.—(LIGNUM VITÆ.) Fig. 173.

Leaflets in 2 pairs, obovate or oval, obtuse.

Habitat. West Indies.
Quality. Acrid, stimulant, emmenagogue.
Uses. Chronic rheumatism, gout, scrophula, syphilis, painful menstruation ; leaves used as a substitute for soap ; hard and heavy timber called lignum vitæ.

2. G. *sanctum* Linnæus.

Leaflets in 5 or 7 pairs, oval, obtuse, mucronulate ; petioles and twigs somewhat downy.

Habitat. Porto Rico.
Quality and *Uses.* As in the last.

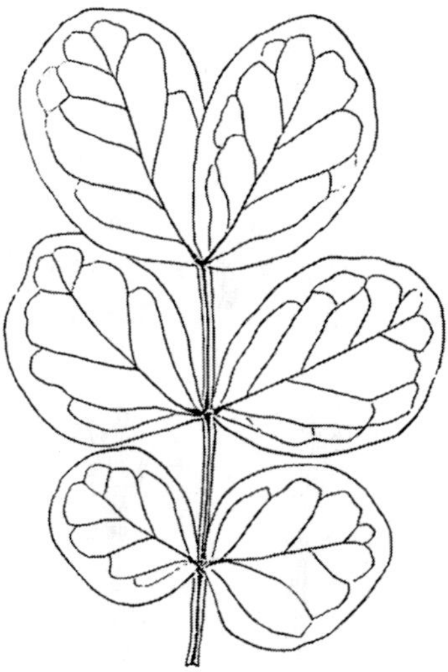

173

THE GERANIAL ALLIANCE (V. K., p. 484.)

Natural Orders of Geranials.

Flaxworts (*Linaceæ.*) Fruit beakless. Seeds exalbuminous. Leaves jointless.

Oxalids (*Oxalidaceæ.*) Fruit beakless. Seeds albuminous. Leaves jointed.

Cranes-bills (*Geraniaceæ.*) Fruit beaked.

Fig. 173.—Leaf of Guaiacum officinale.

Natural Order, Flaxworts; *Linaceæ* (V. K., p. 485.)

Prevailing Quality. Purgative, emollient.

LINUM. *Linnæus.*

Sepals 3-5, persistent. *Petals* 3-5, slightly united at the base. *Stamens* slightly monadelphous. *Capsule* 3-5-celled, each cell subdivided by a spurious partition.

1. L. *usitatissimum* Linnæus.—(COMMON FLAX.) Fig. 174.

Stem simple, erect; leaves alternate, lanceolate; flowers blue; sepals ovate, acuminate, without glands.

Habitat. Hedges and woods of Europe.
Quality. Seeds, called Linseed, emollient, demulcent.
Uses. To allay inflammation; in tea, poultices. Oil largely employed by painters. The crushed seeds are the oil-cake of farmers.

2. L. *catharticum* Linnæus.—(PURGING FLAX.) Fig. 175.

174

a

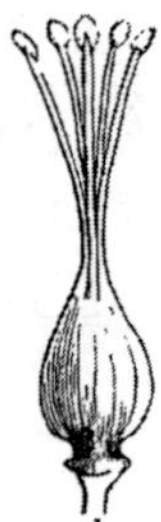

b

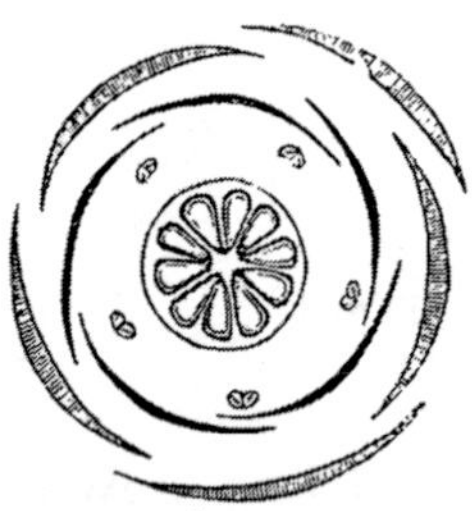

c

175

a b

Stem branched; leaves opposite, rough-edged, the lowest obovate, the others lanceolate; flowers white; sepals fringed with glands.

Habitat. Damp meadows and fields.
Quality. Bitter, purgative, diuretic. *Uses.* In rheumatism.

Fig. 174.—*a*, Linum usitatissimum; *b*, its pistil; *c*, diagram of its flower; 175. Linum catharticum; *a*, a flower seen in front; *b*, do. from behind.

Natural Order, Oxalids; *Oxalidaceæ* (V. K., p. 488.)

Prevailing Quality. Acidity.

OXALIS. Linnæus.

Stamens 10. *Fruit* capsular.

1. O. *Acetosella* Linnæus.—(WOOD SORREL. SHAMROCK.) Fig. 176.

Stemless; leaflets obcordate, downy; peduncle longer than the petiole, 1-flowered, with a pair of bracts above the middle.

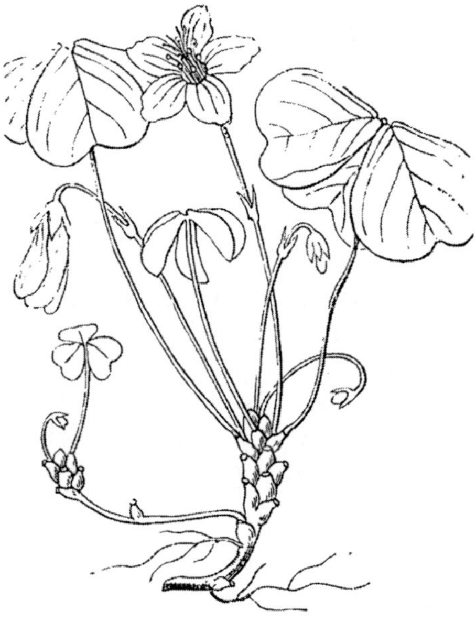

176

Habitat. Dry woods, very common.
Quality. Acid; refrigerant. *Uses.* Leaves used in salad; infusion in fevers.

2. O. *crenata* Jacquin.—(OCA.)

Tuberous; stem erect, leafy; leaflets obovate; peduncle 5-6-flowered, longer than the leaves; petals crenate.

Habitat. Mountains of South America.
Quality. Extremely acid, nutritious.
Uses. Tubers, after exposure to light, sweeten and become a useful food.

Natural Order, Cranes-bills; *Geraniaceæ* (V. K., p. 496.)

Prevailing Quality. Aromatic, resinous.

GERANIUM. *Linnæus.*

Flowers regular. *Stamens* 10, all fertile.

1. G. *Robertianum* Linnæus.—(HERB ROBERT.)

Leaves in 3-5 divisions; lobes trifid, pinnatifid; petals entire, twice as long as the aristate calyx.

Fig. 176.—Oxalis Acetosella.

Habitat. A common weed.
Quality. Astringent, aromatic.
Uses. In nephritic disorders.

2. G. *maculatum* Linnæus.—(ALUM-ROOT.)

Stem somewhat angular, erect, dichotomous, hairy backwards; leaves in 3-5 divisions, cut, toothed, those next the root on long stalks, the uppermost opposite and sessile; petals entire.

Habitat. United States.
Quality. Root extremely astringent.
Uses. Gargles; diarrhœa, chronic dysentery, cholera infantum, hemorrhages, gleet, gonorrhœa, &c.

ERODIUM. *L'Heritier.*

Flowers regular. *Stamens* 10, half being sterile.

1. E. *moschatum* Willdenow.

Stem procumbent; leaves pinnatifid, the segments on short stalks, ovate, unequally cut, serrated; peduncles many-flowered, covered with glandular pubescence.

Habitat. South of Europe.
Quality. Astringent, aromatic, musky.

THE SILENAL ALLIANCE (V. K., p. 495.)

Natural Orders of Silenals.

Silenads, or Cloveworts (*Caryophyllaceæ.*) Polypetalous, symmetrical. Leaves opposite, without stipules.

Purslanes (*Portulacaceæ.*) Polypetalous, unsymmetrical. (Sepals 2).

Buckwheats (*Polygonaceæ.*) Apetalous. Leaves alternate, with ochreate stipules.

Natural Order, **Silenads**, or Cloveworts; * *Caryophyllaceæ* (V. K., p. 496.)

Prevailing Quality. Saponaceous; poisonous.

AGROSTEMMA. *Linnæus.*

Calyx with 5 long leafy teeth, naked. *Stamens* 10. *Styles* 5. *Capsule* completely 1-celled.

1. A. *Githago* Linnæus.—(CORN-COCKEL.)

Petals blunt; segments of the calyx longer than both the tube and the petals.

Habitat. Corn-fields; annual.
Quality and *Uses.* Seeds contain saponine, a poisonous principle; they render flour unwholesome when ground among corn.

VACCARIA. *De Candolle.*

Calyx 5-toothed, naked, angular, winged. *Stamens* 10. *Styles* 2. *Capsule* 4-toothed.

1. V. *vulgaris* Host. *Saponaria Vaccaria* Linn.—(SOAPWORT.)

Stem erect, smooth; leaves lanceolate, connate at the base; flowers loosely corymbose.

* It seems desirable to change the name of *Cloveworts*, because of the identity in name of the garden Clove, a type of this order, and the Clove of the shops, which belongs to Myrtle-blooms.

Habitat. Europe, in fields and on hills ; an annual.
Quality and *Uses.* Roots contain saponine ; herbage said to increase the milk of cows.

Natural Order, **Purslanes** ; *Portulacaceæ* (V. K., p. 500.)

Prevailing Quality. Insipidity.

PORTULACA. *Linnæus.*

Calyx bifid, deciduous, leaving behind a circular base. *Stamens* 8-15, distinct. *Stigmas* 6. *Capsule* circumscissile.

1. P. *oleracea* Linnæus.—(PURSLANE.) Fig. 177.

Stem and branches fleshy, prostrate ; leaves wedge-shaped, succulent ; flowers solitary, sessile, yellow ; sepals bluntly keeled.

177

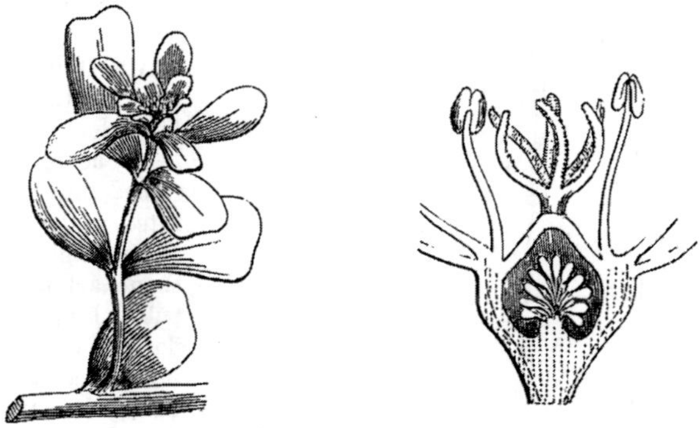

a *b*

Habitat. Cultivated in gardens.
Quality. Antiscorbutic, refrigerant.
Uses. Leaves a common potherb, and salad.

Natural Order, **Buckwheats** ; *Polygonaceæ* (V. K., p. 502.)

Prevailing Quality. Purgative, astringent.

POLYGONUM. *Linnæus.*

Calyx 4-5-cleft, more or less coloured. *Stamens* variable in number, in 2 rows, generally with glands at the base. *Styles* more or less united at the base. *Nut* invested by the permanent calyx. *Embryo* lateral.

1. P. *Hydropiper* Linnæus.—(WATER PEPPER.) Fig. 179.

Leaves lanceolate ; ochreæ with short fringes ; spikes filiform, lax, pendulous ; flowers hexandrous, with glandular dots.

Fig. 177.—*a*, a twig of Portulaca oleracea, natural size ; *b*, section of the flower of Portulaca oleracea, magnified.

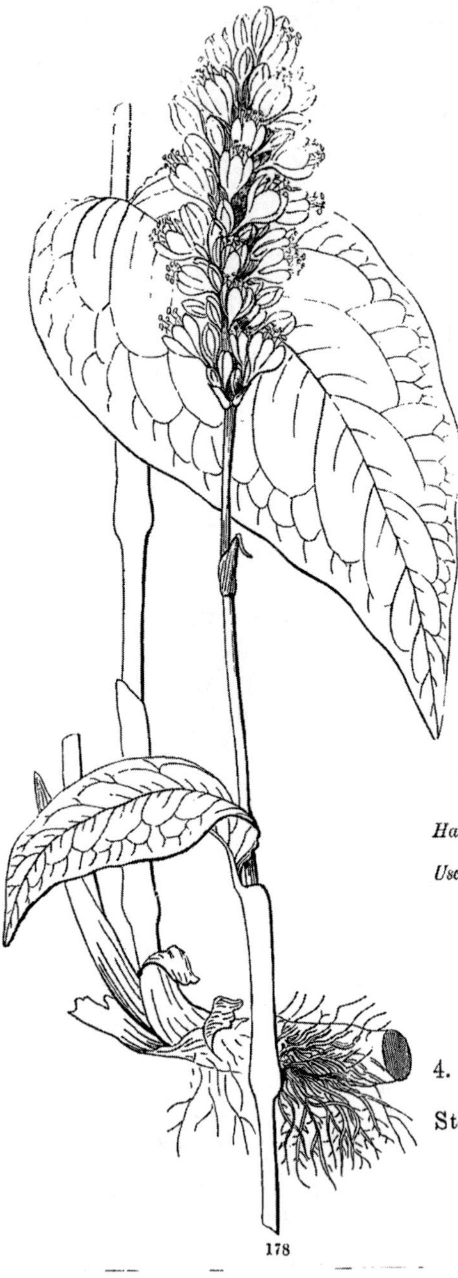

Habitat. Ditches.
Quality. Very acrid, diuretic.
Uses. Leaves are vesicant, but they require to be fresh.

2. P. *Bistorta* Linnæus.—(BISTORT.) Fig. 178.
Leaves oblong-ovate, rather cordate and wavy; petioles winged; flowers in a dense terminal spike.

Habitat. Meadows.
Quality. Powerfully astringent; tonic.
Uses. Decoction in leucorrhœa and gleet, spongy gums, relaxed sore throat, hemorrhages, fluxes; intermittents.

3. P. *amphibium* Linnæus.
Stems creeping, root-like; leaves stalked, generally floating, ovate-lanceolate, more or less acute, single-ribbed, minutely serrated, smooth, bright green, slightly heart-shaped at the base; stipules abrupt, smooth; flowers in dense, ovate, bracteated heads; stamens 5; styles 2, united at base; stigmas globular, red.

Habitat. Ponds, ditches, and wet places.
Uses. The root-like stems bear some resemblance to Sarsaparilla, and are substituted for the foreign drug by the herbalists of Nancy; it is said that it resembles true Sarsaparilla in its properties, and that the apothecaries and druggists of Lorraine use it in preference.

4. P. *aviculare* Linnæus.—(KNOT GRASS). Fig. 180.
Stems prostrate, or nearly so, leafy up to the ends; leaves lanceolate or oblong; ochreæ bifid, 6-nerved; flowers axillary; nuts rugose.

178

Fig. 178.—Polygonum Bistorta.

Habitat. Waste places, every where.
Quality and *Uses.* The small nuts reputed to be emetic and purgative.

179

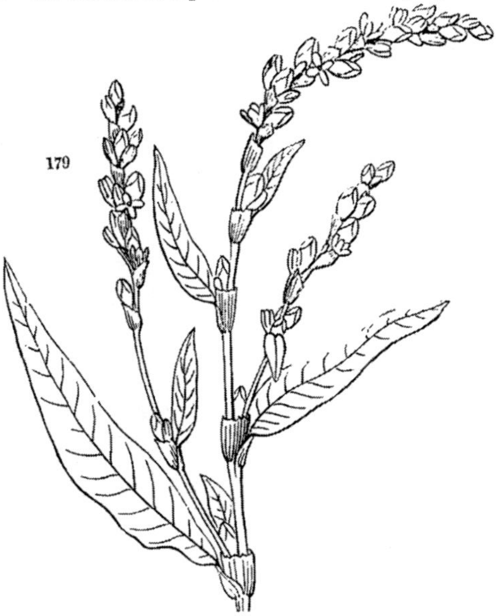

180

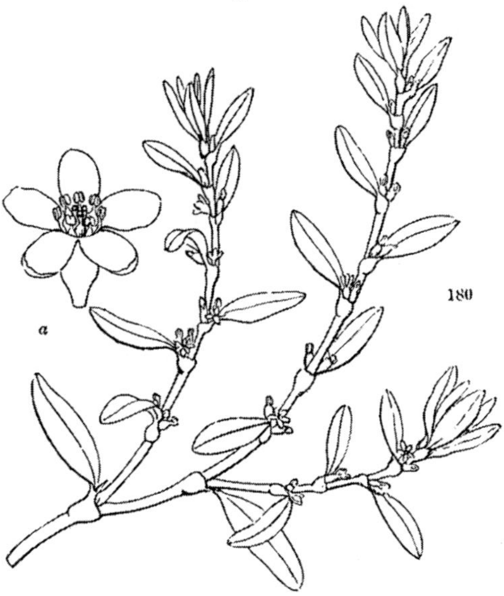

FAGOPYRUM. *Tournefort.*

As Polygonum ; but embryo central, and flowers in panicled racemes.

Fig. 179.—Polygonum Hydropiper ; 180. P. aviculare ; *a*, a flower magnified.

1. F. *esculentum* Meisner.—(Buckwheat.) Fig. 181.

Leaves sagittate, cordate, acuminate; racemes corymbose; flowers white.

Habitat. Commonly cultivated.
Quality. Nuts filled with starchy matter; grateful to pheasants.
Uses. The flour largely consumed as food. Herbage ploughed in as a green-crop.

181

a b

Rumex. *Linnæus.*

Sepals 6, the 3 inner larger, growing, converging, and finally concealing the nut. *Stamens* 6. *Stigmas* pencilled. *Embryo* lateral.

1. R. *alpinus* Linnæus.—(Monk's Rhubarb.)

Radical leaves roundish, cordate, obtuse; inner sepals cordate, membranous, entire, or slightly toothed, without a callus.

Habitat. Alps of Europe.
Quality. Root yellow, fleshy, astringent, purgative.
Uses. As rhubarb formerly, but not so powerful.

2. R. *Hydrolapathum* Hudson.—(Water Dock.) Fig. 182.

Leaves lanceolate, narrowed at the base, with the petiole flat on the upper side; inner sepals ovate-triangular, entire, or slightly toothed, all calliferous; racemes panicled, leafless.

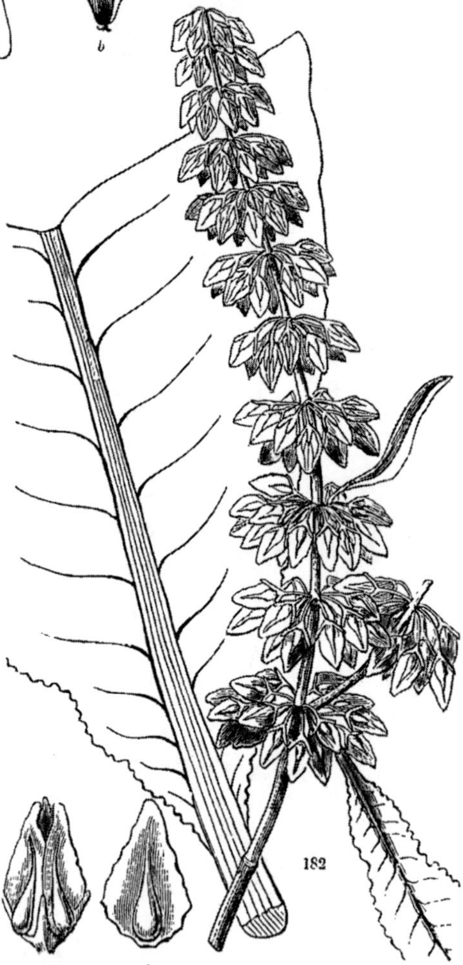

182

a b

Fig. 181.—*a*, flower of Fagopyrum esculentum, magnified; *b*, fruit of do.; 182. Rumex Hydrolapathum; *a*, ripe sepals concealing the fruit; *b*, one of the sepals.

Habitat. Ditches and ponds.
Quality. Root acrid, bitter, astringent, antiscorbutic.
Uses. Scurvy, skin diseases, rheumatism; root a dentifrice, and an astringent gargle.

3. R. *Acetosa* Linnæus. — (COMMON SORREL.) Fig. 183.

Leaves sagittate or hastate, veiny; flowers diœcious; inner sepals roundish, cordate, not calliferous, with a deflexed scale at the base.

Habitat. Woods and pastures.
Quality. Acid, astringent, slightly nutritive, refrigerant.
Uses. A pot-herb and salad; cooling drinks.

RHEUM. *Linnæus.*

Sepals 6, withering. *Stamens* 9. *Stigmas* discoid. *Nut* naked, 3-winged. *Embryo* central. Fig. 184.

1. R. *palmatum* Linnæus.

Leaves half palmate; petiole terete.

184

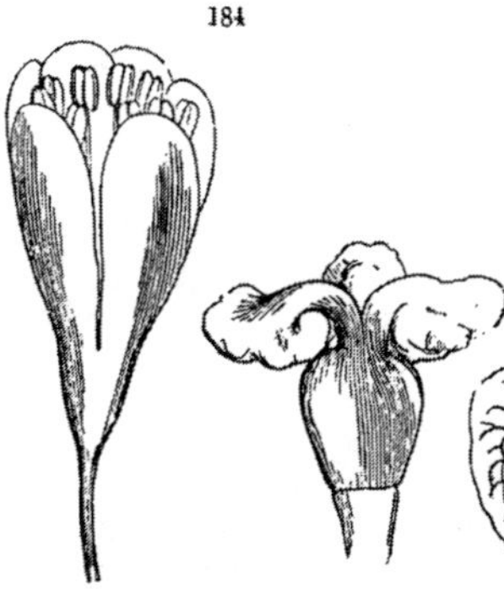

a b

b 183 a

Fig. 183.—Rumex Acetosa; *a*, ripe calyx; *b*, one of the sepals; 184. *a*, flower of Rheum, much magnified; *b*, its pistil, do.

Habitat. Chinese Tartary.
Quality. This and the following astringent, tonic, purgative.
Uses. As a purgative in diarrhœa, dyspepsia; as an external application in healing indolent ulcers. Supposed to be Russian Rhubarb.

2. R. *undulatum* Linnæus.

Leaves oval, obtuse, extremely wavy, when young covered with short white hairs; petioles crimson, semi-cylindrical, with elevated edges.

Habitat. Siberia.
Quality and *Uses.* As in the last. Supposed to be Russian Rhubarb.

3. R. *Emodi* Wallich.

Leaves roundish, cordate, entire, rather wavy, very rough; with angular rough petioles.

Habitat. Himalayah.
Quality and *Uses.* As in the last, but more astringent, and less aromatic. Its root is Himalayan Rhubarb.

Coccoloba. *Linnæus.*

Calyx becoming pulpy and investing the bony nut. *Embryo* central.

1. C. *uvifera* Linnæus.—(Seaside Grape.)

Leaves orbicular, cordate, coriaceous, shining, entire; a small tree.

Habitat. Sea-coast in West Indies.
Quality. Astringent, sub-acid.
Uses. Fruit succulent, eatable; extract found extremely astringent. Jamaica Kino.

THE CHENOPODAL ALLIANCE (V. K., p. 505.)

Natural Orders of Chenopodals.

Nyctagos (*Nyctaginaceæ.*) Calyx tubular, becoming bony at the base.
Phytolaccads (*Phytolaccaceæ.*) Sepals distinct. Carpels several.
Chenopods (*Chenopodiaceæ.*) Sepals distinct. Carpel single.

Natural Order, Nyctagos; *Nyctaginaceæ* (V. K., p. 506.)

Prevailing Quality. Purgative.

Mirabilis. *Linnæus.*

Involucel resembling a calyx, 1-flowered. *Calyx* petaloid, funnel-shaped. *Stamens* 5. *Stigma* capitate. *Fruit* a corrugated false nut, formed out of the hardened base of the calyx.

1. M. *Jalapa* Linnæus.—(Marvel of Peru.) Fig. 185.

Leaves ovate, cordate, smooth; flowers clustered; tube of calyx twice as long as the limb, the segments of which are nearly entire.

Habitat. Tropics of both hemispheres.
Quality. Roots drastic, purgative, when old; inert when young.

185

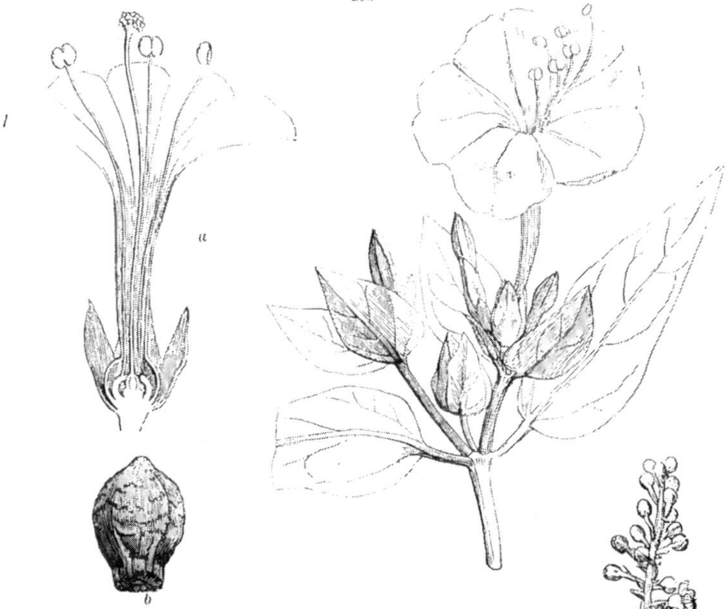

Natural Order, Phytolaccads; *Phytolaccaceæ* (V. K., p. 508.)

Prevailing Quality. Emetic, narcotic.

PHYTOLACCA. *Linnæus.*

Calyx 5-parted. *Stamens* 8-10. *Stigmas* 8-10. *Berry* 8-10-celled, each cell one-seeded.

1. P. *decandra* Linnæus.—(POCAN BUSH.) Fig. 186. Flowers with 10 stamens and 10 styles.

Habitat. United States.
Quality. Purgative, emetic, narcotic; leaves acrid.
Uses. Like those of Guaiacum; against chronic rheumatism and syphilitic pains. Young tender blanched shoots eatable when boiled; pulp of berries employed in the adulteration of wine. Root an emetic, approaching nearly to Ipecacuanha. Bigelow says, that "from abundant experience, he is satisfied that, when properly prepared, it operates in the same doses and with the same certainty" as that drug. Its exhibition sometimes attended by slight narcotic symptoms. Externally applied it excites a sense of heat and smarting; cures psora, and tœnia capitis.

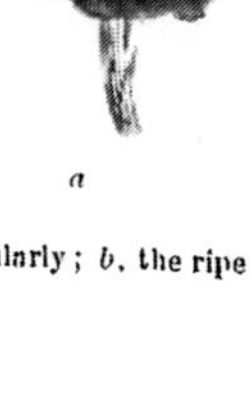

186

Fig. 185.—Mirabilis Jalapa; *a*, a flower cut perpendicularly; *b*, the ripe fruit; 186. Flowers of Phytolacca decandra; *a*, the ripe fruit.

Natural Order, Chenopods; *Chenopodiaceæ* (V. K., p. 512.)

Prevailing Quality. Insipid ; rarely aromatic and stimulating.

Beta. *Linnæus.*

Calyx 5-cleft. *Stamens* 5, inserted on a fleshy ring, surrounding the ovary. *Fruit* adhering to the calyx, and collected in clusters of 2 or 3.

1. B. *vulgaris* Linnæus.—(Garden Beet.) Fig. 187.

Root large, fleshy, succulent ; radical leaves ovate, obtuse, somewhat cordate.

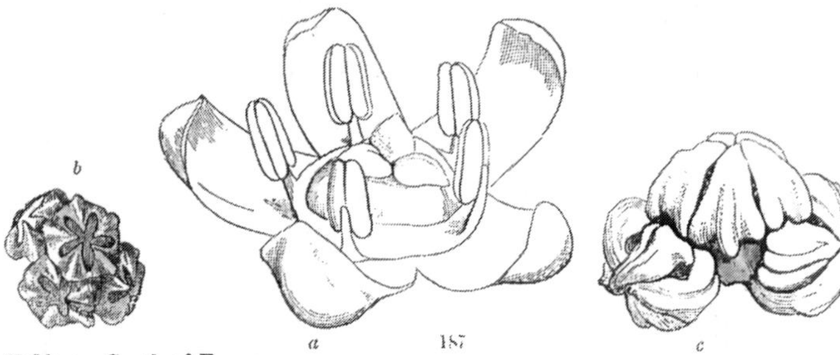

Habitat. South of Europe.
Quality. Roots sugary, nutritious.
Uses. As food for man and cattle ; leaves like cabbage, but earthy and unpleasant.

Spinacia. *Linnæus.*

Flowers polygamo-diœcious. ♂ *Calyx* 4-parted. *Stamens* 4. ♀ *Calyx* 2-3-cleft ; *styles* 4. *Fruit* connate with the hardened calyx.

1. S. *oleracea* Linnæus.—(Spinach.)

Leaves hastate, oblong-ovate.

Habitat. Levant.
Quality. Insipid, nutritious.
Uses. Leaves a common pot-herb.

Atriplex. *Linnæus.*

Flowers polygamo-monœcious. ♂ or ☿ *Calyx* 3-5-parted. *Stamens* 3-5. ♀ *Calyx* compressed, 2-lobed, or 2-parted.

1. A. *hortensis* Linnæus.—(Garden Orach.)

An annual; leaves cordate, triangular, the upper rather hastate; calyx of fruit roundish ovate, netted, entire.

Habitat. Commonly cultivated.
Quality. Leaves insipid, nutritious. Seeds emetic ?
Uses. Leaves an old-fashioned pot-herb.

Salsola. *Linnæus.*

Sepals 5 ; with a transverse appendage at the back when ripe. *Stamens* 5.

1. S. *Kali* Linnæus.

Leaves subulate, spiny at the point ; ripe sepals cartilaginous, with acuminate segments as long as the round spreading appendages.

Habitat. Salt marshes of Europe.
Quality. Saline.
Uses. A common source of soda.

Fig. 187.—*a*, flower of Beta vulgaris much magnified ; *b*, its fruit, natural size ; *c*, the same magnified.

SALICORNIA. *Linnæus.*

Calyx fleshy, entire, sunk in an excavation of the rachis. *Stamens* 1 or 2.

1. S. *annua* Smith.—(SALTWORT.) Fig. 188.
Stem herbaceous; calyxes placed in a triangle.

Habitat. Salt marshes.
Quality. Saline.
Uses. Shoots, when young, pickled, and sold under the false name of Samphire. A source of soda.

CHENOPODIUM. *Linnæus.*

Flowers hermaphrodite. *Calyx* 5-cleft, without appendages. *Stamens* 5. *Utricle* depressed.

1. C. *olidum* Curtis.—(STINKING GOOSE-FOOT.)
Leaves rhomboid-ovate, entire, hoary with meal; racemes leafless; seeds shining, finely dotted.

Habitat. Waste ground, especially at the foot of walls.
Quality. Nauseously fetid.
Uses. As an antispasmodic and emmenagogue; a popular remedy in much repute.

2. C. *Botrys* Linnæus. *Ambrina* Spach.—(JERUSALEM OAK.) Fig. 189.
Leaves pinnatifid, sinuous, obtuse, covered with viscid glands, racemes axillary and terminal, naked.

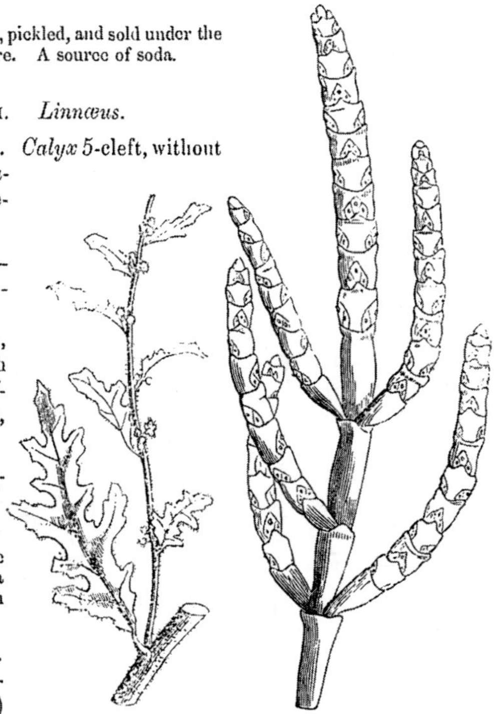

189 188

Habitat. South of Europe.
Quality. Fragrant, expectorant; anthelmintic.
Uses. Catarrh, and humoral asthma.

3. C. *anthelminticum* Linnæus.—(WORMSEED.)
Leaves smoothish, oblong, narrowed to the base, acute, unequally sinuate and serrate; racemes spicate, naked, axillary, and terminal.

Habitat. United States.
Quality. Strong scented, somewhat aromatic.
Uses. Seeds yield Wormseed oil; both employed in the expulsion of worms.

Fig. 188.—Salicornia annua; 189. Chenopodium Botrys.

THE PIPERAL ALLIANCE (V. K., p. 515.)

Natural Orders of Piperals.

Pepperworts (*Piperaceæ.*) Ovule erect. Leaves usually alternate.

Chloranths (*Chloranthaceæ.*) Ovule suspended. Leaves opposite, with intermediate stipules.

Natural Order, **Pepperworts**; *Piperaceæ* (V. K., p. 515.)

Prevailing Quality. Pungent, stimulant, aromatic.

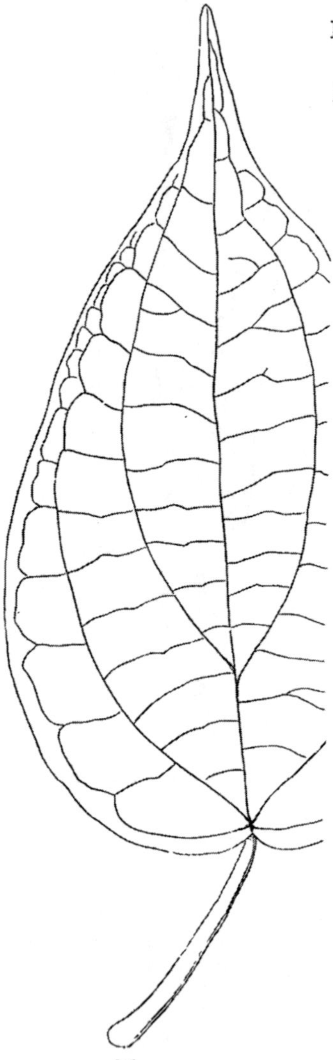

190

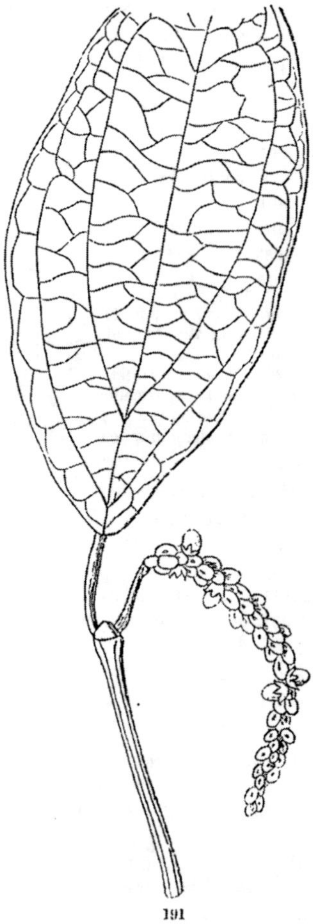

191

Fig. 190.—Leaf of Piper nigrum; 191. P. trioicum in flower.

PIPER. *Miquel.*

Woody. *Spikes* solitary, opposite the leaves. *Flowers* sessile, polygamo-dioecious. *Bracts* oblong, sessile, decurrent.

1. P. *nigrum* Linnæus.—(BLACK PEPPER. WHITE PEPPER.) Fig. 190. Lower leaves roundish ovate, nearly equilateral, pale beneath, dotted when young; amenta ⚥ or ♀, filiform, pendulous.

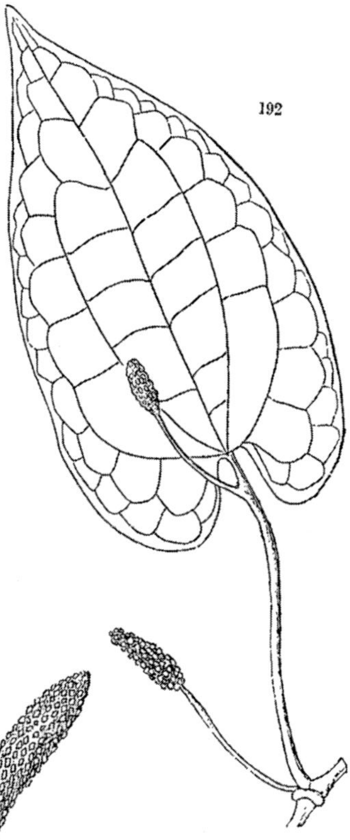

192

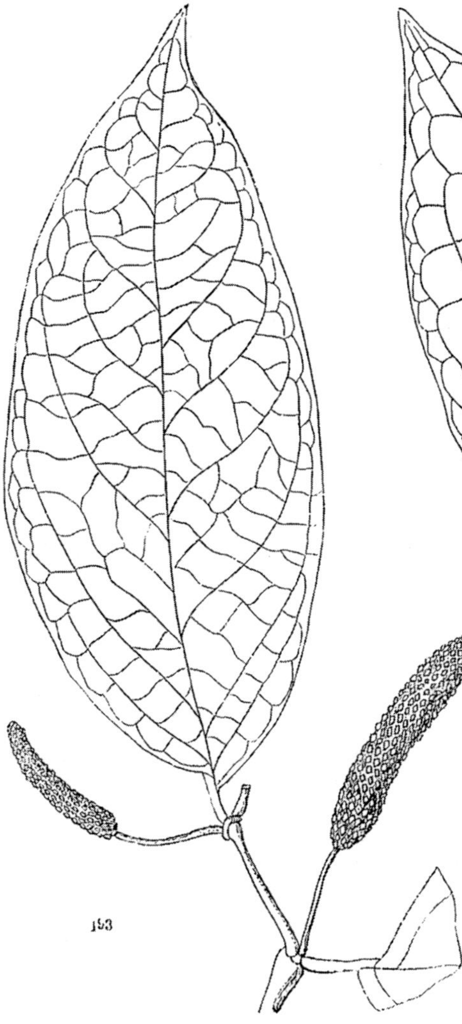

193

Habitat. Tropical Asia.

Quality. Acrid, aromatic; rubefacient, vesicant, stimulant.

Uses. In cookery; intermittent fevers, relaxed uvula, gonorrhœa, paralysis of the tongue.

2. P. *trioicum* Roxburgh. Fig. 191. Lower leaves cordate, obliquely elliptical, acuminate, the uppermost much narrower; amenta polygamous, ♂ filiform, ♀ stiffer and shorter.

Fig. 192.—Leaf and flowers of Chavica Roxburghii; 193. Of Chavica officinarum.

Habitat. East Indies.
Quality and *Uses.* Like the last, but more pungent.

Chavica. *Miquel.*

Woody. *Spikes* solitary, opposite the leaves. *Flowers* sessile, diœcious. *Fruit* sessile. *Bracts* stalked, quadrangular, peltate. *Style* 0.

1. C. *Roxburghii* Miquel.—(Common Long Pepper.) Fig. 192.

Rather hairy; lower leaves roundish ovate, 7-nerved; female spikes cylindrical, about as long as their stalk.

Habitat. Bengal.
Quality and *Uses.* Yields the common Long Pepper of the shops; see C. officinarum.

2. C. *officinarum* Miquel. *Piper longum* Linnæus.—(Java Long Pepper.) Fig. 193.

Smooth; lower leaves ovate-cordate, 3-5 nerved; female spikes cylindrical, short, rather narrowed at the point.

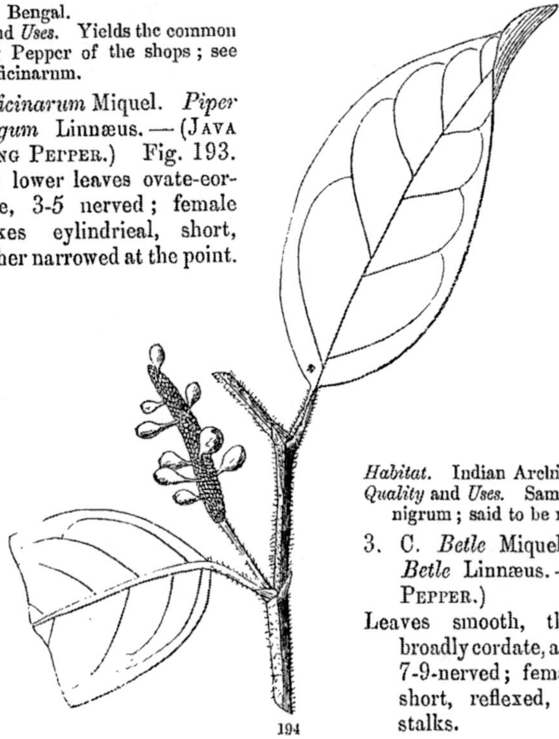

194

Habitat. Indian Archipelago.
Quality and *Uses.* Same as Piper nigrum; said to be more acrid.

3. C. *Betle* Miquel. *Piper Betle* Linnæus.—(Betle Pepper.)

Leaves smooth, the lower broadly cordate, acuminate, 7-9-nerved; female spikes short, reflexed, on long stalks.

Habitat. All over the East Indies.
Quality and *Uses.* The leaf wrapped round the Areca, with a little quicklime, is chewed by Oriental nations as a stimulating narcotic and astringent: but it is uncertain what its precise quality is.

Cubeba. *Miquel.*

Woody. *Spikes* solitary, opposite the leaves. *Flowers* diœcious. *Fruit* stalked. *Bracts* sessile.

1. C. *officinalis* Miquel. *Piper Cubeba* Linnæus.—(Cubebs.)

Leaves coriaceous, smooth, rather large; fruit globose, shorter than their stalks.

Fig. 194.—Cubeba canina.

Habitat. Java.
Quality. Acrid, aromatic, stomachic, stimulant of the urinogenital apparatus, aphrodisiac, diuretic.
Uses. Gonorrhœa, gleet, leucorrhœa ; dyspepsia.

2. C. *canina* Miquel.—(CUBEBS.) Fig. 194.

Leaves membranous, hairy ; fruit roundish-ovate, almost longer than their stalks.

Habitat. Java.
Quality and *Uses.* According to Blume, this furnishes part of the Cubebs of the shops.

ARTANTHE. *Miquel.*

Woody. *Spikes* solitary, opposite the leaves. *Flowers* ☿. *Style* 0. *Bracts* peltate or cucullate.

1. A. *elongata* Miquel. *Piper angustifolium* Ruiz and Pavon. — (MATICO PLANT.) Fig. 195.

Leaves short-stalked, harsh, oblong-lanceolate, acuminate, tessellated on the upper side, because of the sunken veins.

Habitat. Peru.
Quality. Powerful styptic.
Uses. Diseases of genital organs and rectum ; hæmorrhages ; also as Cubebs.

MACROPIPER. *Miquel.*

Woody. *Spikes* clustered, axillary. *Flowers* diœcious.

1. M. *methysticum* Miquel. *Piper methysticum.*—(KAWA, or AVA PLANT.)

Leaves membranous, equal-sided, broadly ovate, roundish, deeply cordate, with 9-10 nerves.

Habitat. Society Islands.
Quality. Stimulating narcotic, sudorific, aromatic.
Uses. Chronic rheumatism ; venereal diseases.

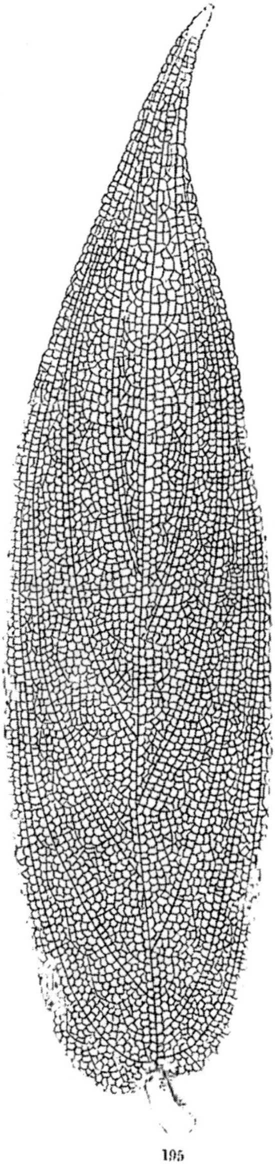

195

Fig. 195.—Leaf of Artanthe elongata.

Natural Order, Chloranths; *Chloranthaceæ* (V. K., p. 519.)

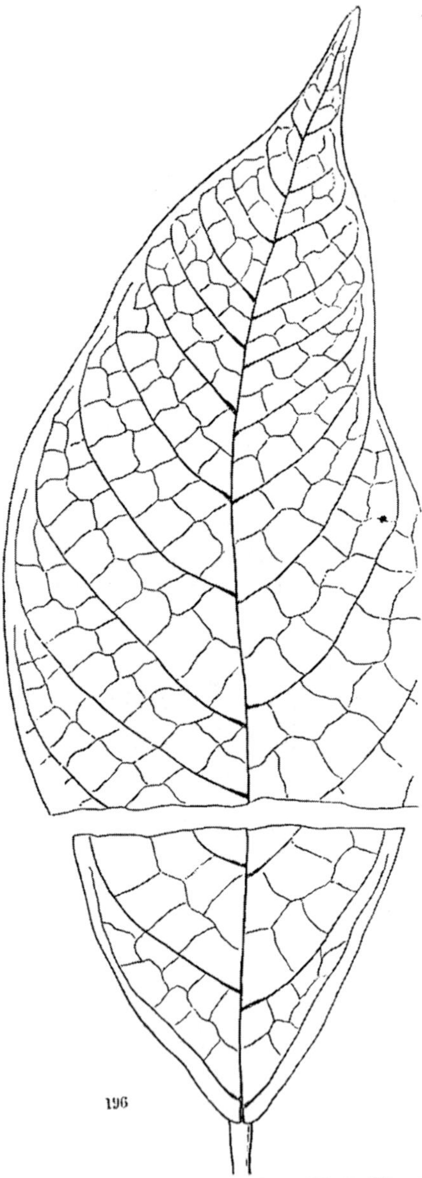

196

Fig. 196.—Leaf of Theobroma Cacao.

Prevailing Quality. Aromatic, stimulant.

CHLORANTHUS. *Swartz.*

Bracts scale-like. *Anthers* lobed, 2-or-4-celled. *Stigma* sessile. A *drupe.*

1. C. *officinalis* Blume.

Leaves oblong, acuminate, thin, shining, with glandular serratures; bracts dotted with glands.

Habitat. Java.
Quality. Aromatic, fragrant, powerfully stimulant.
Uses. Spasms, typhoid fevers, malignant small-pox.

2. C. *inconspicuus* Swartz.—(The CHU-LAN.)

Leaves thin, obovate, crenate or dentate, with blunt glandular serratures; bracts without glands.

Habitat. China.
Quality. Like the last.
Uses. Spikes of flowers scent Tea in China.

THE MALVAL ALLIANCE (V. K., p. 359.)

Natural Orders of Malvals.

Byttneriads (*Byttneriaceæ.*) Stamens monadelphous, partly sterile.

Mallow-worts (*Malvaceæ.*) Stamens columnar, all perfect.

Lindenblooms (*Tiliaceæ.*) Stamens separate.

Natural Order, Byttneriads; *Byttneriaceæ* (V. K., p. 363.)

Prevailing Quality. Uncertain.

THEOBROMA. *Jussieu.*

Sepals 5. *Petals* 5, arched at the base, and extended into a strap. *Fruit* fleshy, 5-celled. *Seeds* lying in buttery pulp.

1. T. *Cacao* Linnæus.—(COCOA or CHOCOLATE TREE.) Fig. 196.

Leaves quite entire, ovate-oblong, acuminate, smooth and the same colour on both sides.

Habitat. West Indies.
Quality. Somewhat astringent, oily, nutritious.
Uses. Forms chocolate with Vanilla, &c.

GUAZUMA. *Plumier.*

Sepals 5, more or less combined. *Petals* 5, 2-horned. *Capsule* woody, warted, valveless, 5-celled, pierced by many perforations.

1. G. *ulmifolia* Lamarck.

Leaves when full-grown oblong, unequally toothed, smooth on each side.

Habitat. West Indies.
Quality. Mucilaginous, sudorific.
Uses. Sweet succulent fruit, eaten in Brazil; bark in cutaneous diseases.

Natural Order, Mallow-worts; *Malvaceæ* (V. K., p. 369.)

Prevailing Quality. Mucilaginous.

ALTHÆA. *Linnæus.*

Calyx 5-cleft, surrounded by a 6-9-cleft involucel. *Styles* numerous. *Fruit* as in Malva.

1. A. *officinalis* Linnæus.—(MARSH MALLOW. GUIMAUVE *Fr.*)

Leaves soft and hoary on each side, the lower 5-lobed, the upper 3-lobed.

Habitat. Meadows in Europe.
Quality. Demulcent, pectoral.
Uses. As Malva sylvestris; injections in difficult parturition.

2. A. *rosea* Cavanilles.—(HOLLYHOCK.)

Stem tall, straight, hairy; leaves cordate, 5-7-angled, crenate, rugose; flowers axillary, sessile, or forming terminal spikes; petals hairy at base.

Habitat. South of Europe.
Quality. Flowers mucilaginous, demulcent. Leaves dye blue.
Uses. As the last.

ABELMOSCHUS. *Medicus.*

Involucel of from 8-15 bracts. *Carpels* many-seeded. *Seeds* smooth, or hairy along a dorsal line. *Corolla* spreading flat.

1. A. *esculentus* Wight and Arnott. *Hibiscus esculentus* Linn.—(OCHRO. GOBBO.)

Stem unarmed; leaves cordate, 5-lobed, toothed; bracts 10, deciduous; calyxes bursting laterally.

Habitat. Both Indies.
Quality. Fruit extremely mucilaginous.
Uses. In thickening soups; leaves as poultices.

2. A. *moschatus* W. and A. *Hibiscus Abelmoschus* Linn.—(MUSK-SEED.) Fig. 197.

197

Leaves rather peltate, cordate, 7-angled, acuminate, serrated; stem hispid; bracts 8-9; capsule bristly.

Habitat. Tropical America and Asia.
Quality. Seeds musky, stimulating, cordial, stomachic.
Uses. Tincture against serpent bites; seeds said to be roasted along with coffee by the Arabs.

198

MALVA. *Linnæus.*

Calyx 5-cleft, in a 3-leaved involucel. *Styles* numerous, connate downwards. *Fruit* orbicular, many-celled, with a convex centre.

1. M. *sylvestris* Linnæus.—(COMMON MALLOW.) Fig. 198.

Leaves angular, 5-7-lobed; peduncles in clusters, erect after flowering; valves of the fruit netted.

Habitat. Hedges and roadsides.
Quality. Emollient, demulcent.
Uses. Decoction in irritation of pulmonary and urinary organs, tenesmus; poultices in external inflammation.

GOSSYPIUM. *Linnæus.*

Bracts 3, leafy, connate at the base, and toothed. *Carpels* many-seeded. *Seeds* buried in wool.

1. G. *herbaceum* Linnæus.—(COTTON.)

Leaves 5-lobed, palmate, with a single gland on the under side, lobes rounded, mucronate; bracts serrated; stem smooth.

Habitat. East Indies.
Quality and *Uses.* Cotton-wool, applied to burns, allays pain and irritation; it is the great manufacturing substance from which proceeds a vast majority of modern linen fabrics; but it is weak and perishable.

2. G. *barbadense* Linnæus. Yields American cotton.

Fig. 197.—Seed of Abelmoschus moschatus, magnified; 198. Malva sylvestris.

Natural Order, **Lindenblooms**; *Tiliaceæ* (V. K., p. 371.)

Prevailing Quality. Mucilaginous, subastringent.

CORCHORUS. *Linnæus.*

Sepals 5, deciduous. *Petals* 5. *Stamens* 00. *Capsule* with 2-5 septiferous valves. *Seeds* in 2 rows.

1. C. *olitorius* Linnæus.

Leaves ovate-oblong, serrated ; the lower serratures usually extended into a bristle ; capsule oblong, taper, smooth.

Habitat. Tropical countries.
Quality. Mucilaginous, insipid, nutritious.
Uses. A pot-herb.

TILIA. *Linnæus.*

Flowers growing from a thin membranous bract. *Sepals* 5, deciduous.

1. T. *europæa* Linnæus.—(LIME or LINDEN TREE.) Fig. 199.

Leaves roundish-cordate, oblique, hairy beneath, especially at the axils of the leaves.

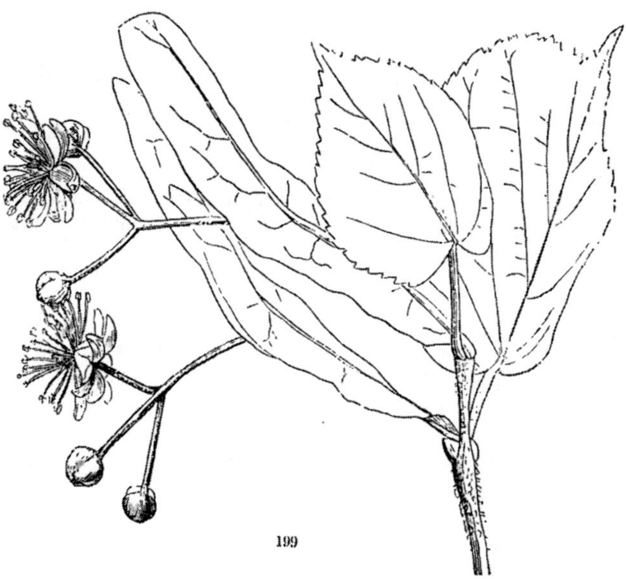

199

Habitat. Woods of Northern Europe.
Quality. Fibrous tunic very tough ; flowers sudorific, emollient ; bracts astringent.
Uses. Inner bark makes Russian mats ; infusion of flowers in vertigo and spasms, and against cough.

Fig. 199.—Tilia europæa in flower.

THE CISTAL ALLIANCE (V. K., p. 348.)

Natural Orders of Cistals.

Rock-roses (*Cistaceæ*). Flowers $\sqrt[3]{}$ or $\sqrt[5]{}$. Stamens 00.

Crucifers (*Brassicaceæ*). Flowers $\sqrt[4]{}$. Stamens tetradynamous.

Capparids (*Capparidaceæ*). Flowers $\sqrt[4]{}$. Stamens 00.

Natural Order, **Rock-roses**; *Cistaceæ* (V. K., p. 349.)

Prevailing Quality. Resinous, fragrant, stimulant.

CISTUS. *Linnæus.*

Sepals 5, the 2 outer unequal or deficient. *Capsule* 5-10 valved.

1. C. *creticus* Linnæus.—(LADANUM BUSH.)

Leaves ovate, obtuse, hairy, blunt, much wrinkled; flowers purple; style as long as the stamens.

Habitat. Candia, &c.
Quality. Resinous juice stimulant.
Uses. In plaisters; in perfumery, and pastiles; as an expectorant.

Natural Order, **Crucifers**; *Brassicaceæ* (V. K., p. 351.)

Prevailing Quality. Pungent, antiscorbutic.

BRASSICA. *Linnæus.*

Silique terete; each valve with one straight dorsal rib, and no lateral veins. *Seeds* globose in one row. *Embryo* conduplicate O>>.

1. B. *oleracea* Linnæus.—(CABBAGE.)

Leaves glaucous, never hispid; siliques and calyx both erect.

Habitat. Sea-coast of Europe, on cliffs.
Quality. Antiscorbutic, nutritious.
Uses. A common pot-herb.

2. B. *Rapa* Linnæus.—(TURNIP.)

Leaves bright green, hispid; root fleshy, succulent.

Habitat. France and the south of Europe.
Quality and *Uses.* Like the last.

3. B. *Napus* Linnæus.—(RAPE.)

Leaves glaucous, never hispid; siliques spreading.

Habitat. North of Europe?
Quality. Like the last.
Uses. Seeds yield rape-oil; the crushed residuum rape-cake; much grown as a green-crop.

SINAPIS. *Linnæus.*

Silique terete; the valves with 3 or 5 straight strong ribs. *Seeds* globose in a single row. *Embryo* O>>.

1. S. *alba* Linnæus.—(WHITE MUSTARD.)

Nearly smooth ; leaves pinnated ; valves of the silique 5-nerved, terminated by a sword-shaped horn.

Habitat. Fields.
Quality. As in S. nigra, but milder.
Uses. Seeds in torpor of digestive organs ; young leaves as salad.

2. S. *nigra* Linnæus.—(BLACK MUSTARD.)

Rough with hispid hairs ; leaves lyrate ; silique short, bluntly quadrangular, pressed close to the axis, without a sword-shaped point.

Habitat. Fields.
Quality. Acrid, stimulant, volatile, pungent, diuretic ; oil purgative.
Uses. Seeds produce vomiting ; employed in dyspepsia, dropsy, intermittents, and for forming sinapisms ; distilled water has been used against itch.

RAPHANUS. *Linnæus.*

Silique indehiscent, spongy, somewhat jointed. *Embryo* O>>.

1. R. *sativus* Linnæus.—(RADISH.)

Siliques terete, acuminate, hardly longer than their stalks ; roots fleshy, annual.

Habitat. Commonly cultivated.
Quality. Roots sub-acrid, succulent and tender when young.
Uses. A well-known esculent vegetable.

CRAMBE. *Linnæus.*

Silicle indehiscent, of 2 unequal joints, of which the upper is globose, the lower obsolete. *Embryo* O>>.

1. C. *maritima* Linnæus.—(SEA KALE.) Fig. 200.

Leaves roundish, sinuated, wavy, toothed, cæsious ; the longer filaments forked.

Habitat. Sea-coast, in clay.
Quality. Antiscorbutic, nutritive, acrid when old.
Uses. The blanched sprouts a favourite esculent vegetable.

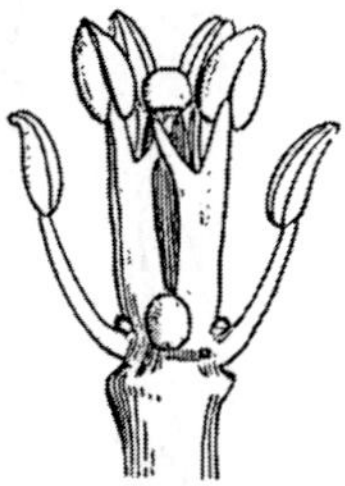
200

CARDAMINE. *Linnæus.*

Silique compressed, with flat nerveless valves. *Stigma* capitate. *Seeds* in one row. *Embryo* accumbent O=.

1. C. *pratensis* Linnæus.—(CUCKOO-FLOWER.)

Leaves pinnate ; leaflets roundish-ovate, of the upper leaves linear, entire ; petals obovate, 3 times as long as the calyx.

Habitat. Meadows.
Quality. Flowers stimulant, diaphoretic, diuretic, nervine.
Uses. Epilepsy, cholera, spasmodic asthma.

NASTURTIUM. *Brown.*

Silique short, cylindrical ; *valves* convex, with scarcely any nerves. *Stigma* capitate. *Seeds* in 2 irregular rows. *Embryo* O=.

Fig. 200.—The stamens of Crambe maritima magnified.

1. N. *officinale* Brown.—(WATERCRESS.) Fig. 201.

Leaves pinnated ; leaflets repand, the lateral elliptical, the terminal ovate, rather cordate.

Habitat. Running streams.
Quality. Pungent, antiscorbutic.
Uses. A favourite salad.

COCHLEARIA. *Linnæus.*

Silicle globose, with very convex valves. *Seeds* numerous. *Embryo* O=.

1. C. *officinalis* Linnæus.—(SCURVY-GRASS.) Fig. 202.

Radical leaves broadly ovate, rather cordate ; the upper amplexicaul ; valves of the silicle one-ribbed. An annual.

Habitat. Sea-coast of Europe.
Quality. Stimulant, aperient, diuretic, antiscorbutic.
Uses. Visceral obstructions ; salads.

c a b 201

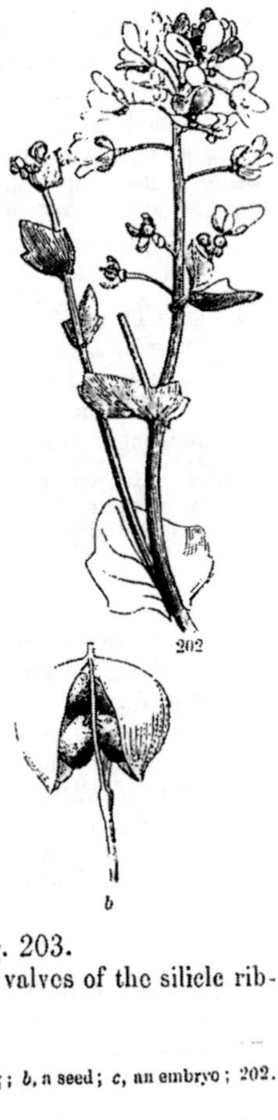

202 b

2. C. *Armoracia* Linnæus.—(HORSE-RADISH.) Fig. 203.

Radical leaves cordate, or obovate-oblong, crenate ; valves of the silicle ribless. A tap-rooted perennial.

Fig. 201. Nasturtium officinale ; *a*, the silique in the act of dehiscing ; *b*, a seed ; *c*, an embryo ; 202. Cochlearia officinalis ; *b*, its silicle in the act of dehiscing.

203

Habitat. Meadows of Europe.
Quality. Pungent, acrid, stimulant, vesicant.
Uses. A condiment; masticatory; hoarseness; in infusion in cases of poisoning.

LEPIDIUM. *Linnæus.*

Silicle roundish, with compressed valves keeled at the back. *Seeds* 1 in each cell. *Embryo* incumbent O||, with multifid cotyledons.

Fig. 203.—Colchlearia Armoracia in flower.

1. L. *sativum* Linnæus.—(GARDEN-CRESS.)
Lower leaves stalked, irregularly cut, lobed, pinnate; silicles roundish, winged, emarginate, pressed close to the rachis.

Habitat. Waste places.
Quality. Pungent, antiscorbutic.
Uses. A common vegetable, esculent when only the first leaves after germination have been formed.

Natural Order, Capparids; *Capparidaceæ* (V. K., p. 357.)

Prevailing Quality. Acridity.

CAPPARIS. *Linnæus.*

Calyx 4-parted. *Petals* 4. *Carpophore* slender. *Stamens* 00. *Berry* with a rind.

1. C. *spinosa* Linnæus.—(CAPER BUSH.) Fig. 204.
Leaves roundish, blunt, or emarginate; stipules spiny, curved.

Habitat. South of Europe.
Quality. Antiscorbutic, stimulant, aperient.
Uses. Young buds as "Capers," a well-known pickle; bark of root diuretic?

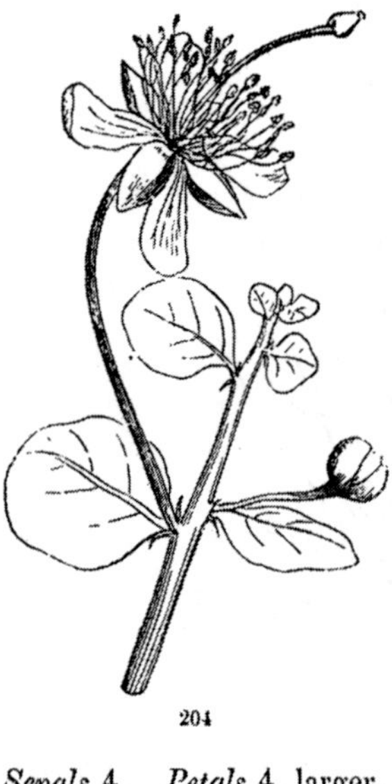

204

POLANISIA. *Rafinesque.*

Sepals 4, spreading. *Petals* 4. *Stamens* 8-32. *Silique* sessile, or with a short stalk.

1. P. *viscosa* De Candolle. *Cleome dodecandra* and *icosandra* Linnæus.
Hairy, glandular; leaflets 3-5, obovate, cuneate, or oblong; stamens 8-16; pod sessile, striated, glandular.

Habitat. East Indies.
Quality. Leaves pungent, like Mustard.
Uses. Leaves to form sinapisms; root as a vermifuge.

CRATÆVA. *Linnæus.*

Sepals 4. *Petals* 4, larger. *Stamens* 8-28. *Berry* stipitate, with a thin rind, pulpy internally.

1. C. *gynandra* Linnæus.—(GARLIC PEAR.)
Leaves ovate, acute; stamens 20-24, inserted on a cylindrical stipe, longer than the petals.

Habitat. Jamaica.
Quality. Bark of root blisters like Cantharides.

Fig. 204.—Capparis spinosa.

THE RANAL ALLIANCE (V. K., p. 416.)

Natural Orders of Ranals.

Magnoliads (*Magnoliaceæ.*) Carpels distinct. Stipules convolute. Flowers $^{5}/$.

Anonads (*Anonaceæ.*) Carpels distinct. Stipules 0. Corolla valvate. Flowers $^{3}/$.

Crowfoots (*Ranunculaceæ.*) Carpels distinct. Stipules 0. Corolla imbricated.

Poppyworts (*Papaveraceæ.*) Carpels consolidated. Placentæ parietal.

Natural Order, **Magnoliads**; *Magnoliaceæ* (V. K., p. 417.)

Prevailing Quality. Bitter, tonic.

205

DRYMIS. *Forster.*

Carpels crowded, berried, many-seeded. *Connective* broad, and disjoining the anther-cells.

1. D. *Winteri* Forster.—(WINTER'S BARK TREE.) Fig. 205.

Leaves dotted, oblong, obtuse, very glaucous beneath; flowers corymbose; sepals 2-3.

Habitat. Mountainous parts of South America.
Quality. Bark stimulant, aromatic, tonic.
Uses. As Cinnamon; in scurvy.

ILLICIUM. *Linnæus.*

Carpels whorled, one-sided, opening near the upper end. *Sepals* 3-6 petaloid.

1. I. *anisatum* Linnæus.—(STAR ANISE PLANT.)

Leaves evergreen, smooth, dotted; petals about 30, yellowish, the outer oblong, the inner subulate.

Habitat China and Japan.
Quality. Fruit aromatic, smelling like Anise, carminative.
Uses. In preparation of liqueurs.

Fig. 205. Small branch in flower of Drymis Winteri.

MAGNOLIA. *Linnæus.*

Carpels spiked, arranged in cones, opening at the dorsal suture. *Seeds* pendulous by a long umbilical cord.

1. M. *glauca* Linnæus.—(SWAMP SASSAFRAS.)

Leaves thin, elliptical, obtuse, glaucous beneath ; petals ovate concave.

Habitat. United States.
Quality. Bark and fruit bitter, aromatic, febrifugal.
Uses. As Cinchona ; in chronic rheumatism.

LIRIODENDRON. *Linnæus.*

Carpels spiked, arranged in cones, indehiscent, winged at the end.

1. L. *tulipifera* Linnæus.—(TULIP TREE.)

Leaves truncate, 4-lobed.

Habitat. United States.
Quality and *Uses.* As in the last.

Natural Order, Anonads; *Anonaceæ* (V. K., p. 420.)

Prevailing Quality. Aromatic.

XYLOPIA. *Linnæus.*

Fruits on a convex receptacle, dry, indehiscent, cylindrical or moniliform. *Seeds* several, adhering to the pericarp.

1. X. *aromatica* Blume. *Habzelia aromatica* A. De Candolle.—(AFRICAN PEPPER.)

Leaves ovate-lanceolate, acute, smooth, glaucous beneath ; fruits taper, moniliform.

Habitat. Sierra Leone.
Quality. Aromatic, pungent, stimulant.
Uses. As pepper ; the fruit is the Piper Æthiopicum of authors.

2. X. *glabra* Linnæus.—(BITTER-WOOD.)

Leaves oblong-ovate, smooth, dotted ; fruit oblong, stipitate.

Habitat. West Indies.
Quality. All the parts aromatic and intensely bitter.

3. X. *grandiflora* St. Hilaire. *Uvaria febrifuga* Martius.

Leaves elliptical, lanceolate, acute, pubescent above, downy beneath ; inner petals 3-cornered, each with 2 auricles.

Habitat. Brazil.
Quality. Fruit aromatic, tonic.
Uses. A valuable remedy for fevers in Brazil.

Natural Order, Crowfoots; *Ranunculaceæ* (V. K., p. 425.)

Prevailing Quality. Acridity, causticity.

CLEMATIS. *Linnæus.*

Sepals 4, valvate or induplicate. *Petals* 0. *Carpels* one-seeded *achænia.*

1. C. *erecta* Linnæus.

Erect; leaves pinnate; leaflets ovate acuminate; sepals smooth, downy at the edge.

Habitat. Europe.
Quality. Leaves extremely acrid and epispastic.
Uses. Unsafe vesicants; in cachectic diseases.

2. C. *Vitalba* Linnæus.—(TRAVELLER'S JOY.) Fig. 206.

Stem climbing, woody; leaflets 5, cordate, unequally cut, finely hairy; carpels with long feathery tails.

206

207

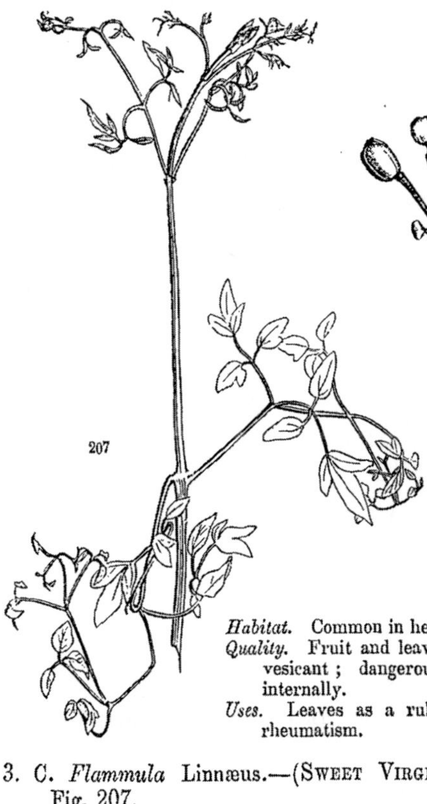

b

a

Habitat. Common in hedgerows.
Quality. Fruit and leaves acrid and vesicant; dangerous if taken internally.
Uses. Leaves as a rubefacient in rheumatism.

3. C. *Flammula* Linnæus.—(SWEET VIRGIN'S-BOWER.) Fig. 207.

Climbing; leaves bipinnate; sepals oblong, obtuse.

Habitat. South of Europe.
Quality. Leaves powerfully epispastic; flowers deliciously fragrant.
Uses. As the last.

ANEMONE. *Linnæus.*

Sepals ∛, imbricated, not distinguishable from the petals. *Achænia* soft, woolly, tailed, or tailless.

Fig. 206.—Clematis Vitalba; *a*, a portion of the inflorescence; *b*, an achænium; 207. Young shoot of Clematis Flammula.

1. A. *nemorosa* Linnæus.—(Wood Anemone.) Fig. 208.

Rhizome creeping ; leaves ternate ; leaflets three-lobed cut ; bracts exactly like the leaves ; flowers solitary, erect, white.

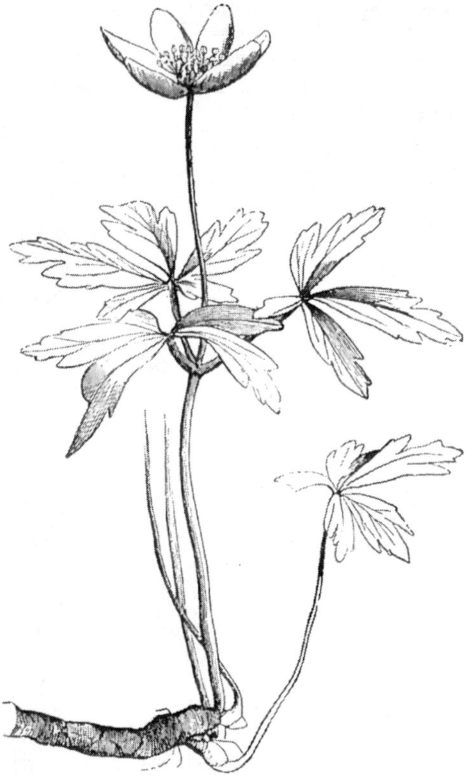

208

Habitat. Woods of the north of Europe.

Quality. Rhizome acrid, vesicant.

Uses. Has been recommended, along with other species, in rheumatism, and in obstinate cases of tænia.

Ranunculus. *Linnæus.*

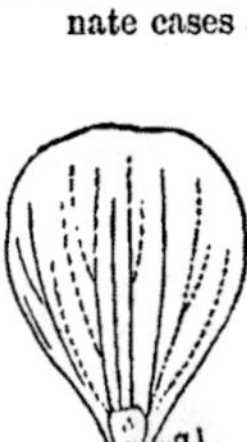

209

Sepals 3-5. *Petals* with a gland at their base. *Achænia* mucronate. Fig 209.

1. R. *Ficaria* Linnæus.—(Pilewort.) Fig. 210.

Leaves roundish cordate, shining, the upper angular sepals 3.

Habitat. Plantations, lawns and meadows.

Quality. Subacrid, antiscorbutic.

Uses. Leaves sometimes used as a potherb ; boiling renders them insipid.

Fig. 208.—Anemone nemorosa; 209. A petal of Ranunculus; the distinctive mark of the genus exhibited by *gl*, the gland.

2. R. *Flammul* Linnæus. Fig. 211.
Leaves lanceolate or linear ; stem many-flowered ; achænia smooth, with a short point.

210 211

Habitat. Ditches and wet meadows.
Quality. Epispastic, poisonous.
Uses. Distilled water an active and useful emetic.

3. R. *acris* Linnæus.—(Upright Crowfoot.)
Leaves palmated, with rather rhomboid cut divisions, the upper 3-parted ; peduncles terete ; carpels lenticular, with a beak much shorter than the achænium ; receptacle smooth.

Habitat. Meadows.
Quality. As a rubefacient and epispastic.
Uses. A powerful acrid.

Fig. 210.—Ranunculus Ficaria ; 211. R. Flammula.

4. R. *sceleratus* Linnæus. Fig. 212.
Succulent, smooth, annual; lower leaves palmate, crenated, upper 3-parted; head of achænia like a spike; achænia not keeled, finely wrinkled.

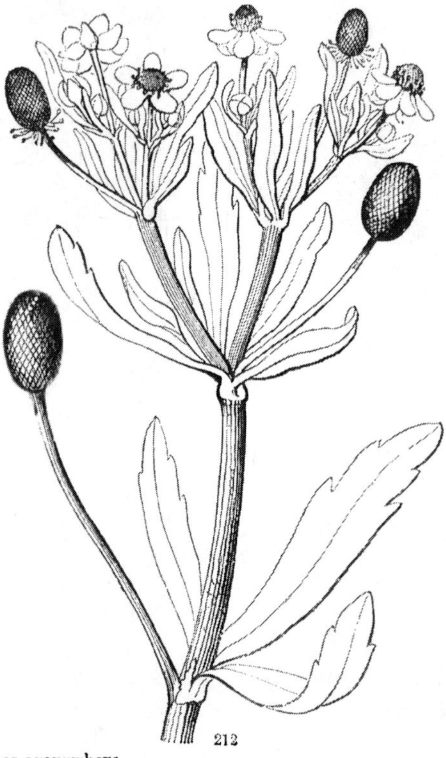

212

Habitat. Ditches everywhere.
Quality. Acrid, corrosive, very dangerous if taken internally; water an antidote.

HELLEBORUS. *Linnæus.*

Sepals 5, petaloid, permanent. *Petals* small, tubular, 2-lipped. *Follicles* sessile, many-seeded.

1. H. *niger* Linnæus.—(BLACK HELLEBORE. CHRISTMAS ROSE.) Fig. 213.
Scape 1-2-flowered, with 2-3 oval bracts; flowers whitish.

Habitat. Shady woods of Central Europe.
Quality. Drastic purgative, emmenagogue; a narcotico-acrid poison.
Uses. Mania, epilepsy, dropsy, chronic skin diseases.

2. H. *orientalis* De Candolle; *officinalis* Sibthorp.—(ORIENTAL HELLEBORE.)
Scape 3-5-flowered, with lanceolate finely serrated bracts; sepals acute; flowers whitish.

Habitat. Asia Minor.
Quality and *Uses.* As H. niger. The true Black Hellebore of the ancients.

Fig. 212.—Ranunculus sceleratus.

3. H. *fœtidus* Linnæus.—(STINKING HELLEBORE.)
Scape many-flowered, leafy; bracts oval; stem leafy; flowers green.

Habitat. Woods and waste places.
Quality. Leaves emetic, purgative, poisonous.
Uses. As an emmenagogue; against the Ascaris lumbricoides.

NIGELLA. *Linnæus.*

Sepals 5, petaloid. *Petals* minute, unguiculate, with a scale at the base. *Follicles* 5 united into a spuriously 10-celled capsule.

1. N. *sativa* Linnæus.—(BLACK CUMMIN.) Fig. 214.
Involucre 0; capsules rough; seeds wrinkled transversely.

Habitat. Corn-fields and gardens in Europe.
Quality. Seeds pungent, aromatic, strong-smelling.
Uses. As a condiment instead of pepper; pounded, as a sternutatory.

214

a b

213

DELPHINIUM. *Linnæus.*

Sepals 5, the upper spurred. *Petals* 4; the 2 upper having spurs plunged in the sepaline spur. *Follicles* 1-5, many-seeded.

Fig. 213.—Helleborus niger; 214. Nigella sativa; *a*, a branchlet; *b*, a seed much magnified.

1. D. *Consolida* Linnæus. —(BRANCHING LARKSPUR.) Fig. 215.

Stem straggling, much branched; leaves finely cut; petals united into 1; carpel solitary.

Habitat. Corn-fields in Europe.
Quality. Seeds acrid, emetic.
Uses. Tincture in asthma; in preparation of cosmetics.

2. D. *Staphisagria* Linnæus.—(STAVESACRE.) Fig. 216.

Leaves palmate, broad, 5-cleft, with pallid stains; petals all distinct, beardless.

Habitat. South of Europe.
Quality. Seeds emetic, purgative; acrid, narcotic.
Uses. To destroy pediculi, against worms, and itch.

ACONITUM. *Linnæus.*

Sepals 5, the upper galeate. *Petals* 2, hammer-headed. *Follicles* 3-5.

1. A. *Napellus* Linnæus. — (MONK'S - HOOD.) Fig. 217.

Roots clustered, fusiform; flowers purple, in erect racemes; young carpels divaricating.

215 a

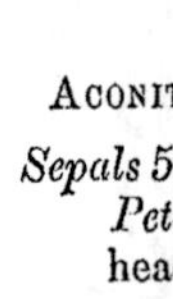

b

216

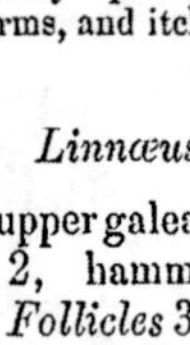

Fig. 215.—*a*, branch of Delphinium Consolida; *b*, one of its seeds magnified; 216. Seed of D. Staphisagria magnified.

Habitat. Europe.
Quality. Root produces numbness and tingling; an acrid narcotic poison.
Uses. Neuralgia, rheumatism, dropsy, hypertrophy of the heart. Yields Aconitina, "the most virulent poison known."—*Pereira.*

2. A. *Cammarum* Linnæus. A. *paniculatum* Störke.

Flowers white and blue, in straggling panicles; young carpels incurved.

Habitat. Switzerland and Germany.
Quality and *Uses.* Like those of A. Napellus, but extremely feeble.

3. A. *ferox* Wallich.

Flowers purple, in rather loose panicles; helmet semicircular; young carpels very downy; lobes of the leaves much acuminated and divaricating.

Habitat. Nepal.
Quality and *Uses.* Like those of A. Napellus; but still more virulent.

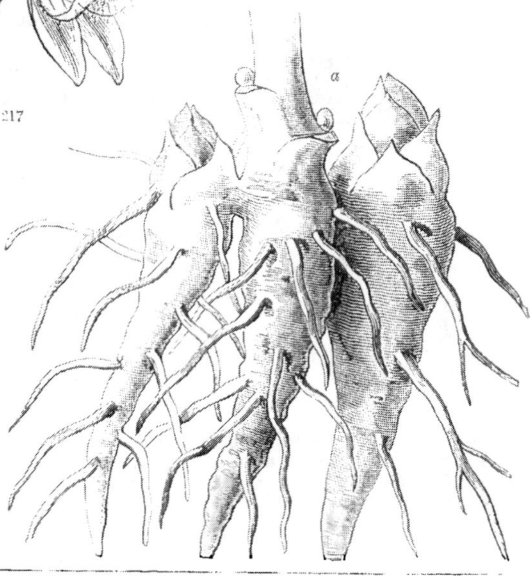

N.B.—Dr. Fleming's experiments prove that the roots are more active immediately after the period of flowering than at any other time, and that the leaves lose their power when the seeds begin to form. The seeds themselves are comparatively weak.

Fig. 217.—Flowers of Aconitum Napellus; *a*, one of its roots.

4. A. *Lycoctonum* Linnæus.—(WOLFSBANE.) Fig. 218.
Flowers yellow ; petals with a filiform circinate spur.

Habitat. Switzerland.
Quality. Scentless, bitter, subacrid.
Uses. Similar to those of A. Napellus, but much more feeble.

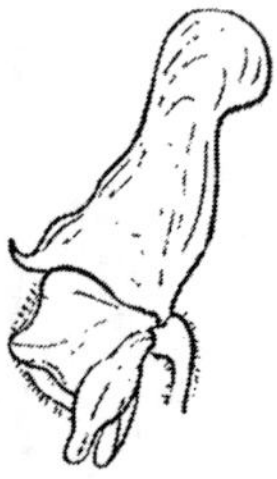

218

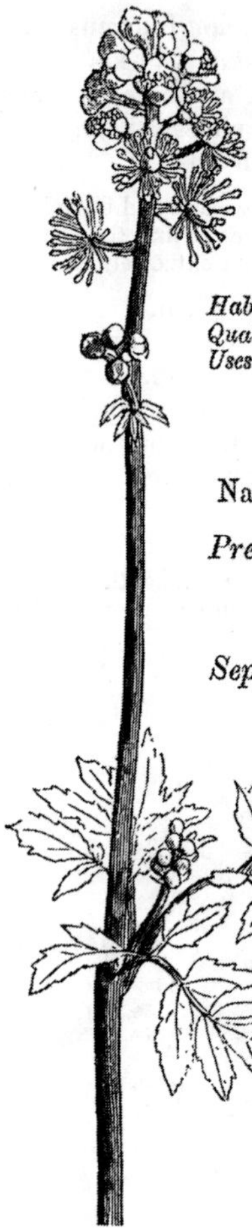

ACTÆA. *Linnæus.*

Sepals 4, deciduous. *Petals* 4. *Carpel* single, baccate.

1. A. *spicata* Linnæus. — (BANEBERRY.) Fig. 219.

Leaves ternate, twice pinnate ; leaflets ovate-oblong, incised ; racemes ovate.

Habitat. North of Europe.
Quality. Poisonous, antispasmodic, astringent.
Uses. Roots in catarrh.

Natural Order, **Poppyworts** ; *Papaveraceæ* (V.K., p. 430.)

Prevailing Quality. Narcotic, poisonous.

PAPAVER. *Linnæus.*

Sepals 2, deciduous. *Petals* 4. *Stigma* with from 4 to 20 rays.

1. P. *somniferum* Linnæus. — (OPIUM POPPY.)

Leaves oblong, amplexicaul, glaucous, smooth ; capsule smooth.

Habitat. Syria.
Quality. A stimulating narcotic ; aphrodisiac ?
Uses. In fevers, inflammatory diseases, cholera, insanity, delirium tremens, convulsive diseases, venereal disorders, &c., &c. Seeds (of a variety) called maw seeds, demulcent.

a

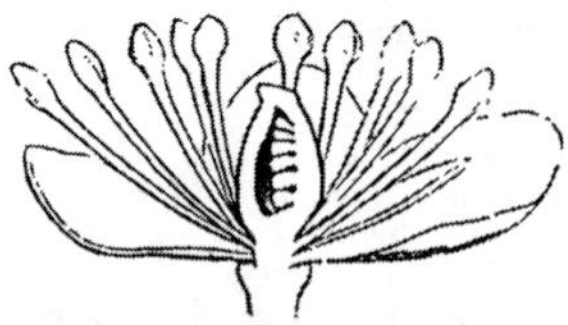

219

N.B.—From the wounded half-ripe capsules flows the juice which concretes into opium. From the dried capsules the decoction, syrup, and extract of Poppies are prepared. Dr. Pereira justly observes, that these capsules or "heads" would be more active if gathered before ripeness: when full grown and just when the first change of colour is perceptible should be the best time to collect them. The seeds are not narcotic, but yield a bland oil similar to that obtained from Olives.

Fig. 218.—Flower of Aconitum Lycoctonum; 219. Actæa spicata; *a*, a perpendicular section of a flower.

2. P. *Rhœas* Linnæus.—(CORN POPPY.) Fig. 220.

Leaves hispid, pinnate and bipinnate; capsule smooth, obovate, roundish.

Habitat. Corn-fields.
Quality. Sub-narcotic.
Uses. Petals as a colouring ingredient.

CHELIDONIUM. *Linnæus.*

Sepals 2, deciduous. *Petals* 4. *Capsules* siliquose, 2-valved; opening from the base to the apex.

1. C. *majus* Linnæus.—(CELANDINE.) Fig. 221.

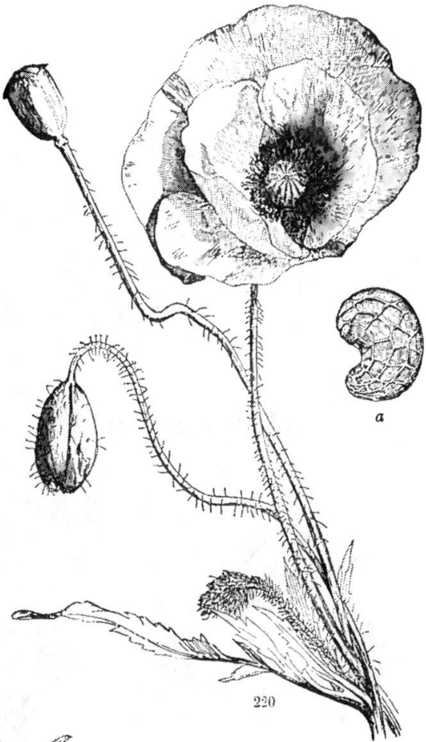

Flowers in umbels; calyx nearly smooth; filaments broadest upwards.

Habitat. Plantations and waste places.
Quality. Strong-smelling, acrid, narcotic, emetic, purgative, poisonous.
Uses. Juice against warts, and opacities of the cornea.

ARGEMONE. *Tournefort.*

Petals 4-6. *Stigmas* 4-7, radiating, concave. *Capsule* obovate, opening by valves at the point.

Fig. 220.—Papaver Rhœas; *a*, one of its seeds; 221. Chelidonium majus.

1. A. *mexicana* Linnæus. Fig. 222.

Leaves sessile, repand, sinuated, spiny, variegated with white; flowers yellow.

222

Habitat. Common in tropical countries, derived from Mexico.
Quality. Narcotic, purgative, diuretic.
Uses. Seeds instead of opium, and of ipecacuanha; juice in ophthalmia, and as an application to chancres.

SANGUINARIA. *Linnæus.*

Petals 8-12. *Stigma* 2. *Capsule* oblong, ventricose, with 2 deciduous valves and a persistent many-seeded frame.

1. S. *canadensis* Linnæus.—(PUCCOON. BLOOD ROOT.) Fig. 223.

Rhizome fleshy, with red juice; leaves solitary, radical, roundish, deeply cordate, with about 7 toothed angles; flower solitary, radical.

Habitat. United States.
Quality. Acrid, narcotic, emetic.
Uses. Rhizome in typhoid pneumonia, catarrh, pertussis, croup, rheumatism, jaundice, &c. Over-dose dangerous.

Fig. 222.—Argemone mexicana.

THE GUTTIFERAL ALLIANCE (V. K., p. 392.)

Natural Orders of Guttiferals.

Theads (*Ternströmiaceæ.*) Leaves alternate.

Guttifers (*Clusiaceæ.*) Leaves opposite. Seeds few. Petals equilateral.

Tutsans (*Hypericaceæ.*) Leaves opposite. Seeds 00. Petals oblique.

Natural Order, **Theads**; *Ternströmiaceæ* (V. K., p. 396.)

Prevailing Quality. Stimulating, subnarcotic.

223

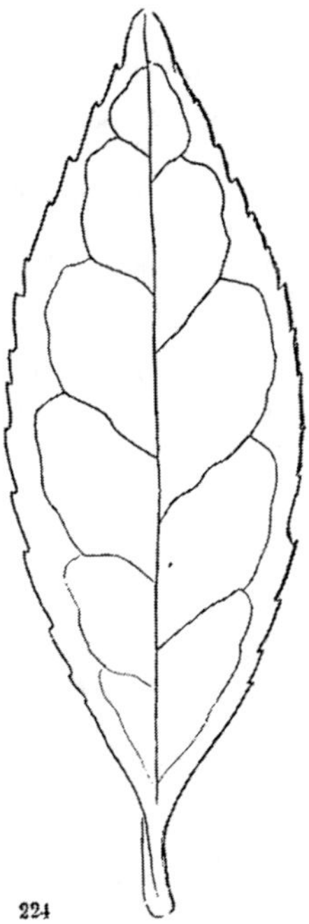

224

THEA. *Linnæus.*

Sepals 5-6. *Petals* 6-9, in two rows. *Stamens* nearly distinct. *Capsule* 3-valved, septicidal.

1. T. *Bohea* Linnæus.—(The TEA PLANT.) Fig. 224.

Fig. 223.—Plant in flower of Sanguinaria canadensis; 224. Leaf of Thea Bohea.

Leaves coriaceous, flat, oblong-lanceolate or somewhat obovate, slightly toothed.

Habitat. Southerly districts of China. *Quality* and *Uses.* As in the last.

2. T. *viridis* Linnæus.—(The TEA PLANT.) Fig. 225.

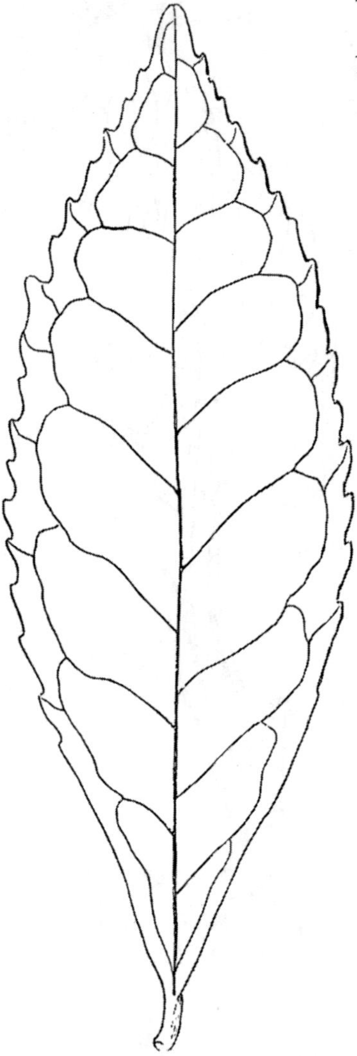

225

Leaves thin, lanceolate, rather wavy, coarsely toothed.

Habitat. Northerly districts of China.
Quality. Astringent, antisoporific, sedative, diluent, diaphoretic, diuretic.
Uses. A well-known beverage is prepared from the leaves. Black tea is the leaf more fermented than green tea.

Natural Order, **Guttifers**; *Clusiaceæ* (V. K., p. 400.)

Prevailing Quality. Acrid, purgative.

GARCINIA. *Linnæus.*

Flowers polygamous. *Stamens* more or less united ; *Anthers* opening longitudinally. *Fruit* succulent, 4-10-celled.

1. G. *Mangostana* Linnæus.—(MANGOSTEEN.)

Leaves rather rhomboidal, obtuse ; male flowers fascicled ; females solitary, terminal ; fruit 6-10-celled, about the size of an orange.

Habitat. Indian Archipelago.
Quality. Refrigerant, nutritious, laxative.
Uses. The most delicious of known fruits.

HEBRADENDRON. *Graham.*

Flowers polygamous. *Stamens* more or less united; *anthers* circumscissile. *Fruit* succulent, 4-10-celled.

1. H. *pictorium* Christison.—(MYSORE GAMBOGE TREE.) Fig. 226.

Leaves oval, acute at each end; flowers axillary, solitary; fruit slightly furrowed when ripe ; seeds 4.

Habitat. Mysore and Malabar.
Quality. Gum resin hydragogue, and drastic ; acrid.
Uses. Constipation, apoplexy, dropsy, tapeworm cases.

Fig. 225.—Leaf of Thea viridis.

226

2. H. *gambogioides* Graham. *Cambogia Gutta* Linn.—(CEYLON GAMBOGE TREE.)

Leaves oval or slightly obovate, obtusely acuminate; flowers axillary, aggregate; fruit globose; seeds 4.

Habitat. Ceylon. *Quality* and *Uses.* As the last.

Fig. 226.—Hebradendron pictorium; from the *Pharmaceutical Journal*, Vol. vi., p. 69.

Natural Order, Tutsans; *Hypericaceæ* (V. K., p. 405.)

Prevailing Quality. Astringent, subacrid.

ANDROSÆMUM. *Allioni.*

Capsule baccate, indehiscent.

1. A. *officinale* Allioni.—(TUTSAN.)

Branches quadrangular; leaves ovate or cordate, sessile.

Habitat. Woods.
Quality and *Uses.* An ancient and powerful vulnerary; bruised leaves employed.

HYPERICUM. *Linnæus.*

Capsule membranous, 3-5 valved.

1. H. *perforatum* Linnæus.—(ST. JOHN'S WORT.) Fig. 227.

Leaves opposite, ovate, obtuse, strongly marked with clear transparent dots; flowers in terminal panicles.

Habitat. Groves, hedges, and waysides.
Quality. Astringent.
Uses. Gargles and lotions.

VISMIA. *Vandelli.*

Berry membranous. *Styles* 5; *stigmas* peltate.

1. V. *guianensis* Persoon.—(AMERICAN GAMBOGE.)

Stem 4-cornered; leaves ovate-lanceolate, acuminate, dilated at the base, short-stalked, rufous beneath, smooth above.

Habitat. Guiana.
Quality. Resin acrid, purgative.
Uses. As those of Gamboge.

THE FICOIDAL ALLIANCE (V. K., p. 523.)

Prevailing Quality. Mucilaginous, insipid.

TETRAGONIA. *Linnæus.*

(*Natural Order.* Aizoons.)

Calyx 4-cleft, adherent; by degrees gaining several horn-like processes. *Nut* bony, adherent, winged or horned.

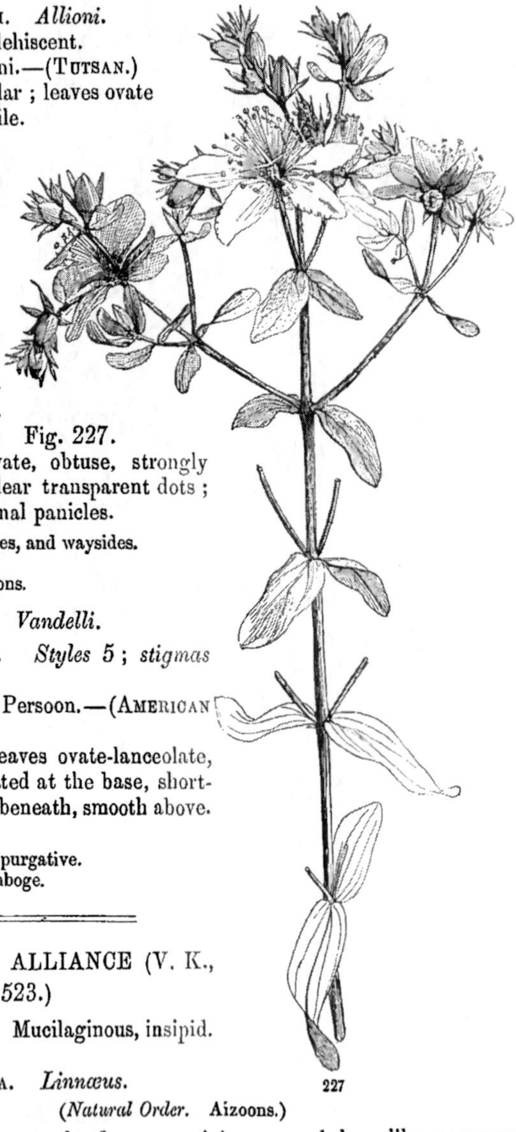

227

Fig. 227.—Hypericum perforatum.

228

1. T. *expansa* Aiton.—(NEW ZEALAND SPINACH.) Fig. 228. A trailing succulent annual; leaves stalked, ovate, rhomboid; fruit 4-horned, 6-8-seeded.

Habitat. New Zealand.
Quality and *Uses.* Cultivated in gardens for its mucilaginous, insipid, succulent foliage; used instead of Spinach.

MELLOCA. *Lindley.*

(*Natural Order.* Basellads.)

Calyx membranous, 5-parted, free, stellate. *Ovary* 1-celled, 1-seeded.
1. M. *tuberosa* Lindley. *Ullucus tuberosus.*—(The MELLOCO.) Fig. 229. Leaves reniform, cordate; flowers in short axillary spikes.

a

229

Habitat. Peru.
Quality. Mucilaginous, amylaceous, nutritious.
Uses. Tubers employed like potatoes; leaves insipid, as Spinach.

THE DAPHNAL ALLIANCE (V. K., p. 529.)

Natural Orders of Daphnals.

Daphnads (*Thymelaceæ.*) Anther-valves straight.

Laurels (*Lauraceæ.*) Anther-valves recurved.

Fig. 228.—Fruit of Tetragonia expansa; 229. Melloca tuberosa; *a*, an expanded flower.

Natural Order, Daphnads; *Thymelaceæ* (V. K., p. 530.)
Prevailing Quality. Acridity.

DAPHNE. *Linnæus.*

Calyx long, tubular, naked at the mouth. *Stamens* 8. *Hypogynous* scales 0. *Fruit* a drupe.

1. D. *Mezereum* Linnæus.—(MEZEREON.) Fig. 230.
Leaves deciduous; flowers lateral, sessile.

230

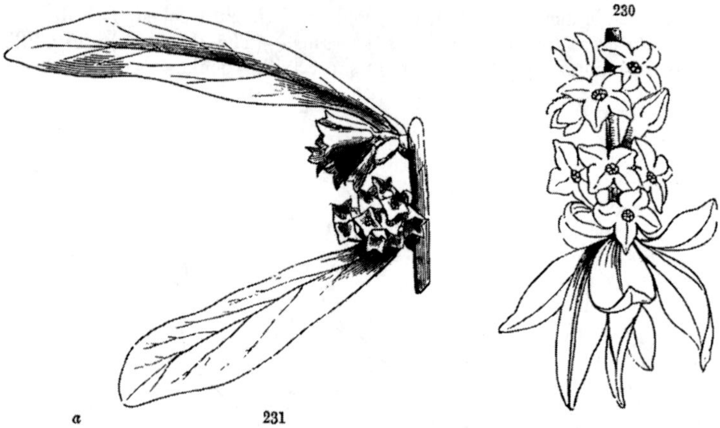

a 231

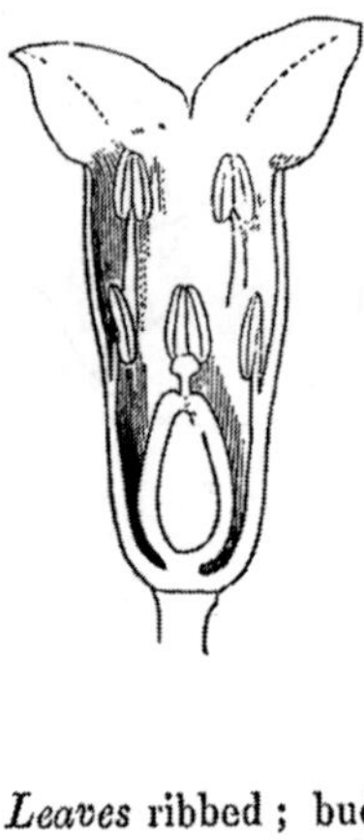

Habitat. Woods; indigenous.
Quality. Bark acrid, laxative, poisonous.
Uses. Venereal, rheumatic, scrofulous and chronic cutaneous diseases; a masticatory.

2. D. *Laureola* Linnæus.—(SPURGE LAUREL.) Fig. 231.
Leaves evergreen; flowers in axillary clusters.

Habitat. Woods; indigenous.
Quality and *Uses.* As in the last; fruits the food of birds; poisonous to other animals.

Natural Order, Laurels; *Lauraceæ* (V. K., p. 535.)
Prevailing Quality. Aromatic, stimulant.

CINNAMOMUM. *Nees.*

Leaves ribbed; buds not scaly. *Fertile* stamens, 9 in 3 rows, with 4-celled anthers, of which the inner only are extrorse; the inner ones with 2 glands at their base. *Calyx* coriaceous.

1. C. *zeylanicum* Nees. *Laurus Cinnamomum* Linnæus.—(CINNAMON.)
Leaves ovate or ovate-oblong, with a blunt point, 3-nerved or triple-nerved.

Habitat. Ceylon.
Quality. A spicy stimulating aromatic, astringent, tonic.
Uses. The same as that of other spices; diarrhœa, low fevers, flatulence, colic.

Fig. 230.—Daphne Mezereum; 231. D. Laureola; *a*, perpendicular section of a flower.

2. C. *Cassia* Blume.—(Cinnamon Cassia. Chinese Cinnamon.)
Leaves long-oblong, acute at each end, triple-nerved; the ribs vanishing below the point.

Habitat. China.
Quality and *Uses.* As in Cinnamon, less sweet, and more astringent.

Camphora. *Nees.*

Leaves ribbed; buds scaly. *Calyx* thin. Otherwise as Cinnamomum.
1. C. *officinarum* Nees. *Laurus Camphora* Linnæus.—(Camphor Tree.)
Leaves triple-nerved, shining on the upper side, with a porous gland in the axil of the veins on the under side.

Habitat. Island of Formosa, China, Japan.
Quality. Acrid, anodyne, vascular excitant, sudorific, poisonous, anaphrodisiac.
Uses. Typhoid fevers, internal inflammations, measles, scarlatina, spasms, strangury, satyriasis, nymphomania, onanism, cholera.

Nectandra. *Rottboell.*

Leaves veiny. Fertile *stamens* 9, with 4-celled subsessile anthers, of which the inner only are extrorse. *Calyx* rotate.
1. N. *Rodiæi* Schomburgk.—(Bibiri or Greenheart Tree.)
Leaves nearly opposite, oblong-elliptical, shortly acuminate, coriaceous, smooth, shining and obscurely netted on the upper side; panicles few-flowered, axillary, much shorter than the leaves, finely downy; anthers all thick, oblong, without glands. *Bentham.*

Habitat. Guiana.
Quality. Powerfully tonic and febrifugal.
Uses. Yields the alkaloid Beebcerine, of great value in intermittents.

Sassafras. *Nees.*

Leaves veiny. *Anthers* all introrse, 4-celled. Fertile *stamens* 9.
1. S. *officinale* Nees. *Laurus Sassafras* Linnæus.—(Sassafras Tree.)
Leaves thin, oblong, entire or 2-3-lobed.

Habitat. United States.
Quality. Wood and bark stimulant, sudorific, subacrid.
Uses. Rheumatism, venereal diseases.

Laurus. *Linnæus.*

Leaves veiny. *Anthers* all introrse, 2-celled. Fertile *stamens* 12.
1. L. *nobilis* Linnæus.—(Sweet Bay.)
Leaves lanceolate or oblong-lanceolate, acute, smooth, wavy, and rather toothed, purple-ribbed, with a hairy pore at the axil of the under veins.

Habitat. South of Europe.
Quality. Aromatic, stimulant, narcotic.
Uses. Leaves used by confectioners to flavour creams; dyspepsia, flatulence.

THE RHAMNAL ALLIANCE (V. K., p. 576.)

Natural Orders of Rhamnals.

Elmworts (*Ulmaceæ.*) Flowers apetalous.

Rhamnads (*Rhamnaceæ.*) Flowers polypetalous. Calyx valvate.

Spindle Trees (*Celastraceæ.*) Flowers polypetalous. Calyx imbricated.

Sapotads (*Sapotaceæ.*) Flowers monopetalous. Ovules ascending.

Storaxworts (*Styracaceæ.*) Flowers monopetalous. Ovules suspended.

Natural Order, Elmworts; *Ulmaceæ* (V. K., p. 580.)

Prevailing Quality. Mucilaginous, astringent.

PLANERA. *Michaux.*

Fruit globose, membranous, wingless, indehiscent.

1. P. *Abelicea* Römer and Schultes.

Leaves elliptical, equally serrate, unequal-sided, tomentose and discoloured on the under side.

Habitat. Candia.

Quality and *Uses.* Wood aromatic; formerly official, under the name of Pseudo-santalum creticum.

CELTIS. *Linnæus.*

Fruit a fleshy, globose, 1-celled drupe.

1. C. *australis* Linnæus.—(NETTLE TREE.) Fig. 232.

Leaves oblong-lanceolate, acuminate, finely serrated, rough above, soft with down beneath, unequal-sided.

Habitat. South of Europe.

Quality. Astringent, sub-aromatic.

Uses. Branches boiled and the infusion used in dysentery. and blennorrhœa.

232

a

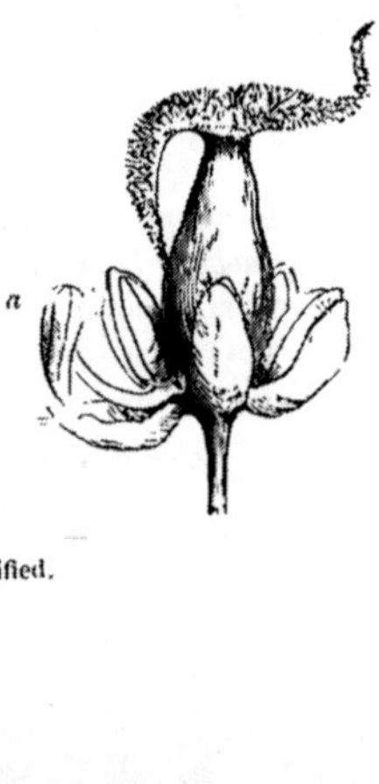

Fig. 232.—Celtis australis; *a*, a flower magnified.

ULMUS. *Linnæus.*

Fruit thin, 2-winged, veiny, indehiscent.

1. U. *montana* Smith.—(WITCH-ELM.) Fig. 233.

Leaves obovate, cuspidate, doubly and coarsely serrate; branches cinereous, smooth.

Habitat. Great Britain, &c.
Quality. Inner bark mucilaginous, bitter, astringent, diaphoretic, diuretic.
Uses. Cutaneous eruptions, ichthyosis; a substitute for Sarsaparilla; wood the basis of some soft snuffs.

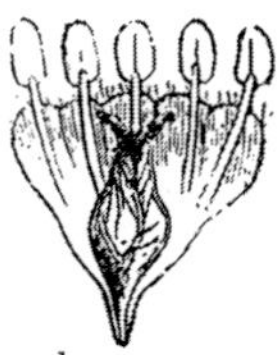

b

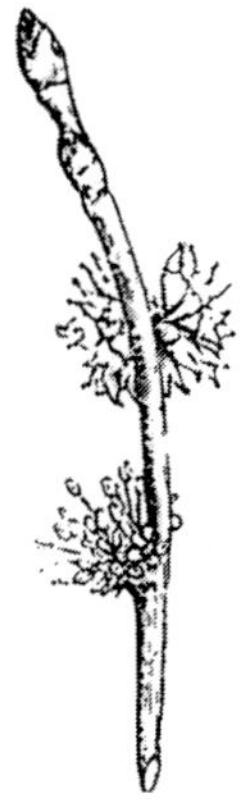

a

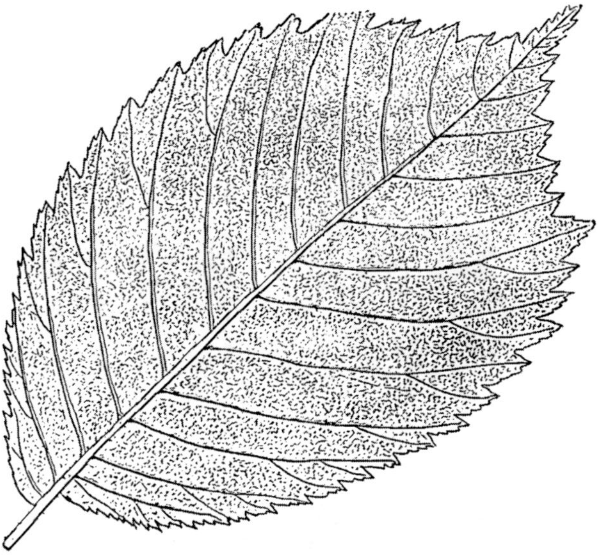

233

Natural Order, **Rhamnads**; *Rhamnaceæ* (V. K., p. 581.)

Prevailing Quality. Purgative.

RHAMNUS. *Linnæus.*

Calyx campanulate, 4-5-cleft, circumscissile after flowering. *Fruit* globose, rather dry, or spongy, or succulent, with 2 to 4 stones.

Fig. 233.—Leaf of Ulmus montana; *a*, its flowers, natural size; *b*, a flower cut open and magnified.

1. R. *catharticus* Linnæus.—(BUCKTHORN.) Fig. 234.
Leaves deciduous, ovate, crenate ; flowers fascicled ; branches spiny.

234

Habitat. Hedges and plantations.
Quality and *Uses.* Berries hydragogue, cathartic ; inconvenient to use, because of the sickness and thirst that accompany their exhibition.

2. R. *Frangula* Linnæus.—(BLACK ALDER.) Fig. 235.
Leaves oval, entire, with 10-12 lateral pinnated veins, smooth as well as the calyx.

Habitat, Quality, and *Uses.* Like the last.

3. R. *infectorius* Linnæus.—(DYERS' BUCKTHORN.)
Leaves ovate-lanceolate, serrulate, nearly smooth ; branches procumbent, spiny.

Habitat. South of Europe.
Quality. Berries purgative.
Uses. Unripe fruit, called yellow berries, yields a brilliant yellow dye; it stains yellow morocco.

Fig. 234.—Rhamnus catharticus in flower and fruit ; *a*, a flower magnified.

ZIZYPHUS. *Tournefort.*

Calyx rotate, 5-cleft, circumscissile after flowering. *Fruit* oblong, drupaceous, with a single 2-celled stone.

1. Z. *vulgaris* Lamarck.—(JUJUBE.) Leaves ovate, retuse, rather toothed, smooth ; branches with no hooks, or very few, in pairs, of which one is recurved ; drupes oblong.

Habitat. Levant.
Quality. Fruit subacid, pleasant.
Uses. Employed in the preparation of *Pâte de Jujube,* and in hoarseness and sore-throat.

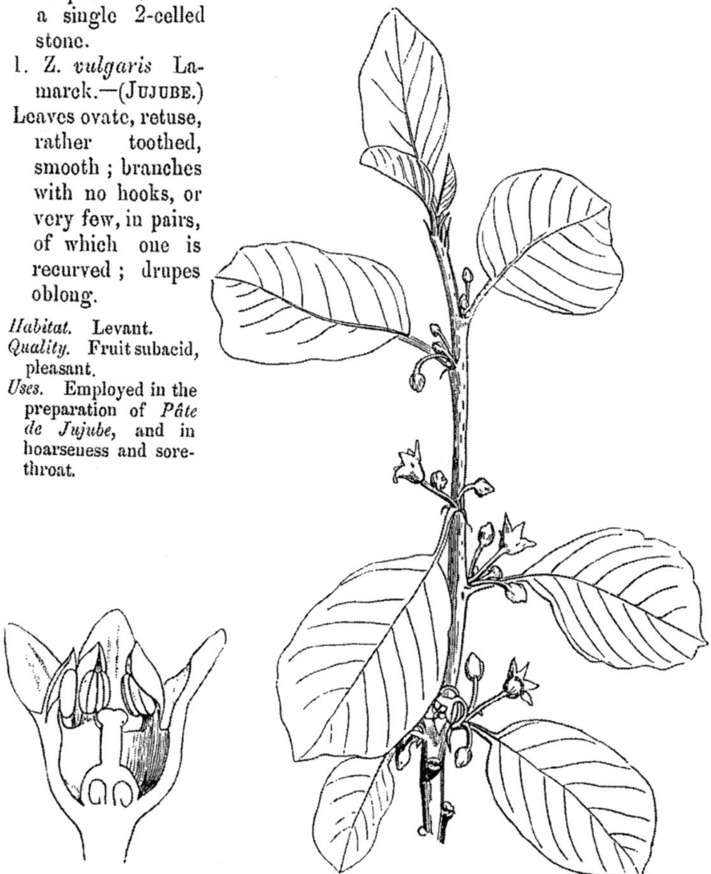

a

235

Natural Order, **Spindle Trees** ; *Celastraceæ* (V. K., p. 586.)

Prevailing Quality. Uncertain.

CATHA. *Forskähl.*

Stamens 5, inserted below the edge of a disk. *Ovary* 3-celled. *Capsule* 3-cornered. *Calyx* rotate.

Fig. 235.—Rhamnus Frangula in flower ; *a*, perpendicular section of a flower magnified.

1. C. *edulis* Forskähl.—(Arabian Tea.)
Erect, smooth; leaves elliptical, obtusely serrated; cymes axillary, dichotomous.

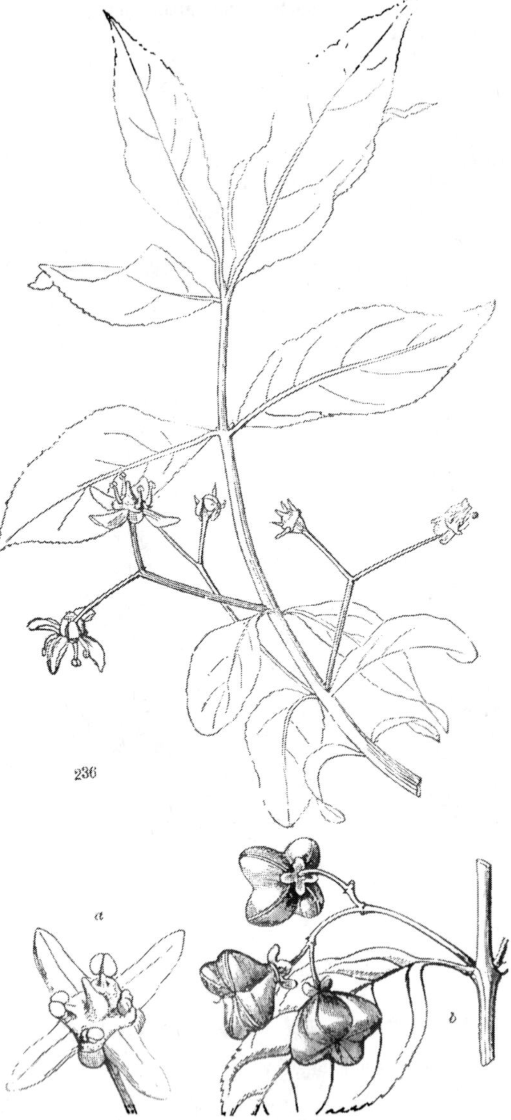

Habitat. Arabia.
Quality. Leaves stimulant, antisoporific, narcotic.
Uses. Employed by the Arabs instead of green tea to produce wakefulness.

Fig. 236.—Euonymus europæus; *a*, a flower magnified; *b*, a cluster of fruit.

Euonymus. *Linnæus.*

Stamens 5, standing on as many glands. *Ovary* 4-5-celled. *Capsule* lobed. *Calyx* rotate.

1. E. *europæus* Linnæus.—(Spindle Tree.) Fig. 236.

Branches smooth; leaves lanceolate-ovate, finely serrated; peduncles bearing about 3 flowers; petals oblong, acute.

Habitat. Hedges and woods; indigenous.
Quality. Seeds acrid, nauseous, purgative, emetic.
Uses. Ointment prepared from them kills pediculi.

Natural Order, Sapotads; *Sapotaceæ* (V. K., p. 590.)

Prevailing Quality. Astringent, febrifugal, lactescent.

Achras. *Linnæus.*

Some of the *stamens* sterile, lanceolate, alternate with the lobes of the corolla. *Fruit* a berry, containing erect, nut-like, shining seeds with a broad scar occupying all the inner angle.

1. A. *Sapota* Linnæus.—(Sapodilla Plum.)

Leaves elliptical, acute at each end; petioles and calyx covered with ferruginous down; flowers $\frac{6}{}\!\!\surd$.

Habitat. West Indies.
Quality and *Uses.* Fruit subacid, sweet, esteemed in the West Indies; bark astringent, febrifugal, equivalent to cinchona.

Bumelia. *Gærtner.*

Two sterile *stamens* between each lobe of the corolla; fertile 5, opposite the lobes, with a pair of scales at the back of each. *Fruit* 1-seeded, somewhat drupaceous.

1. B. *lycioides* Gærtner.

Leaves obovate, obtuse, with very distinct areolate nervures; flowers fascicled; corolla about twice as long as the calyx.

Habitat. United States.
Quality and *Uses.* Fruit austere, sweetish; recommended in diarrhœa.

Isonandra. *Wight.*

Stamens all fertile, twice as numerous as the lobes of the corolla.

1. I. *Gutta* Hooker.—(Gutta Percha Tree.) Fig. 237.

Leaves on long stalks, obovate-oblong, with a short point, golden beneath; flowers axillary, fascicled; stamens 12.

Habitat. Malay Archipelago.
Quality and *Uses.* Yields the gum resin called Gutta Percha, now in extensive use for various economical purposes.

237

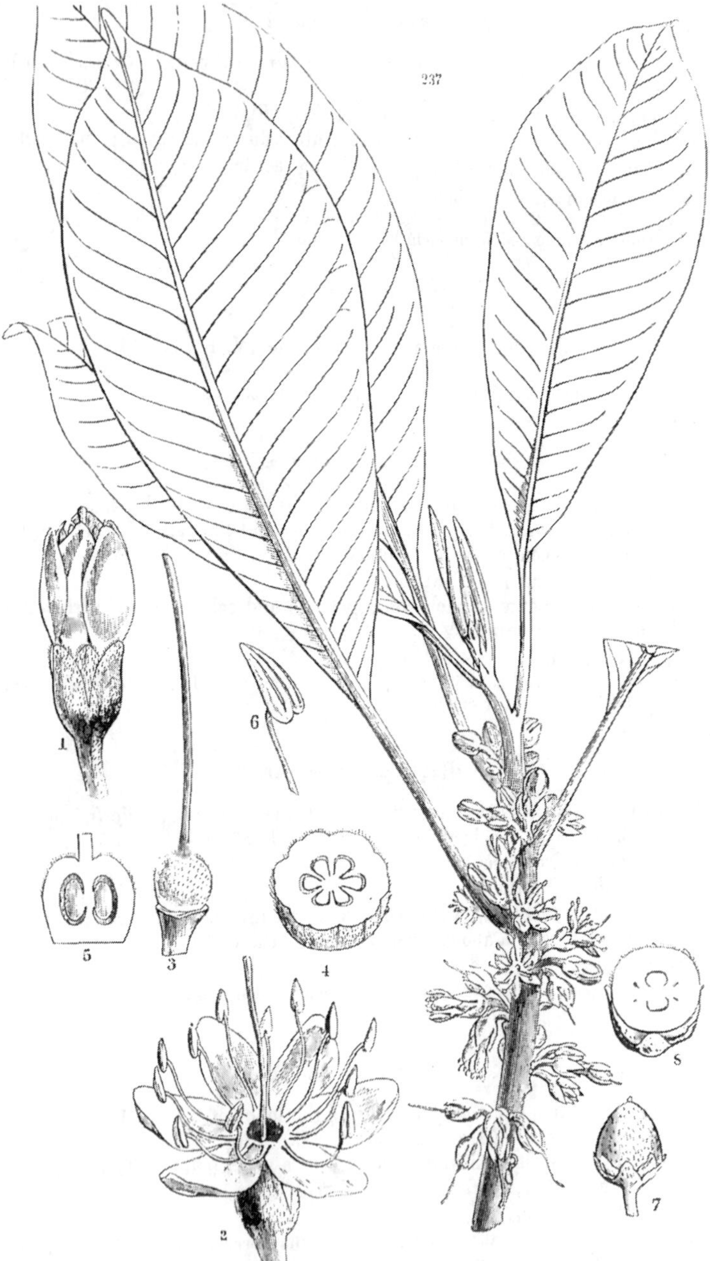

Natural Order, Storaxworts; *Styracaceæ* (V. K., p. 592.)

Prevailing Quality. Stimulant, fragrant.

STYRAX. *Linnæus.*

Calyx campanulate, slightly toothed, or entire. *Corolla* hoary, 5-parted. *Stamens* monadelphous. *Drupe* globular, downy, seated in the permanent calyx.

1. S. *officinale* Linnæus.—(STORAX TREE.) Fig. 238.

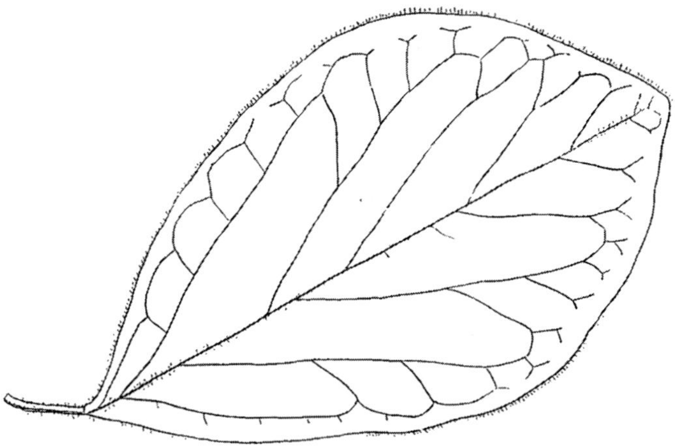

238

Leaves downy beneath, oval, obovate, smooth on the upper side; racemes few-flowered.

Habitat. Coasts of the Mediterranean.
Quality. Resin balsamic, stimulating, expectorant; detergent.
Uses. Chronic bronchial affections; foul ulcers.

2. S. *Benzoin* Dryander.—(BENJAMIN TREE). Fig. 239.

Leaves downy and white beneath, oblong, acuminate; racemes compound, many-flowered, rather shorter than the leaves.

Habitat. Indian Archipelago.
Quality. As the last; resin very fragrant.
Uses. Chronic pulmonary affections, fumigations.

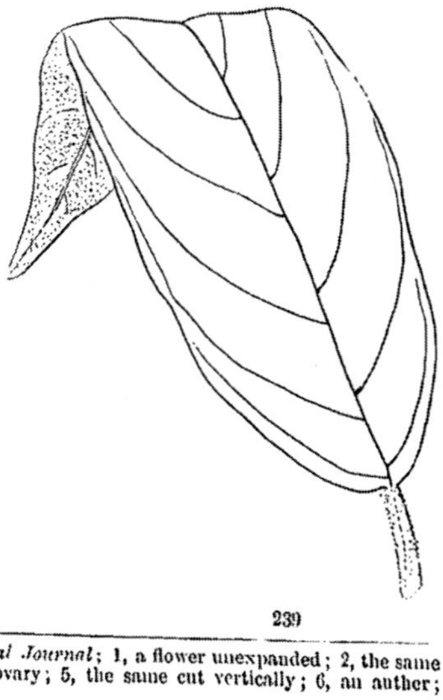

239

Fig. 237.—Isonandra Gutta, from the *Pharmaceutical Journal*; 1, a flower unexpanded; 2, the same full-blown; 3, the pistil; 4, a cross section of the ovary; 5, the same cut vertically; 6, an anther; 7, an unripe fruit; 8, its cross section.

Fig. 238.—Styrax officinale; 239. Styrax Benzoin.

THE ROSAL ALLIANCE (V. K., p. 539.)

Natural Orders of Rosals.

Leguminous Plants (*Fabaceæ.*) Carpel solitary. A Legume.

Almondworts (*Drupaceæ.*) Carpel solitary. A Drupe. Petals 5.

Appleworts (*Pomaceæ.*) Carpels several, adherent.

Roseworts (*Rosaceæ.*) Carpels several, free.

Sanguisorbs (*Sanguisorbaceæ.*) Carpel 1 or 2. Petals 0. Calyx-tube hardened.

Natural Order, **Leguminous Plants**; *Fabaceæ* (V. K., p. 544.)

Prevailing Quality. Deleterious.

---

* Papilionaceous Genera.

Lathyrus. *Linnæus.*

Calyx 5-toothed. *Stamens* diadelphous. *Style* dilated upwards. *Pod* continuous, 2-valved. *Leaves* with tendrils.

1. L. *Cicera* Linnæus.

Leaflets in 1 or 2 pairs; peduncles 1-flowered, shorter than the leaf; ripe pod with 2 narrow wings on the upper edge; seeds angular; flowers red.

Habitat. Fields throughout Europe.
Quality. Ripe seeds narcotic, poisonous, spoiling flour in which they are ground.

2. L. *Aphaca* Linnæus. Fig. 240.

Leaflets 0; petioles filiform; stipules large, ovate, auriculate.

Habitat. Indigenous.
Quality. Ripe seeds narcotic.
Uses. Green seeds and pods eaten as a potherb.

240

3. L. *tuberosus* Linnæus.

Leaves pinnated, of one pair; stem angular, wingless; peduncles many-flowered; the upper calycine teeth short, triangular.

Habitat. Europe, in chalky or heavy land.
Quality. Tubers amylaceous, eatable.
Uses. Employed as food in Holland; sometimes called Dutch Mice.

Fig. 240.—Stipules and filiform petiole of Lathyrus Aphaca.

ERVUM. *Linnæus.*

Calyx 5-toothed, with sharp linear divisions. *Style* hairy all round. *Pod* oblong, continuous, 2-4-seeded. *Leaves* with tendrils.

1. E. *Lens* Linnæus.—(LENTIL.)

Leaflets about 8, oblong, smoothish; stipules lanceolate, ciliated; peduncles 2-3-flowered, as long as the leaf; pod broad, short, somewhat truncate, finely netted, smooth, 2-seeded.

Habitat. Hedges in Europe.
Quality. Seeds amylaceous, nutritious; subnarcotic in large doses; difficult of digestion.
Uses. A food in southern countries; the base of Revalenta, a flour so called.

FABA. *Tournefort.*

Calyx 5-toothed. *Stamens* diadelphous. *Style* filiform. *Pod* coriaceous, tumid, spongy inside. *Seeds* with a large scar.

1. F. *vulgaris* Mœnch. *Vicia Faba* Linnæus.—(COMMON BEAN.)

Leaflets thick, 2-5, oval, mucronate; teeth of calyx linear.

Habitat. Borders of the Caspian Sea.
Quality. Roots diuretic; seeds nutritious when young, but somewhat poisonous when ripe.

GLYCYRRHIZA. *Tournefort.*

Calyx tubular, 5-cleft, 2-lipped. *Stamens* diadelphous. *Pod* ovate or oblong, 1-4-seeded. *Leaves* pinnated, with an odd one.

1. G. *glabra* Linnæus.—(LIQUORICE.) Fig. 241.

Leaflets ovate, rather blunt; stipules 0; spikes stalked, longer than the leaf; pods smooth, 3-4-seeded.

Habitat. South of Europe.
Quality. Emollient, demulcent, nutritious.
Uses. Root, or its extract, in coughs, and in the preparation of pills.

241

Fig. 241.—Glycyrrhiza glabra.

Colutea. *Linnæus.*

Calyx 5-toothed. *Stamens* diadelphous. *Pod* stipitate, membranous, inflated, indehiscent or nearly so.

1. C. *arborescens* Linnæus.—(Bladder Senna.)

Leaflets elliptical, retuse ; peduncles with about 6 flowers.

Habitat. South of Europe.
Quality. Leaves purgative.
Uses. Employed in adulterating blunt-leaved Senna.

Coronilla. *Linnæus.*

Calyx campanulate, with the 2 upper of its 5 teeth close together. *Claws* of the petals longer than the calyx. *Stamens* diadelphous. *Pod* jointed.

1. C. *Emerus* Linnæus.—(Scorpion Senna.)

Shrubby, smooth ; stipules minute ; leaflets 5-7, obovate ; peduncles 3-flowered.

Habitat. South of Europe.
Quality. Leaves purgative, drastic ; inconvenient on account of their griping effects.

Spartium. *Linnæus.*

Calyx membranous, spathaceous, 2-lipped. *Stamens* monadelphous. *Pod* compressed, many-seeded.

1. S. *junceum* Linnæus.—(Spanish Broom.)

Branches smooth, spongy ; leaves few, simple, lanceolate ; flowers large, yellow, in terminal racemes.

Habitat. South of Europe.
Quality. Seeds emetic, purgative, diuretic, tonic.
Uses. Dropsy.

Phaseolus. *Linnæus.*

Calyx campanulate, 2-lipped. *Stamens* diadelphous ; these, the keel and the style, rolled spirally.

1. P. *multiflorus* Willdenow.—(Scarlet Runner.)

Roots tuberous ; stem twining ; leaflets ovate-acuminate ; racemes stalked, longer than the leaves.

Habitat. East Indies.
Quality. Green fruit nutritious, digestible ; roots narcotic, dangerous.
Uses. Young pods eaten boiled as Kidney Beans ; ripe seeds eaten stewed under the name of Haricot Beans.

Astragalus. *Linnæus.*

Calyx 5-toothed. *Stamens* diadelphous. *Keel* obtuse. *Pod* continuous, 2-celled by the expansion of the dorsal suture. *Leaves* pinnate, with an odd one.

1. A. *gummifer* Labillardière.—(White Tragacanth.) Fig. 242.

A spiny bush ; leaflets smooth, in 4-6 pairs, oblong-linear ; flowers 3-5, axillary, sessile ; calyxes woolly, 5-cleft.

Habitat. Koordistan.
Quality. Gum emollient, demulcent; nutritive.
Uses. Irritation of mucous membranes, gonorrhœa, as a vehicle for calomel.

2. A. *strobiliferus* Lindley.—(RED TRAGACANTH.) Fig. 243.
A spiny bush; leaflets woolly, in 3 pairs, aristate; flowers in sessile axillary ovate cones; calyx feathery.

242

243

Habitat. Koordistan.
Quality and *Uses.* As the last, but of inferior quality.

3. A. *verus* Olivier.
Usually said to be the source of gum Tragacanth, is very little different from No. 1.

MUCUNA. *Adanson.*

Calyx campanulate; the upper of its two lips broad and entire. *Standard* shorter than the wings and keel. *Stamens* diadelphous, alternately imperfect. *Pod* oblong, torose, 2-celled, covered with brittle stinging hairs.

1. M. *pruriens* De Candolle.—(COWAGE PLANT.)
Flowers racemose; leaflets 3, acuminate; the middle rhomboid, hairy on the under side.

Habitat. East Indies.
Quality. Root diuretic; hairs of the pod brittle and productive of unbearable itching.
Uses. Hairs in worm cases.

TEPHROSIA. *Persoon.*

Calyx nearly equal, 5-toothed. *Standard* broad, downy on the outside. *Stamens* irregularly monadelphous or diadelphous. *Pod* sessile, compressed, many-seeded, with flat valves.

1. T. *Apollinea* De Candolle.—(EGYPTIAN INDIGO.) Fig. 244.
Spreading, with close-pressed hairs; leaflets in 4 or 5 pairs, obovate, silky beneath.

Fig. 242.—Astragalus gummifer; 243. A. strobiliferus.

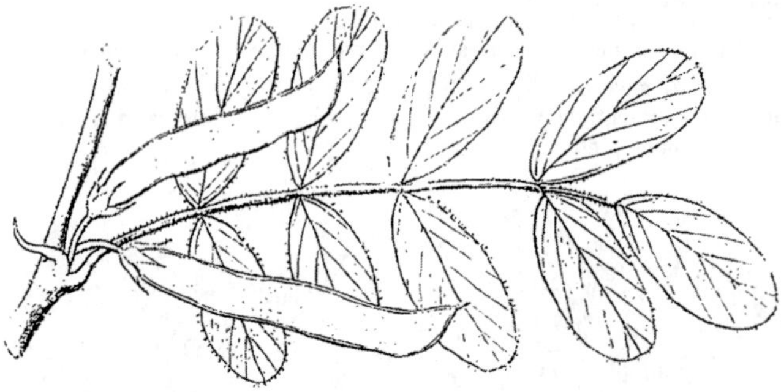

244

245

Habitat. Egypt.
Quality. Narcotic ; yields a fine blue dye.
Uses. Leaves often mixed, sometimes largely, with Alexandrian Senna. It is commonly cultivated for its indigo in Nubia.

CYTISUS. *De Candolle.*

Calyx 2-lipped, the upper lip generally entire. *Standard* broad. *Stamens* monadelphous. *Pod* compressed, many-seeded.

1. C. *scoparius* Link. — (BROOM.) Fig. 245.

Branches angular, smooth ; leaves small, the upper simple, the lower trifoliate ; flowers axillary, solitary, stalked ; pods hairy at the edges.

Habitat. Heaths of all Europe.
Quality. Emetic, purgative, diuretic.
Uses. Broomtops in dropsy.

2. C. *Laburnum* Linnæus.—(LABURNUM.)

A tree ; leaflets 3, ovate-lanceolate, downy beneath ; racemes pendulous, many-flowered.

Habitat. Alps of Europe.
Quality. Seeds narcotic, dangerous.
Uses. Seeds are a frequent cause of accidents among children. This is owing to the presence of an active principle called Cytisine.

3. C. *alpinus* Miller.—(The SCOTCH LABURNUM of Gardens.)

Perhaps a variety of the last ; has, no doubt, similar properties.

Fig. 244.—Tephrosia Apollinea ; 245. Cytisus scoparius.

TRIGONELLA. *Linnæus.*

Calyx campanulate, 5-cleft. *Carina* very small; the wings and standard presenting the appearance of a tripetalous corolla. *Pod* many-seeded, cylindrical, acuminate.

1. T. *Fœnum Græcum* Linnæus.—(FENUGREEK.)

Leaflets 3, obovate, obscurely toothed; stipules falcate, entire; flowers sessile; pods netted lengthwise, with a falcate beak.

Habitat. South of France.
Quality. Seeds emollient.
Uses. Poultices of the flour employed in veterinary practice.

INDIGOFERA. *Linnæus.*

Calyx 5-cleft. *Keel* bent back with elasticity. *Stamens* diadelphous. *Style* filiform, smooth. *Pod* taper, many-seeded. *Seeds* truncate, usually separated by cellular diaphragms.

1. I. *tinctoria* Linnæus.—(EAST INDIAN INDIGO.)

Leaflets in 4-5-pairs, oval, nearly smooth beneath; racemes axillary, shorter than the leaf; pods torulose, curved, deflexed.

Habitat. East Indies.
Quality. Emetic, cathartic, narcotic.
Uses. Epilepsy, infantine convulsions, chorea, hysterics; said to be an effectual cure for the "yaws" in the West Indies.

PTEROCARPUS. *Linnæus.*

Calyx 5-toothed. *Stamens* 10, variously united. *Pod* irregular, nearly orbicular, surrounded by a wing, often varicose, 1-seeded.

1. P. *santalinus* Linnæus.—(RED SANDAL-WOOD.)

Stamens 1 and 9; leaflets 3-5, alternate, rather round, retuse, smooth; racemes axillary; petals crenate. A tree.

Habitat. East Indies.
Quality. Astringent.
Uses. A mere colouring agent.

2. P. *erinaceus* Lamarck.—(AFRICAN KINO TREE.)

Leaflets alternate, elliptical, obtuse, smooth above, covered beneath with brown down; pod with a short lateral point. A tree.

Habitat. Gambia.
Quality. Astringent.
Uses. Chronic diarrhœa, pyrosis, leucorrhœa; gargles, injections.

3. P. *Marsupium* Roxburgh.—(INDIAN KINO TREE.)

Stamens monadelphous; leaflets 5-7, alternate, elliptical, rather emarginate, leathery, smooth; branches and calyxes smooth; panicle terminal; pod half circular, oblique, smooth. A tree.

Habitat. Malabar.
Quality. Astringent.
Uses. In restraining mucous discharges.

MYROSPERMUM. *Jacquin.*

Calyx campanulate, 5-toothed. *Standard* much larger than the other petals. *Stamens* free. *Pod* stipitate, samaroid.

Leaflets coriaceous, oval, obtuse, smooth; pod-wing very thick on one side, not veiny on the other.

1. M. *peruiferum* De Candolle.—(QUINQUINO.)

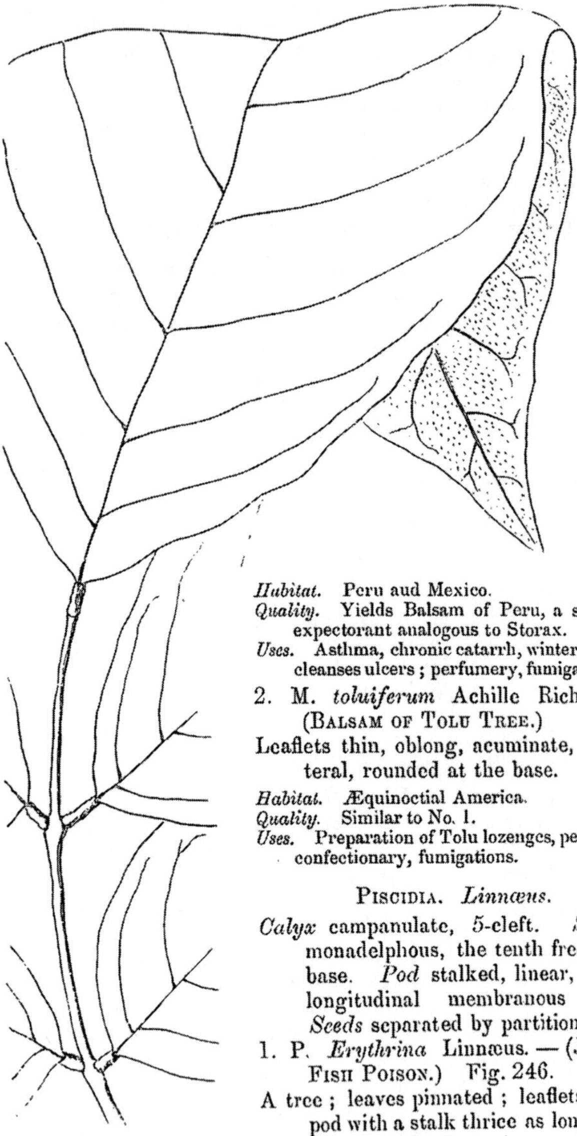

Habitat. Peru and Mexico.
Quality. Yields Balsam of Peru, a stimulant expectorant analogous to Storax.
Uses. Asthma, chronic catarrh, winter coughs; cleanses ulcers; perfumery, fumigations.

2. M. *toluiferum* Achille Richard. — (BALSAM OF TOLU TREE.)

Leaflets thin, oblong, acuminate, equilateral, rounded at the base.

Habitat. Æquinoctial America.
Quality. Similar to No. 1.
Uses. Preparation of Tolu lozenges, perfumery, confectionary, fumigations.

PISCIDIA. *Linnæus.*

Calyx campanulate, 5-cleft. *Stamens* monadelphous, the tenth free at the base. *Pod* stalked, linear, with 4 longitudinal membranous wings. *Seeds* separated by partitions.

1. P. *Erythrina* Linnæus. — (JAMAICA FISH POISON.) Fig. 246.

A tree; leaves pinnated; leaflets ovate; pod with a stalk thrice as long as the calyx, and broken wings.

246

Habitat. West Indies. *Quality.* Bark of root a powerful narcotic.
Uses. Substitute for opium; as a poison for stupifying fish.

Fig. 246.—Leaf of Piscidia Erythrina.

* * Cæsalpineous Genera.

Cassia. *Linnæus.*

Sepals 5, somewhat unequal. *Petals* 5, unequal. *Stamens* 10, unequal: 3 long, 4 short, 3 abortive.

1. C. *elongata* Lemaire.—(Tinnivelly Senna.)

Leaflets lanceolate, rather downy beneath, with fine close-pressed hairs; petiole without gland; pods oblong, membranous.

Habitat. Arabia.
Quality. Leaves purgative, scarcely tonic.
Uses. Constipation, worms, &c.

2. C. *acutifolia* Delile.—(Alexandrian Senna.) Fig. 247.

Leaflets ovate, with long spreading hairs near the midrib on the under side; petiole without gland; pods roundish-oblong, membranous.

247 *a*

Habitat. Egypt and Nubia.
Quality and *Uses.* As in the last.

3. C. *obovata* Colladon. C. *Senna* Linnæus.—(Blunt-leaved Senna.)

Leaflets obovate, obtuse; petiole without gland; pods flat, curved, tumid, and crested about the middle.

Habitat. Egypt, India.
Quality and *Uses.* As the last, but less esteemed; probably owing to the extensive adulteration to which this sort is subject.

4. C. *Fistula* Linnæus.—(Pudding Pipe Tree.)

Pods long, cylindrical, with a pulpy chamber for every seed.

Habitat. East Indies.
Quality. Pulp of pods purgative.
Uses. Mixed with other purgatives; called Cassia by pharmaceutical writers.

Tamarindus. *Linnæus.*

Sepals 5, tubular, reflexed: the two lower entirely connate. *Petals* 3. *Stamens* 9-10, monadelphous: 7 sterile. *Pod* many-seeded, filled with pulp.

Fig. 247.—Cassia acutifolia; *a*, an expanded flower.

1. T. *indica* Linnæus.—(TAMARIND TREE.)

A tree; leaves abruptly pinnated, in many pairs; flowers racemose; pods at least six times longer than broad.

Habitat. East Indies.
Quality. Pulp of fruit nutritive, refrigerant, laxative.
Uses. Infusion forms a cooling drink; a constituent of mild laxative medicines.

COPAIFERA. *Linnæus.*

Sepals 4, united at base. *Petals* 0. *Stamens* 10, distinct. *Legume* stipitate, 1-seeded.

1. C. *officinalis* Linnæus.—(COPAIVA TREE.) Fig. 248.

248 *a*

Leaves pinnated; leaflets in 2 or 3 pairs, oblique, ovate, bluntly acuminate, containing oblong, kidney-shaped, and circular oil-cysts of very unequal size.

Habitat. Tropical America.
Quality. The oil stimulant, acting principally upon the mucous membranes, and urino-genital apparatus.
Uses. Gonorrhœa, catarrhus vesicæ, leucorrhœa, &c.

[N.B.—Various species of Copaifera yield Copaivi balsam of various quality.]

CERATONIA. *Linnæus.*

Calyx 5-parted. *Petals* 0. *Stamens* 5. *Stigma* orbicular. *Pod* coriaceous, many-seeded, filled with a spongy pulp.

1. C. *Siliqua* Linnæus.—(CAROB TREE. ALGAROBA. LOCUST TREE. ST. JOHN'S BREAD.)

A large tree; leaves evergreen, abruptly pinnate; leaflets oval, obtuse, flat.

Habitat. Syria.
Quality. Pulp of fruit sweet and nutritious, but laxative.
Uses. A common article of food in the Mediterranean both for man and cattle.

CÆSALPINIA. *Linnæus.*

Sepals 5, unequal, united into a nearly permanent cup, the lowest, largest and arched. *Petals* 5, unguiculate. *Stamens* 10, ascending, all perfect. *Pod* unarmed, compressed.

Fig. 248.—Leaf of Copaifera officinalis; *a*, a magnified view of a portion of a leaf, showing the transparent oil-cysts.

1. C. *coriaria* Willdenow.—(Dividivi Tree.)
Unarmed, smooth; leaves bipinnate, with 5-7 pairs of pinnæ; leaflets in 15-20 pairs, linear, blunt; racemes panicled.

Habitat. Tropical America.
Quality. Pods gathered before ripeness excessively astringent.
Uses. By tanners and dyers.

Hæmatoxylon. *Linnæus.*

Calyx tubular, with 5 deciduous segments. *Petals* 5, scarcely longer than the calyx. *Stamens* 10, hairy; *anthers* glandular. *Pod* compressed, flat, acuminate at each end, indehiscent, 2-seeded, breaking across in the middle.
1. H. *campeachianum* Linnæus.—(Logwood Tree.)
Leaves pinnate, or imperfectly bipinnate; leaflets obcordate; flower-buds red; flowers yellow.

Habitat. Campeachy.
Quality. Bark a mild astringent.
Uses. Old diarrhœas, dysenteries, hemorrhages, leucorrhœa. A dye.

---

* * * Mimoseous Genera.

Acacia. *Linnæus.*

Calyx 4-5-toothed. *Petals* 4-5. *Stamens* 00. *Pod* continuous, juiceless, 2-valved.
1. A. *Verek* Guillemin.—(Gum Arabic Tree.) Fig. 249.
Branches smooth; spines recurved, in threes; petiole unarmed; pinnæ in 3-5 pairs; leaflets in 10-15 pairs, linear, obtuse, glaucous; flowers spiked; pod oblong, thin.

249

Habitat. Senegal.
Quality. Bark astringent, tonic.
Uses. Yields true white Gum Arabic. *Perrottet.*

Fig. 249.—Acacia Verek.

2. A. *arabica* Willdenow. A. *nilotica* Delile.

Spines in pairs ; branches and petioles downy; pinnæ in 4-6 pairs ; leaflets in 10 pairs, oblong-linear, smooth ; a gland on the petiole ; flowers in globose heads ; pod moniliform.

Habitat. Senegal, &c.
Quality. As the last.
Uses. Yields a bad red gum, unknown in commerce. *Perrottet.*

3. A. *vera* Willdenow. *Mimosa nilotica* Linnæus.

Spines in pairs ; branches and leaves smooth ; pinnæ in 2 pairs ; leaflets in 8-10 pairs, oblong-linear ; a gland between the pinnæ ; heads of flowers globose ; pods moniliform.

Habitat. Senegal.
Quality. Pods very astringent.
Uses. Said to yield Gum Arabic, aud a part of Gum Senegal.

4. A. *Adansonii* Guillemin.—(Red Gum Arabic Tree.)

Spines in pairs ; branches and leafstalks downy; pinnæ 4-6 pair ; leaflets 12-16 pair, oblong-linear ; a gland between the upper and lower pinnæ; heads stalked, axillary, in threes or fours ; pod torulose, 10-12-seeded.

Habitat. Senegal.
Quality. Pods and bark powerfully astringent.
Uses. Yields red Gum Arabic. *Perrottet.*

5. A. *Catechu* Willdenow.—(Catechu Tree.)

Stipulary, prickly, straight when young, hooked afterwards ; pinnæ in 10 pairs ; leaflets in 40-50 pairs, with a single depressed gland at the base of the petiole, and 2 or 3 between the upper pinnæ ; spikes cylindrical.

Habitat. East Indies.
Quality. Astringent ; "more powerful than Kino." *Pereira.*
Uses. Relaxed uvula, dyspepsia, diarrhœa, gleets, fluor albus, &c.

Inga. *Plumier.*

Calyx 5-toothed. *Petals* 5, united. *Stamens* 00. *Pod* linear, filled with pulp or farinaceous matter.

1. I. *Sassa* Willdenow.

Pinnæ in 3-4 pairs ; leaflets in 12 pairs, oblong-ovate ; flowers in umbellate panicles ; stamens monadelphous beyond the corolla.

Habitat. Abyssinia.
Quality. Pulp of fruit sweet ; a source of gum.
Uses. Said to produce the Gum Sassa of commerce.

Natural Order, Almondworts; *Drupaceæ* (V. K., p. 557.)

Prevailing Quality. Production of Hydrocyanic acid.

Amygdalus. *Tournefort.*

Drupe with coarsely-furrowed and wrinkled putamen. Young *leaves* conduplicate.

1. A. *communis* Linnæus.—(Almond Tree.) Fig. 250.

Leaves oblong-lanceolate, serrulate ; flowers solitary; drupe downy, with a tough fibrous sarcocarp.

Habitat. Barbary.
Quality. Seeds of the Sweet Almond nutritive and emollient, but indigestible; of the Bitter Almond poisonous, abounding in hydrocyanic acid.
Uses. Sweet Almonds in dessert, confectionary, emulsion; Bitter Almonds when hydrocyanic acid is required, and as a flavouring agent.

2. A. *Persica* Linnæus.—(THE PEACH. THE NECTARINE.)

Leaves oblong-lanceolate, serrulate; flowers solitary; drupe downy or smooth, with a tender succulent sapid sarcocarp.

Habitat. North of India; Persia.
Quality. Nutritious, refrigerant; bark, blossoms, and skin of the seeds poisonous.
Uses. Common at dessert; blossoms a vermifuge; stewed fruit in slight constipation; kernels when bitter are like Bitter Almonds.

250

CERASUS. *Tournefort.*

Drupe smooth, with a polished even putamen. Young *leaves* conduplicate.

1. C. *Laurocerasus* Loiseleur.—(COMMON LAUREL.) Fig. 251.

Leaves ovate-lanceolate, evergreen, convex, remotely serrated, with from 2 to 4 glands on the under side; racemes shorter than the leaves.

Habitat. Trebizond.
Quality. Poisonous, acts like hydrocyanic acid.
Uses. Forms laurel-water, a substitute for hydrocyanic acid in palpitation of the heart, &c. Vapour of bruised leaves destroys small insects.

2. C. *virginiana* Michaux.—(VIRGINIAN BIRD CHERRY.)

Leaves oblong-acuminate, doubly-toothed, smooth, deciduous, with about 4 glands on each petiole; flowers in erect racemes.

Habitat. United States.
Quality. Bark astringent, febrifugal.
Uses. Against intermittents; dysentery.

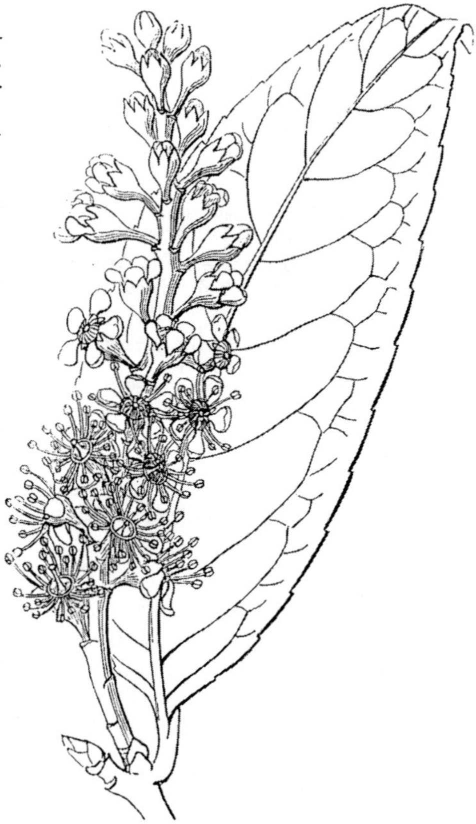

251

Fig. 250.—Flower of Amygdalus communis; 251. Cerasus Laurocerasus.

3. C. *vulgaris* Miller.—(COMMON CHERRY.)

Leaves oblong, serrate, acuminate, deciduous, with long glandular petioles; flowers in clusters; petals white; fruit succulent, sapid.

Habitat. Asia Minor.
Quality. Fruit succulent, sweet, subacid; a favourite article of dessert. Bark yields Cherry-tree gum, a substitute for Tragacanth.

PRUNUS. *Tournefort.*

Drupe smooth, glaucous, succulent, with a smooth putamen. Young *leaves* convolute.

1. P. *spinosa* Linnæus.—(The SLOE.)

Branches spiny; leaves obovate-elliptical or ovate, downy beneath, finely and doubly toothed; peduncles solitary; fruit very austere.

Habitat. Hedges. Indigenous.
Quality. Fruit austere, astringent.
Uses. Leaves dried and mixed with tea; bark in intermittents.

2. P. *Coccomilia* Tenore.

Leaves obovate, smooth on each side, with glandular crenatures; peduncles short, in pairs; fruit ovate-oblong, mucronate.

Habitat. Calabria.
Quality. Bark febrifugal.
Uses. Largely employed against the intermittents of Calabria.

3. P. *domestica* Linnæus.—(The COMMON PLUM.)

Possesses qualities like the last, but much weaker. *Fruit* laxative when stewed. *Bark* exudes a gum analogous to Tragacanth.

4. P. *Armeniaca* Linnæus.—(The APRICOT TREE.)

Leaves cordate, long-stalked, smooth; flowers sessile; fruit sapid, with a downy skin.

Habitat. The Levant, and the Himalayahs.
Quality. Fruit refrigerant, laxative.
Uses. The fruit, whether fresh or dried, is largely employed, and with great success, in the East, against the dangerous fevers of the country.

Natural Order, Applewortš; *Pomaceæ* (V. K., p. 559.)

Prevailing Quality. Austerity.

PYRUS. *Linnæus.*

Fruit 5-celled, with a cartilaginous endocarp, and 2 seeds in each cell.

1. P. *Malus* Linnæus.—(The APPLE TREE.)

Leaves ovate, acute, crenate, woolly beneath, as are the calyx-tube and their own under-side; styles smooth; fruit narrowest next the point.

Habitat. Middle of Europe.
Quality. Fruit agreeable; when cooked, nutritious and digestible.
Uses. A common article of dessert; the more austere varieties form *cyder.*

2. P. *communis* Linnæus.—(The PEAR TREE.) Fig. 252.

Leaves ovate, serrated, smooth on both sides, as well as the wood and buds; fruit narrowest at the base.

Habitat. Middle of Europe.
Quality and *Uses.* As in P. Malus. The fermented juice is *perry.*

3. P. *Aucuparia* Gærtner.—(The MOUNTAIN ASH.)

Leaves pinnated, nearly smooth; leaflets serrated; buds downy; fruit globose.

Habitat. Woods.
Quality and *Uses.* Flowers, bark, and root yield hydrocyanic acid in large quantity. Berries contain malic acid in abundance.

CYDONIA. *Tournefort.*

Fruit 5-celled, with a cartilaginous endocarp; and many mucilaginous seeds in each cell.

1. C. *vulgaris* Persoon.—(The QUINCE TREE.) Fig. 253.

Leaves ovate, entire, deciduous, downy on the under side, as is also the calyx; sepals leafy.

252

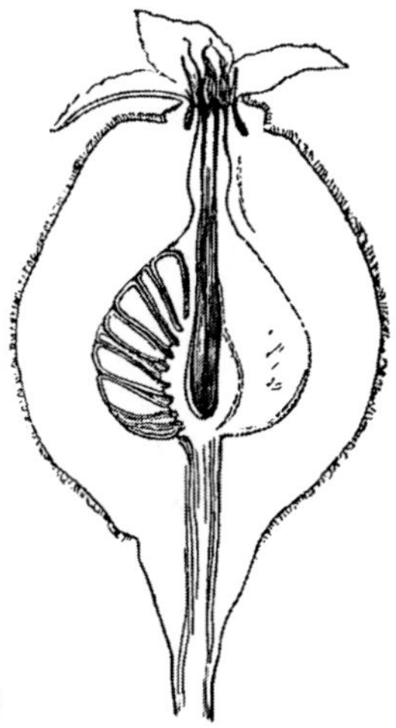
253

Habitat. South of Europe.
Quality. Fruit, when stewed, an agreeable food; mucilage or bassorin of seeds nutritive, demulcent, emollient. Œnanthic ether occurs in the rind.
Uses. Fruit forms a marmalade much esteemed; fermented juice forms a pleasant wine; mucilage of seeds applied externally to cracked lips, cracked nipples, painful hæmorrhoids, erysipelatous affections of the skin.

Natural Order, Roseworts; *Rosaceæ* (V. K., p. 563.)

Prevailing Quality. Astringency.

POTENTILLA. *Linnæus.*

Calyx 4-5-lobed, with as many bracts. *Carpels* 00, dry.

1. P. *reptans* Linnæus.—(CINQUEFOIL. FIVE FINGERS.)

Stem creeping; leaflets 5-nate, obovate, toothed; calyx 5-lobed; flowers axillary, solitary, longer than the leaves.

Habitat. Common in hedges and woods.
Quality and *Uses.* Like those of P. Tormentilla. Also a febrifuge.

Fig. 252.—Flowers of Pyrus communis; 253. Fruit of Cydonia vulgaris cut lengthwise.

2. P. *Tormentilla* Nestler.—(TORMENTIL.) Fig. 254.
Calyx and corolla tetramerous.

254

Habitat. Hedges and heathy downs everywhere.
Quality. Astringent and tonic.
Uses. Chronic diarrhœa, dysentery (especially of cattle); tans in the Orkneys, dyes red in Lapland.

3. P. *anserina* Linnæus.—(SILVERWEED.)
Stem creeping; leaves silvery, with soft hairs, interruptedly pinnate; peduncles solitary.

Habitat. Roadsides.
Quality and *Uses.* Roots extremely astringent, sometimes used for tanning. Distilled water said to be cosmetic.

FRAGARIA. *Linnæus.*

Calyx 5-lobed, with as many bracts. *Carpels* 00, dry, on a convex deciduous succulent torus.

1. F. *vesca* Linnæus.—(WOOD STRAWBERRY.)
Leaves hairy, plaited, thin, pale green; hairs of the flower-stalks close-pressed.

Fig. 254.—Potentilla Tormentilla.

Habitat. Woods of Europe.
Quality. Fruit a delicious article of dessert.

2. F. *virginiana* Miller.—(GARDEN STRAWBERRY.)
Leaves broad, smooth, shining, nearly even, glaucous beneath.

Habitat. United States and Canada.
Quality. Like the last, but not so aromatic, and more sweet.

GEUM. *Linnæus.*

Calyx 5-lobed, with as many bracts. *Carpels* 00, dry, with hardened hooked styles, forming a bur.

1. G. *urbanum* Linnæus.—(AVENS. HERB BENNETT.) Fig. 255.
Radical leaves interruptedly pinnate and lyrate, those of the stem ternate; lower joint of the style much longer than the upper, which is smooth.

Habitat. Hedges and thickets.
Quality. Aromatic, tonic, astringent.
Uses. Diarrhœa, leucorrhœa, dysentery, intermittents; an ingredient in some ales.

2. G. *rivale* Linnæus.—(WATER AVENS.)
Leaves interruptedly pinnate and lyrate, those of the stem ternate; flowers nodding, dull purple; the two joints of the style of equal length.

Habitat. Wet meadows and woods.
Quality and *Uses.* Same as the last. Root in bladder diseases.

3. G. *canadense* Jacquin.—(CHOCOLATE ROOT. BLOOD ROOT.)
This, which is a native of the United States, has some reputation as a mild tonic.

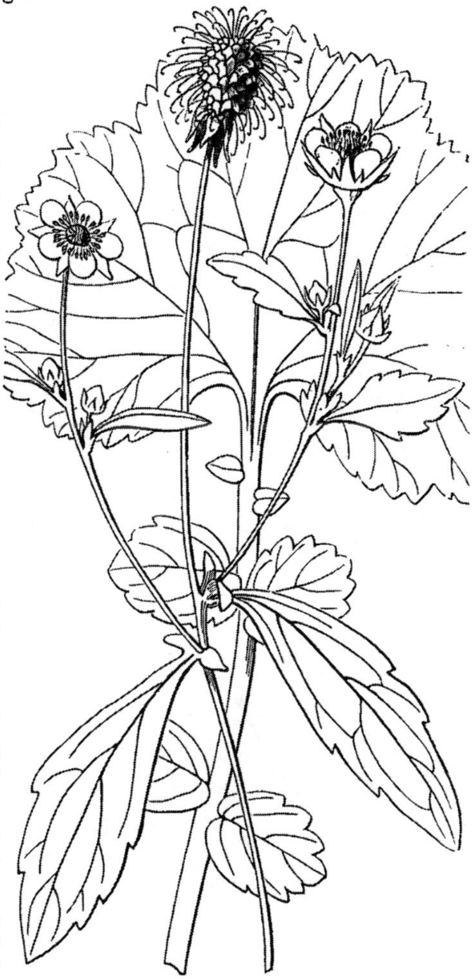
255

Fig. 255.—Geum urbanum.

AGRIMONIA. *Linnæus.*

Calyx 5-cleft, without bracts; tube fleshy, at length tough, covered with hooked bristles, and investing a couple of carpels.

1. A. *Eupatoria* Linnæus.—(AGRIMONY.) Fig. 256.

Leaves interruptedly pinnate, serrate, downy beneath; calyx of the fruit obconic; outer bristles spreading.

Habitat. Fields and roadsides.
Quality. Slightly aromatic, styptic, bitter.
Uses. Decoction used in gargles; dried leaves form a kind of herb-tea; root has been employed as a vermifuge.

COMARUM. *Linnæus.*

Calyx 5-cleft, with as many bracts. *Carpels* 00, dry, on a convex persistent succulent torus.

1. C. *palustre* Linnæus.

Leaves pinnate; leaflets oblong, sharply serrate; flowers dark purple; petals small.

Habitat. Marshes and bogs.
Quality. Said to be a valuable remedy for intermittent fevers.

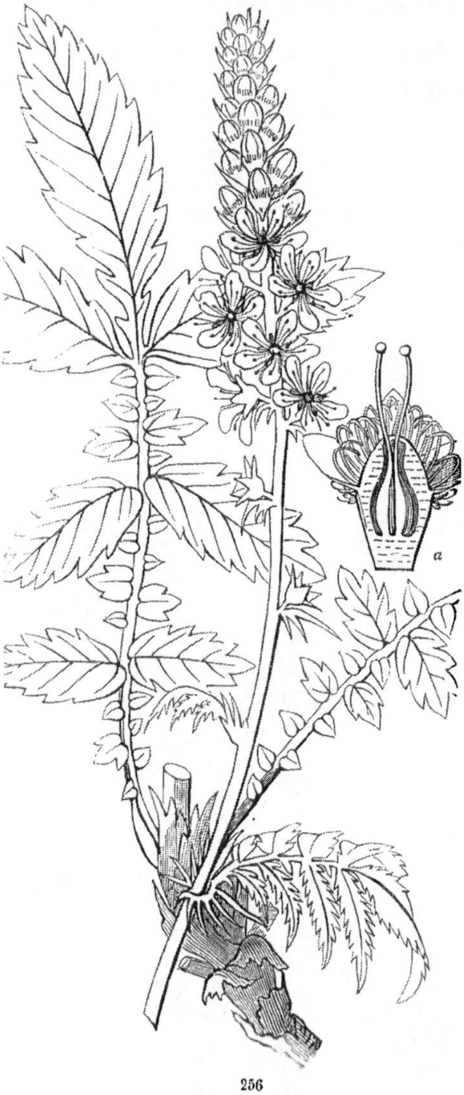

256

b

Fig. 256.—Agrimonia Eupatoria; *a*, perpendicular section of flower; *b*, ripe fruit.

RUBUS. *Linnæus.*

Calyx 5-lobed, without bracts. *Drupes* 00, placed upon a long torus, and adhering to each other.

1. R. *Idæus* Linnæus.—(The RASPBERRY.)

Stem erect ; leaves pinnate, white beneath ; flowers axillary and terminal, corymbose ; calyx woolly.

Habitat. Woods in the west of England.
Quality. Fruit fragrant, subacid, wholesome.
Uses. Forms preserves ; a kind of vinegar ; a useful fever drink.

SPIRÆA. *Linnæus.*

Calyx 5-cleft, without bracts. *Carpels* 1 or more, follicular, many-seeded.

1. S. *Filipendula* Linnæus.—(DROPWORT.)

Leaves interruptedly pinnate ; leaflets all oblong, deeply cut and serrate ; cymes panicled ; follicles hairy.

Habitat. Upland pastures.
Quality. Tonic, fragrant : the tubers rather nutritious.

2. S. *Ulmaria* Linnæus. —(MEADOW SWEET.) Fig. 257.

Leaves interruptedly pinnate ; leaflets ovate, the terminal large, 3-5-lobed ; cymes compound, proliferous ; follicles smooth, contorted.

Habitat. Moist meadows.
Quality. Flowers yield a fragrant distilled water. Roots astringent.

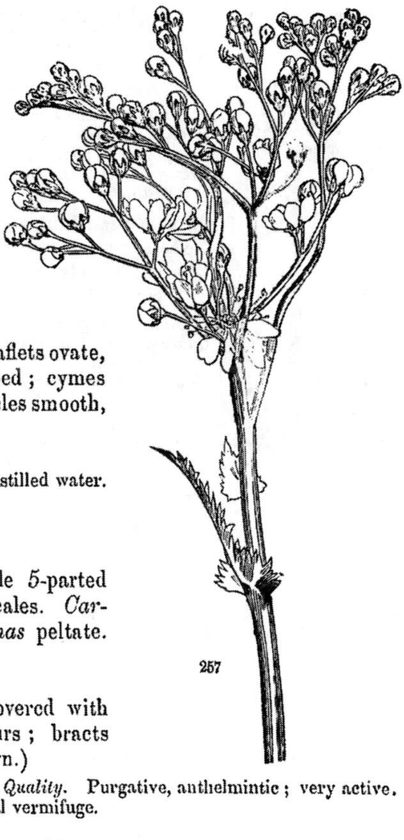

257

BRAYERA. *Kunth.*

Calyx turbinate, with a double 5-parted limb. *Petals* resembling scales. *Carpels* 2, 1-2-seeded. *Stigmas* peltate. *Seeds* solitary, pendulous.

1. B. *anthelmintica* Kunth.

A tree ; peduncles branched, covered with soft hairs; flowers in fours ; bracts roundish. (Leaves unknown.)

Habitat. Abyssinia. *Quality.* Purgative, anthelmintic ; very active.
Uses. Reported to be a very effectual vermifuge.

GILLENIA. *Mœnch.*

Calyx tubular, contracted at the mouth, 5-toothed. *Petals* long, linear. *Stamens* 10-15, short, enclosed. *Carpels* 5, with a filiform style and capitate stigma, partially connate.

Fig. 257.—Spiræa Ulmaria.

1. G. *trifoliata* Mœnch.

Leaves 3-foliolate ; stipules linear, acuminate, entire.

Habitat. United States.

Quality. Root emetic ; employed as Ipecacuanha, but said to be uncertain in its operation.

ROSA. *Linnæus.*

Calyx with a fleshy permanent tube, enclosing 00 bony carpels.

1. R. *gallica* Linnæus.—(FRENCH ROSE.)

Prickles and setæ nearly equal, weak; leaflets stiff, elliptical; flowers erect ; sepals ovate ; fruit bracteate, nearly globose ; sepals compound.

Habitat. Centre of Europe.

Quality. Petals mild astringents and tonics ; also laxative.

Uses. Chiefly used for colouring and flavouring other medicines.

2. R. *centifolia* Linnæus.—(PROVINS ROSE. CABBAGE ROSE.)

Prickles and setæ unequal, the larger falcate ; leaflets oblong, wrinkled, fringed with glands ; flowers nodding ; calyxes viscid; fruit bracteate, oblong.

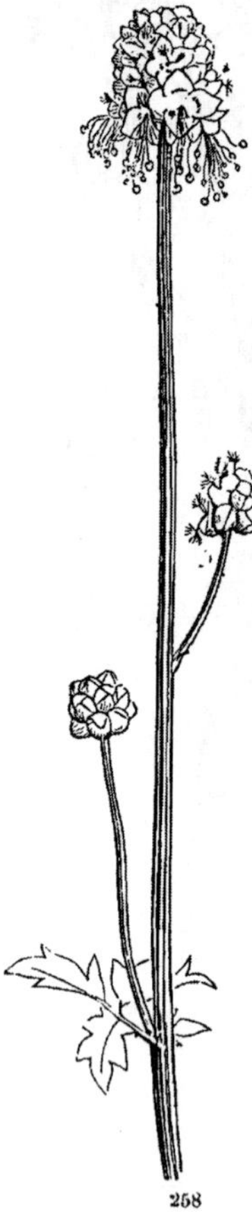

Habitat. The eastern slope of Caucasus.

Quality. Petals laxative, deliciously fragrant ; the odour dangerous to some constitutions.

Uses. Petals form Syrup of Roses, and yield rosewater by distillation.

3. R. *canina* Linnæus.—(DOG ROSE.)

No setæ ; prickles equal, hooked ; leaflets ovate, without glands, with converging serratures, rigid ; sepals deciduous, compound ; root-shoots arched.

Habitat. Hedgerows.

Quality. Pulp of fruit nutritive, slightly refrigerant and astringent.

Uses. Forms Conserve of Heps.

Natural Order, **Sanguisorbs**; *Sanguisorbaceæ* (V. K., p. 561.)

Prevailing Quality. Astringency.

POTERIUM. *Linnæus.*

Flowers polygamous. Three scales at the base of the calyx. *Stamens* 00. *Stigma* pencilled.

1. P. *Sanguisorba* Linnæus.—(BURNET.) Fig. 258.

Leaves pinnate ; leaflets roundish-ovate ; heads of flowers, male at the base, female at the apex.

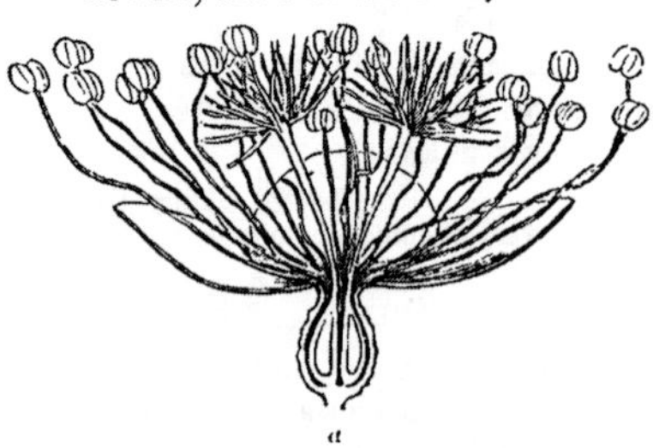

258 *a* *b*

Fig. 258.—Poterium Sanguisorba; *a*, perpendicular section of flower ; *b*, fruit.

Habitat. Meadows. *Quality.* Astringent, tonic.
Uses. Sometimes used as a herb tea. A common sheep food.

THE SAXIFRAGAL ALLIANCE (V. K., p. 566.)

Natural Orders of Saxifragals.

Saxifrages (*Saxifragaceæ.*) Styles distinct. Leaves alternate.

Loosestrifes (*Lythraceæ.*) Styles consolidated. Leaves opposite.

Natural Order, Saxifrages; *Saxifragaceæ* (V. K., p. 567.)

Prevailing Quality. Astringency.

HEUCHERA. *Linnæus.*

Calyx 5-cleft. *Petals* undivided. *Stamens* 5. *Styles* very long.

1. H. *americana* Linnæus.—(ALUM-ROOT.)

Covered with clammy down; leaves roundish, lobed, with mucronate dilated teeth; peduncles dichotomous, straggling.

Habitat. United States. *Quality.* Root strongly styptic.
Uses. Where powerful astringents are needed; forms a wash for wounds and obstinate ulcers.

Natural Order, Loosestrifes; *Lythraceæ* (V. K., p. 574.)

Prevailing Quality. Astringency, acridity.

LYTHRUM. *Linnæus.*

Calyx cylindrical, striated, with a double row of short broad teeth. *Petals* 4-6. *Stamens* twice as many. *Style* filiform. *Stigma* capitate.

1. L. *Salicaria* Linnæus.—(COMMON LOOSESTRIFE.) Fig. 259.

Leaves lanceolate, cordate; flowers spiked, nearly sessile.

Habitat. Ditches and river banks.
Quality. Demulcent, astringent. *Uses.* Diarrhœa, dysentery.

AMMANNIA. *Linnæus.*

Calyx campanulate, with a double row of teeth. *Petals* 0. *Stamens* 4.

1. A. *vesicatoria* Roxburgh.

Stem erect, branched; leaves sessile, lanceolate, tapering to the base; flowers sessile, in close whorls.

Habitat. East Indies.
Quality. The whole plant has a strong muriatic smell; acrid.
Uses. Used in India to raise blisters in rheumatism, &c.

259

Fig. 259.—Flowers of Lythrum Salicaria.

HEIMIA. *Link.*

Calyx campanulate, with 2 bracts at base, with a double row of 12 teeth. *Petals* 6. *Stamens* 12.

1. H. *salicifolia* Link.—(HANCHINOL.)

Leaves in threes or opposite, the upper often alternate ; petals obovate.

Habitat. New Spain.
Quality. A powerful sudorific and diuretic.
Uses. Has a great Mexican reputation in venereal diseases.

THE GENTIANAL ALLIANCE (V. K., p. 594.)

Natural Orders of Gentianals.

Hollyworts (*Aquifoliaceæ.*) Stipules 0. Stigma simple. Placentæ axile.

Dogbanes (*Apocynaceæ.*) Stipules 0. Stigma trochlear.

Loganiads (*Loganiaceæ.*) Stipules interpetiolar.

Gentianworts (*Gentianaceæ.*) Stipules 0. Stigma simple. Placentæ parietal.

Natural Order, **Hollyworts**; *Aquifoliaceæ* (V. K., p. 597.)

Prevailing Quality. Tonic, emetic, diuretic.

PRINOS. *Linnæus.*

Flowers polygamous $\sqrt[6]{}$. *Fruit* with 6 stones.

1. P. *verticillatus* Linnæus.

Leaves deciduous, oval, acuminate, serrated, downy beneath ; male flowers axillary, in small umbels.

Habitat. United States.
Quality. Bark and fruit tonic ; the latter also emetic.
Uses. In cases of great debility, attended by fever.

ILEX. *Linnæus.*

Flowers polygamous $\sqrt[4,5]{}$. *Fruit* with 4-5 stones.

1. I. *Aquifolium* Linnæus.—(The HOLLY TREE.) Fig. 261.

Leaves ovate, acute, wavy, shining, spiny-toothed ; flowers somewhat umbelled.

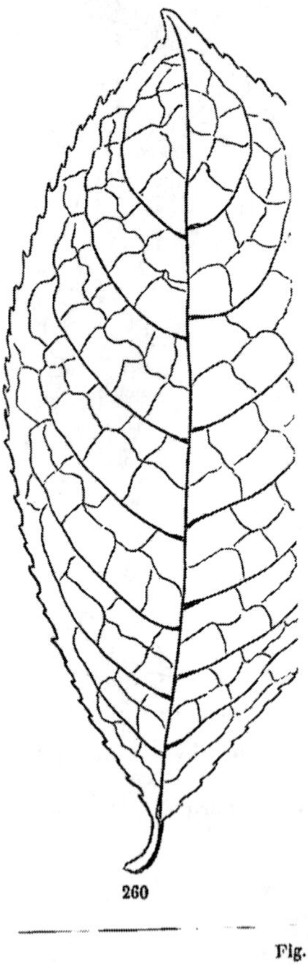

260

Fig. 260.—Ilex paraguayensis.

Habitat. Woods.
Quality. Leaves astringent, tonic ; root and bark emollient, expectorant, diuretic.
Uses. Intermittent fevers ; icterus. Bark yields birdlime.

2. I. *paraguayensis* St. Hilaire. — (PARAGUAY TEA.) Fig. 260.
Perfectly smooth ; leaves lanceolate-oblong, apiculate, serrated ; peduncles axillary, many-parted; stones wrinkled.

Habitat. Paraguay and Brazil.
Quality. Stimulating, narcotic ; diuretic, diaphoretic.
Uses. Forms the Paraguay Tea or Maté, largely used in South America.

3. I. *vomitoria* Aiton.
Leaves oblong, blunt at each end, crenato-serrate, smooth; umbels lateral, nearly sessile.

Habitat. United States.
Quality. A strong decoction acts as a mild emetic.

Natural Order, **Dogbanes**; *Apocynaceæ* (V. K., p. 599.)

Prevailing Quality. Acrid, emetic, drastic, poisonous.

CERBERA. *Linnæus.*

Calyx without glands. *Corolla* hypocrateriform. *Ovules* 2-4 in each ovary. Fruit a *Drupe*.

1. C. *Tanghin* Hooker. *Tanghinia venenifera* Poiret.

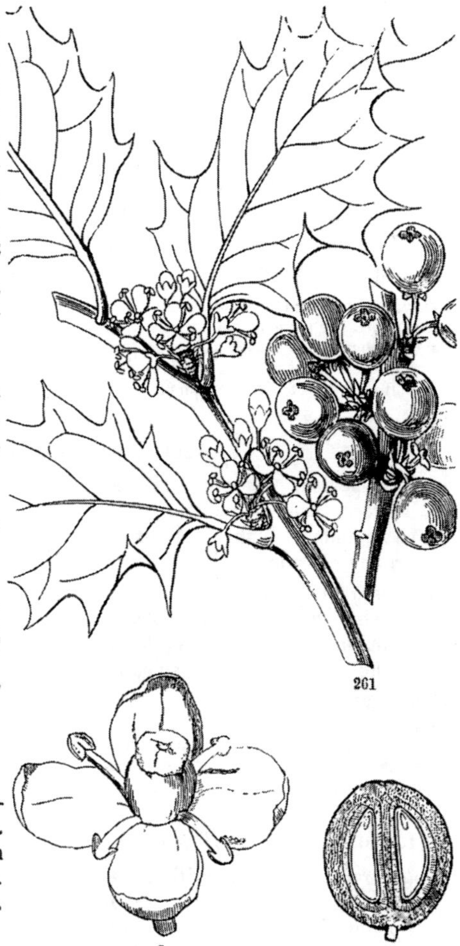

Leaves lanceolate, much tapering to the base, coriaceous, smooth, revolute at the edge ; cymes dichotomous, terminal.

Habitat. Madagascar.
Quality. Kernel a deadly poison. It is asserted that, although not larger than an Almond, one kernel is sufficient to destroy twenty people. It was used in Madagascar as an ordeal, but the practice is now discontinued. The kernel was pounded on a stone with water, and the emulsion thus obtained was distributed among the supposed criminals.

Fig. 261.—Ilex Aquifolium ; *a*, a flower magnified ; *b*, a perpendicular section of the fruit.

2. C. *Manghas* Linnæus.

Leaves lanceolate, tapering to the base, coriaceous, smooth ; cymes dichotomous, terminal.

Habitat. East Indies.
Quality. Kernel emetic ; poisonous, purgative.
Uses. Leaves used in some parts of India as a substitnte for Senna, especially in Java, according to Waiz.

NERIUM. *Linnæus.*

Calyx 5-parted, with numerous glands inside. *Corolla* hypocrateriform, with large lacerated faucial appendages. *Fruit* bifollicular.

1. N. *Oleander* Linnæus.—(The OLEANDER.)

Leaves opposite and ternate, lanceolate, acute.

Habitat. Shores of the Mediterranean.
Quality. All the parts acrid, poisonous, very dangerons.
Uses. Decoction of leaves kills vermin and cures itch ; powdered bark and wood a rat-poison. Cases are recorded of persons having been poisoned by meat roasted npon skewers of Oleander wood.

ALLAMANDA. *Linnæus.*

Calyx 5-parted, without glands. *Flowers* funnel-shaped, with a campanulate limb. *Fruit* a prickly capsule.

1. A. *cathartica* Linnæus.

Leaves whorled or opposite, oblong, acuminate, membranous ; lobes of the calyx acnminate, smooth.

Habitat. West Indies.
Quality. Cathartic, poisonous, emetic.
Uses. An infusion of the leaves in small doses in painters' colic. In over-doses it is violently emetic and purgative.

ICHNOCARPUS. *R. Brown.*

Corolla hypocrateriform, with a callous contracted orifice. *Hypogynous* glands 5, long, capitate. *Stigma* with a long subulate point.

1. I. *frutescens* R. Brown. Fig. 262.

Leaves elliptical, acute at each end, smooth above.

262

Habitat. East Indies, especially the island of Ceylon.
Quality. Purgative, alterative.
Uses. Sometimes nsed in the East Indies as a substitute for Sarsaparilla.

Fig. 262.—Ichnocarpus frutescens in flower, after Burmann's figure.

Natural Order, Loganiads; *Loganiaceæ* (V. K., p. 602.)

Prevailing Quality. Tonic, poisonous.

SPIGELIA. *Linnæus.*

Calyx glandular inside. *Corolla* long, slender, valvate. *Stamens* with long filaments. *Capsule* composed of 2 cocci, circumscissile at the base.

1. S. *marilandica* Linnæus.—(WORM GRASS.)

Perennial; stem simple, quadrangular; leaves sessile, ovate-lanceolate, hairy at the edge; corolla scarlet outside, yellow inside.

Habitat. United States.
Quality. Purgative; a poisonous narcotic.
Uses. A very powerful vermifuge.

2. S. *Anthelmia* Linnæus.

Annual; leaves ovate-oblong, acuminate, the upper in whorls of 4; racemes spicate from the axils of the upper leaves; flowers very small, purplish.

Habitat. Tropical America.
Quality and *Uses.* Like the last; but its properties disappear with keeping.

STRYCHNOS. *Linnæus.*

Corolla hypocrateriform, with valvate lobes. *Stamens* with very short filaments. *Berry* coated with a rind, 1-celled, with discoidal seeds lodged in pulp.

1. S. *Nux-vomica* Linnæus.—(NUX-VOMICA TREE.) Fig. 263.

Cirrhi 0; leaves roundish-oblong, stalked, 3-5-nerved, perfectly smooth; corymbs terminal.

Habitat. East Indies.
Quality. Seeds a most dangerous poison; yield strychnia. Tonic, diuretic.
Uses. Paralysis, nervous affections, dyspepsia, pyrosis, impotence, dysentery.

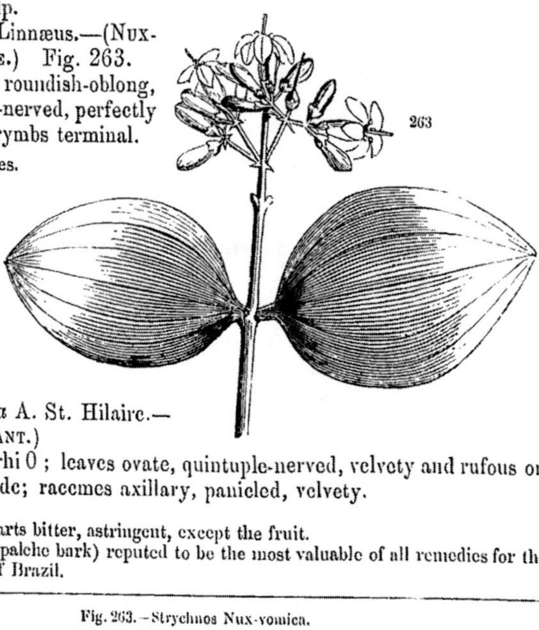
263

2. S. *pseudoquina* A. St. Hilaire.—(COPALCHE PLANT.)

Bark corky; cirrhi 0; leaves ovate, quintuple-nerved, velvety and rufous on the under side; racemes axillary, panicled, velvety.

Habitat. Brazil.
Quality. All the parts bitter, astringent, except the fruit.
Uses. Its bark (Copalche bark) reputed to be the most valuable of all remedies for the intermittents of Brazil.

Fig. 263.—Strychnos Nux-vomica.

3. S. *toxifera* Bentham.—(WOURALY SHRUB.) Fig. 264.

Branches climbing and covered with long rufous hairs, as well as the cirrhi; leaves nearly sessile, oval-oblong, membranous, 3-nerved, acute, covered with long coarse hairs.

Habitat. Demerara.

Quality. A most dangerous poison, acting like Nux vomica; is the base of a formidable substance, called Wooraly, used among the savages of Demerara.

Uses. Bark has been used as an application to foul ulcers.

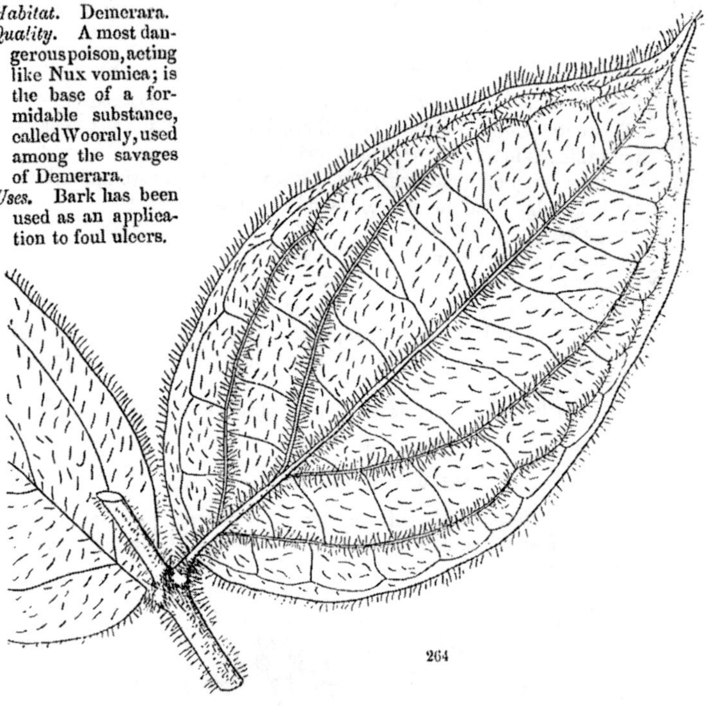

264

Natural Order, Gentianworts; *Gentianaceæ* (V. K., p. 619.)

Prevailing Quality. Pure bitterness.

GENTIANA. *Linnæus.*

Æstivation of the corolla left-handed. *Style* 0. *Flowers* not glandular. *Stigmas* 2, revolute.

1. G. *lutea* Linnæus.—(YELLOW GENTIAN.)

Corolla rotate, yellow; leaves broad, many-nerved, on a stout, erect stem.

Habitat. European Alps.

Quality. A pure bitter; narcotic, deleterious in large doses.

Uses. Dyspepsia; intermittents; against worms. In full doses it is apt to relax the bowels, and it does not always agree with the stomach; in fact it possesses a volatile principle capable of producing nausea and a kind of intoxication. The root contains a good deal of sugar and mucilage, which enables the Swiss to prepare from it a liqueur held in high esteem among that people.

Fig 264.—Leaves of Strychnos toxifera.

2. G. *punctata* Linnæus.

Corolla plaited, campanulate; stigmas naked; capsule sessile; plaits of the corolla triangular; calyx entire; leaves acute.

Habitat. Alps of Europe.
Quality and *Uses.* Quite like the last.

3. G. *purpurea* Linnæus.

Corolla plaited, campanulate; stigmas naked; capsule sessile; plaits of the corolla truncate; leaves smooth at the edge.

Habitat. Alps of Europe.
Quality and *Uses.* As the last. All yield "Gentian-root."

4. G. *Amarella* Linnæus,
5. G. *campestris* Linnæus,
6. G. *pannonica* Murray,

} are other European substitutes for Gentiana lutea.

AGATHOTES. *Don.*

Æstivation of corolla left-handed. *Style* 0. *Corolla* naked at base, with glandular pits, covered in by a fringed scale. *Stamens* monadelphous.

1. A. *Chirayita* Don.—(CHIREETA PLANT.)

Flowers $\sqrt[4]{}$; corolla longer than the calyx; leaves ovate and cordate, smooth.

Habitat. North of India.
Quality. Those of Gentiana lutea.
Uses. Dyspepsia in gouty subjects. "Strengthens the stomach, obviates flatulency and diminishes the tendency to acidity."—*Pereira.* The whole plant is pulled up at the time the flowers begin to decay, and is dried for use. Its febrifugal properties are in high estimation with European practitioners in India, who use it instead of Cinchona when the latter is not to be procured.

ERYTHRÆA. *Renealm.*

Æstivation of corolla left-handed. *Style* distinct, deciduous. *Anthers* having a spiral and projecting connective. *Inflorescence* centrifugal. *Corolla* regular, hypocrateriform.

1. E. *Centaurium* Persoon.—(CENTAURY.) Fig. 265.

Cymes dichotomous; flowers numerous, pink; stigma double; tube of corolla twice as long as the calyx; stem erect.

265

Fig. 265.—Erythræa Centaurium.

Habitat. Dry gravelly and heavy pastures.

Quality and *Uses.* As Gentian. It possesses all the essential properties of the Gentian of the shops, and although not used professionally is a valuable native medicine. In many places it is carefully collected for use in rustic pharmacy.

266

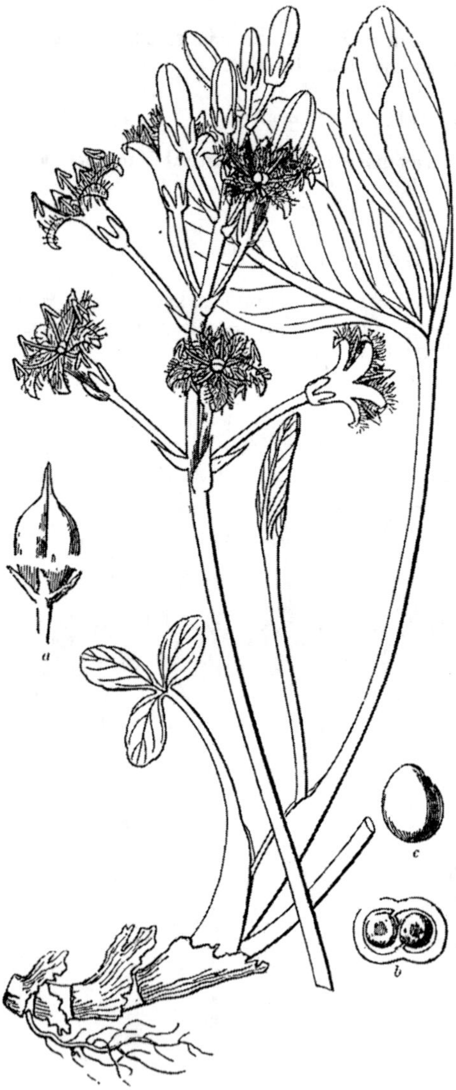

Fig. 266.—Menyanthes trifoliata; *a*, ripe fruit; *b*, a cross section of it; *c*, a seed.

MENYANTHES. *Linnæus.*

Æstivation of corolla induplicative. *Seeds* in the middle of each valve.

1. M. *trifoliata* Linnæus.—(BUCK BEAN.) Fig. 266.

Rhizome creeping; leaflets ternate, entire at the base; corolla bearded.

Habitat. Swamps.

Quality. Tonic, astringent; cathartic, even emetic.

Uses. Where bitter tonics are required, the rhizome, gathered in August, and also the seeds have been used. An occasional substitute for hops. All the plant intensely bitter.

Reckoned one of the most valuable of tonics. Large doses produce vomiting, purging, and frequently powerful diaphoresis. Especially recommended in intermittent and remittent fevers, gout, herpetic complaints, rheumatism, dropsy, scurvy, and worms.

VILLARSIA. *Ventenat.*

1. V. *nymphæoides* Ventenat.

This common wild aquatic possesses properties analogous to those of Menyanthes.

THE CORTUSAL ALLIANCE (V. K., p. 637.)

Natural Order of Cortusals.

Leadworts (*Plumbaginaceæ.*) Stamens opposite petals. Styles 5. Seed 1.

Primworts (*Primulaceæ.*) Stamens opposite petals. Style 1. Seed 00.

Natural Order, **Leadworts**; *Plumbaginaceæ* (V. K., p. 640.)

Prevailing Quality. Acridity, causticity.

PLUMBAGO. *Linnæus.*

Calyx 5-lobed, herbaceous, glandular. *Corolla* hypocrateriform. *Styles* connate.

1. P. *zeylanica* Linnæus.

Stem scrambling, shrubby ; leaves oblong, slightly auriculate ; flowers in long dense spikes ; corolla regular, white.

Habitat. East Indies.
Quality. Extremely acrid.
Uses. Sliced roots and leaves produce blisters as readily as cantharides ; applied in India to incipient buboes ; an infusion in olive oil forms a useful wash for ulcers.

2. P. *scandens* Linnæus.

Stem scrambling, shrubby; leaves oblong-lanceolate, slightly auriculate ; spikes long, terminal, lax ; corolla regular, white.

Habitat. West Indies.
Quality and *Uses.* Like the last.

3. P. *europæa* Linnæus.—(LEADWORT.) Fig. 267.

Branches long, slender, striated ; leaves auriculate, with calcareous powder on the under side ; limb of the corolla slightly irregular ; flowers violet.

Habitat. South of Europe.
Quality. Excessively acrid ; emetic.
Uses. A stimulating wash for old ulcers, but extremely painful ; the leaves chewed against toothache ; forms issues ; raises blisters.

ARMERIA. *Willdenow.*

Calyx membranous, naked. *Styles* distinct, feathery.

1. A. *vulgaris* Willdenow.—(THRIFT.)

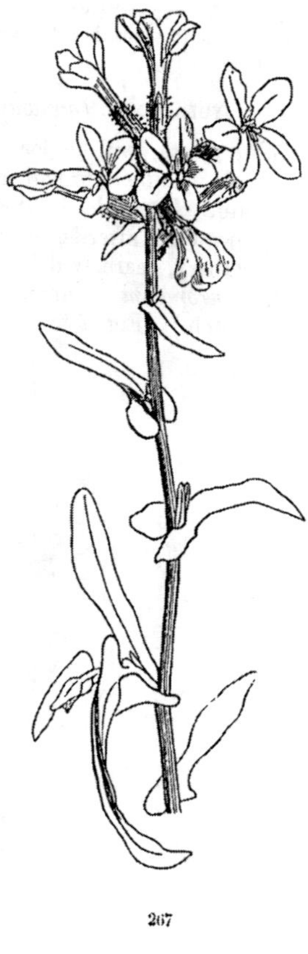

267

Fig. 267.—Plumbago europæa.

Cæspitose; leaves linear, smooth, or nearly so; scapes dwarf, villous; involucral leaves herbaceous at the back.

Habitat. Cliffs. Common in gardens as an edging.
Quality. Flowers active and useful as diuretics.

Natural Order, **Primworts**; *Primulaceæ* (V. K., p. 644.)

Prevailing Quality. Acridity.

PRIMULA. *Linnæus.*

Corolla hypocrateriform, dilated in the orifice. *Filaments* very short. *Capsule* ovate, 5-valved.

1. P. *veris* Linnæus.—(COWSLIP.)

Leaves oblong, rugose; scape longer, bearing an umbel at the end; limb of corolla short, concave.

Habitat. Meadows.
Quality. Flowers sedative.
Uses. When fermented with sugar they form a soporific domestic wine; used for wheys.

CYCLAMEN. *Linnæus.*

Corolla rotate, with a long reflexed limb. *Anthers* prominent, cuspidate. *Peduncles* twisted spirally downwards after flowering. *Fruit* leathery, scarcely dehiscent.

1. C. *europæum* Linnæus.—(SOW-BREAD.) Fig. 268.

268

269

Fig. 268.—Cyclamen europæum; 269. Anagallis arvensis.

Leaves reniform-orbicular, crenated, slightly angular; orifice of the corolla 5-sided.

Habitat. South of Europe.
Quality and *Uses.* Extremely acrid; a drastic purgative; emmenagogue.

ANAGALLIS.

Corolla rotate. *Capsule* circumscissile (a pyxis).

1. A. *arvensis* Linnæus.—(PIMPERNEL.) Fig. 269.

An annual; stems procumbent, angular; leaves opposite, or ternate, ovate, sessile, shorter than the axillary peduncles.

Habitat. Cornfields.
Quality. Acrid.
Uses. Has been prescribed in epilepsy, dropsy, and mania.

THE SOLANAL ALLIANCE (V. K., p. 615.)

Natural Orders of Solanals.

Oliveworts (*Oleaceæ.*) Stamens 2, free.

Nightshades (*Solanaceæ.*) Stamens 5, free. Placentæ axile. Seeds 00.

Asclepiads (*Asclepiadaceæ.*) Stamens and stigma consolidated.

Sebestens (*Cordiaceæ.*) Stamens 5, free. Placentæ axile. Seeds solitary.

Bindweeds (*Convolvulaceæ.*) Stamens 5, free. Placentæ basal.

Natural Order, **Oliveworts**; *Oleaceæ* (V. K., p. 616.)

Prevailing Quality. Bitter, tonic.

OLEA. *Linnæus.*

Calyx 4-toothed. *Corolla* 4-cleft. A *drupe.*

1. O. *europæa* Linnæus.—(The OLIVE TREE.)—Fig. 270.

Leaves lanceolate, opposite, entire, pale beneath; racemes axillary, compound, short.

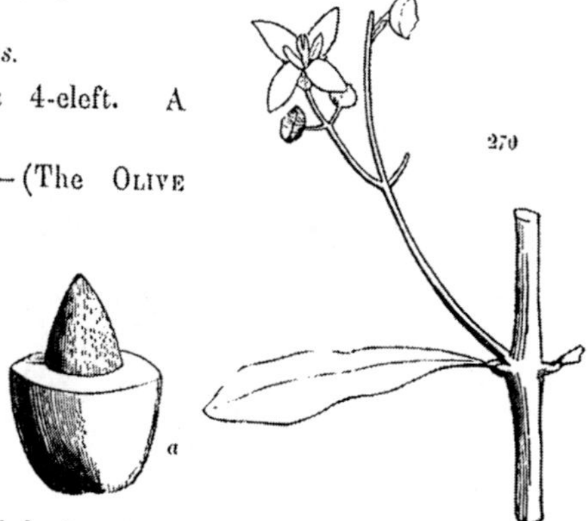

Habitat. South of Europe.
Quality. Oil of pericarp unctuous, scentless, almost tasteless; bark tonic, bitter, astringent.
Uses. Oil in plasters, enemata, salads, &c. Bark a good substitute for cinchona.

Fig. 270.—Olea europæa; *a*, fruit showing the stone.

SYRINGA. *Linnæus.*

Calyx 4-toothed. *Corolla* 4-cleft. A 2-celled *Capsule,* with navicular valves.

1. S. *vulgaris* Linnæus.—(The LILAC.)
Leaves cordate, acuminate.

Habitat. East of Europe. *Quality.* Bitter, tonic.
Uses. Unripe fruit and seeds valuable as a cure for intermittent fevers.

271

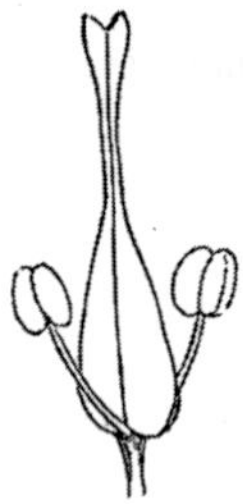

FRAXINUS. *Linnæus.*

Flowers very imperfect. *Calyx* obsolete. *Corolla* 0, or 3-4-parted. A *Samara.*

1. F. *Ornus* Linnæus. *Ornus europæa* Persoon.—(FLOWERING ASH.)
Leaflets in about 3 pairs, lanceolate or elliptical, serrate; flowers complete.

Habitat. South of Europe.
Quality. Branches exude the bitter-sweet laxative substance called Manna.

272

2. F. *excelsior* Linnæus. —(ASH TREE.) Fig. 271.
Leaflets in 3-6 pairs, sessile, lanceolate-oblong, acuminate, serrated; flowers apetalous.

Habitat. Woods.
Quality. Bark tonic; leaves cathartic and diuretic.
Uses. Timber largely used where strength and elasticity are required. Also yields Manna in hot countries.

Natural Order, **Nightshades**; *Solanaceæ* (V. K., p. 618.)

Prevailing Quality. Narcotic.

SOLANUM. *Linnæus.*

Corolla rotate. *Anthers* connivent, opening by a pore. A *Berry.*

1. S. *tuberosum* Linnæus. —(POTATO.)

Fig. 271.—Flower of Fraxinus excelsior, magnified; 272. Solanum Dulcamara.

Leaves interruptedly pinnated, downy; flowers white; tuberous.

Habitat. Chili.
Quality. Leaves and fruit narcotic.
Uses. Tubers abound in starch, which is sometimes used as arrow-root, but it is of very inferior quality, causing flatulence, and disordering the stomach of delicate persons. They are among the most valuable of known food, if perfectly ripe and well cooked.

2. S. *Dulcamara* Linnæus.—(BITTERSWEET.) Fig. 272.
Stem scrambling; leaves ovate-cordate, the upper hastate; corymbs opposite the leaves; flowers violet; berries red.

Habitat. Hedges.
Quality. Diaphoretic, diuretic, acro-narcotic.
Uses. Chronic catarrhs, rheumatism, gout, eczema and psoriasis. The root and young branches, in the form of a decoction, much diluted with milk, have been recommended in scrophulous or glandular obstructions. *Smith.*—The plant is a dangerous narcotic, and its gay tempting berries have occasionally caused serious accidents among children and others who have eaten them. In medicine the plant has been considered serviceable both internally, and used as a wash in lepra, and other cutaneous disorders. It is said to have been advantageously exhibited in asthma.

3. S. *nigrum* Linnæus.—(BLACK NIGHTSHADE.) Fig. 273.
An annual; stem erect, angular; leaves sinuate-toothed, downy; flowers white; berries black, filled with purple pulp, in which the seeds are embedded.

273

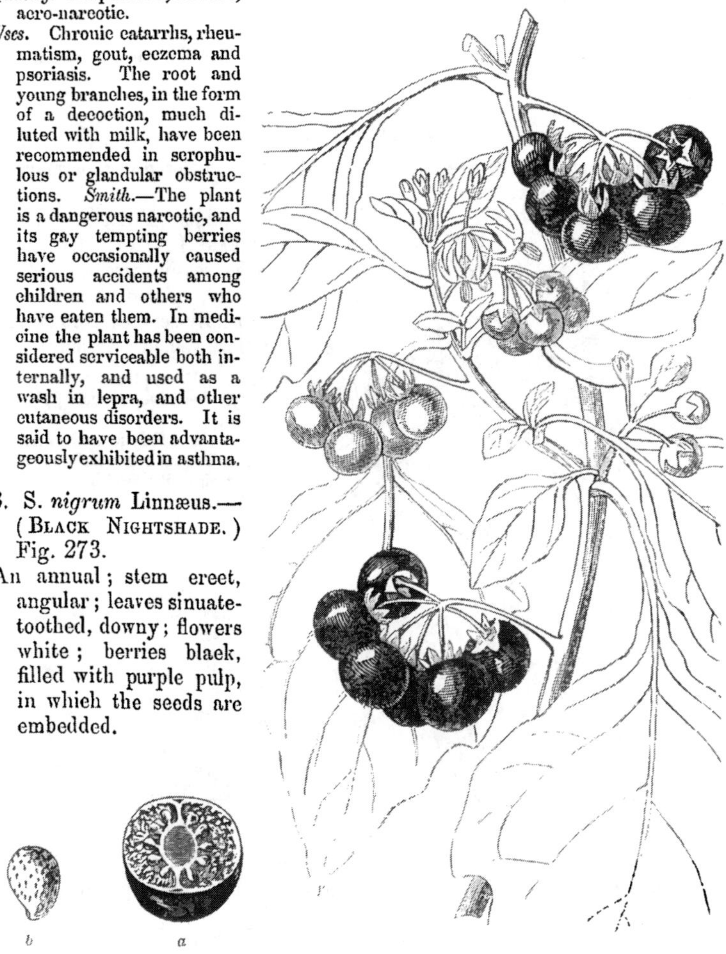

Fig. 273.—Solanum nigrum; *a*, a transverse section of fruit; *b*, seed, both magnified.

Habitat. Waste places all over the world.
Quality. Slightly narcotic.
Uses. As a resolvent; berries eatable without danger; in hot countries, leaves used as spinach.

4. S. *Lycopersicum* Linnæus.—(TOMATO.)
An annual; leaves interruptedly pinnate, hairy; flowers yellow; berries large, ribbed, many-celled.

Habitat. Peru.
Uses. The juice of the fruit subacid, much used as a sauce.

5. S. *Melongena* Linnæus.—(EGG PLANT.)
An annual, gray, with stellate hairs; leaves ovate, angular; flowers purple; fruit oblong, smooth, shining, with a hard skin.

Habitat. East Indies.
Uses. Fruit much used as food, when skilfully cooked. Called, in the countries where it grows, Aubergine, Brinjal, &c.

HYOSCYAMUS. *Linnæus.*

Corolla funnel-shaped, irregular. *Fruit* a *pyxis* enclosed in a permanent hardened calyx.

1. H. *niger* Linnæus.—(HENBANE.) Fig. 274.
Leaves sessile, viscid, amplexicaul, pinnatifid; flowers nearly sessile; corolla yellow, veined with purple.

Habitat. Waste places.
Quality. Sedative, narcotic, poisonous.
Uses. As an anodyne, and antispasmodic; cough, gonorrhœa, toothing, glandular swellings; to dilate the pupil of the eye.

274

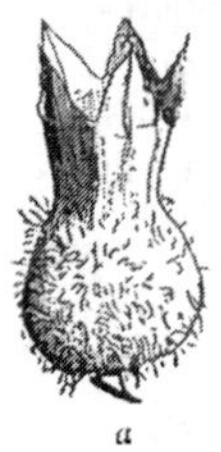

a

b

DATURA. *Linnæus.*

Corolla funnel-shaped, plaited. *Calyx* deciduous, with a permanent circular base. A 4-celled, 4-valved *Capsule*.

1. D. *Stramonium* Linnæus.—(THORN APPLE.) Fig. 275.

Fig. 274.—Hyoscyamus niger; *a*, calyx; *b*, pyxis.

An annual ; leaves ovate, smooth, unequally sinuate-toothed ; capsules erect, spiny ; flowers white.

Habitat. Waste places.
Quality. Like those of Henbane and Belladonna. Seeds produce maniacal delirium.
Uses. To allay pain ; in tic-douloureux, spasmodic asthma, mania, epilepsy.

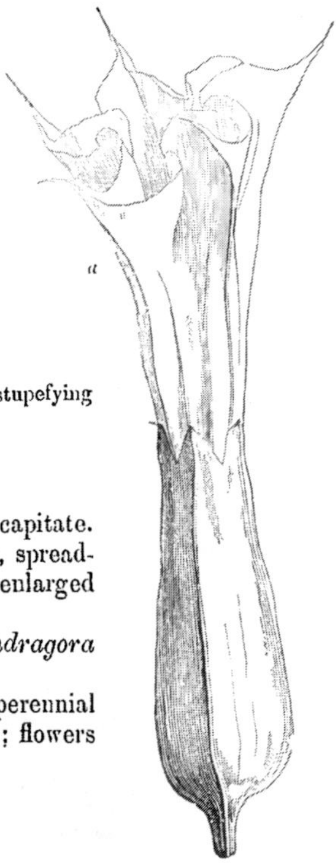
a

2. D. *sanguinea* Ruiz and Pavon.—(FLORIPONDIO.)

A tree ; leaves oblong-acuminate, toothed, angular, hairy ; flowers pendulous, red, with narrow acuminate lobes to the corolla.

Habitat. Peru.
Quality. Seeds produce furious excitement.
Uses. To bring on fits of delirium ; to form a stupefying beverage.

MANDRAGORA. *Tournefort.*

Corolla campanulate, plaited. *Stigma* capitate. *Stamens* inclosed in the tube of the calyx, spreading. *Berry* 2-celled, surrounded by the enlarged calyx.

1. M. *officinarum* Miller. *Atropa Mandragora* Linnæus.—(MANDRAKE.)

A stemless plant, with a large forked fleshy perennial root ; leaves lanceolate, spreading, grey ; flowers hidden among the leaves, pale violet.

Habitat. South of Europe.
Quality. Acro-narcotic, purgative, anæsthetic, aphrodisiac.
Uses. An old ingredient in philtres ; dangerous and disused. " Le professeur Fodevé avait cueilli un pied de Mandragore, et l'avait laissé par inadvertence dans un petit cabinet où il se livrait au travail. Au bout d'un quart d'heure il fut pris de vertiges, de faiblesse, et d'une langueur telle qu'il avait peine à se soutenir. Il ne songeait plus à cette plante, et son premier mouvement fut d'ouvrir la fenêtre. S'étant appuyé sur la plante même, il en

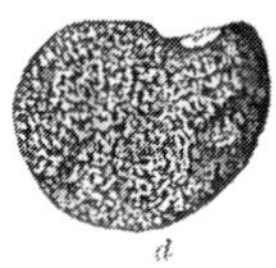
d

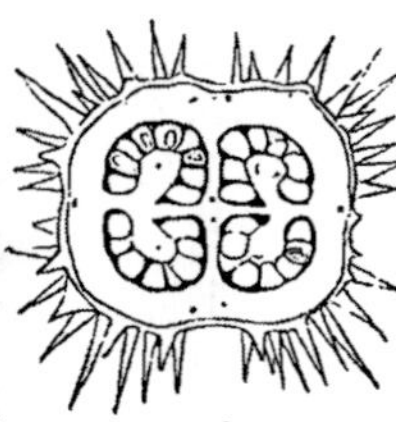
b

c

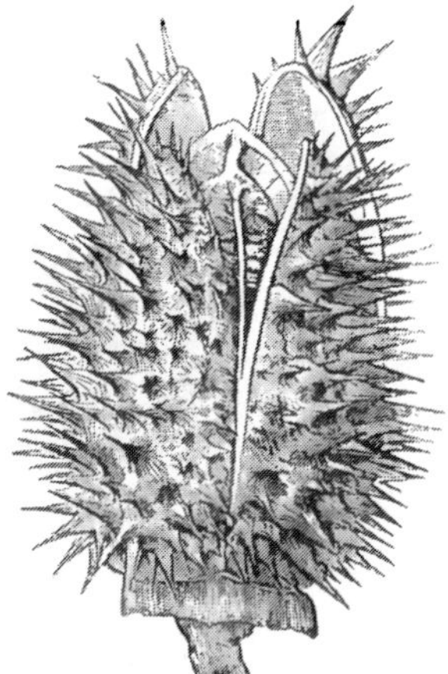
275 c

Fig. 275.—Datura Stramonium ; *a*, flower ; *b*, cross section of ovary ; *c*, ripe fruit ; *d*, magnified view of a seed ; *e*, section of ditto.

sortit uno odeur fortement nauséabonde qui lui fit apercevoir aussitôt la cause de ces accidens." (*Roques, Médecine Légale*, tom. iv. p. 22.) Dr. T. H. Silvester has shown that the root was formerly employed in the same way as chloroform and other anæsthetic agents now are.

ATROPA. *Linnæus.*

Corolla campanulate. *Berry* not filled with pulp.

1. A. *Belladonna* Linnæus.—(DEADLY NIGHT-SHADE.) Fig. 276.

Leaves ovate, entire; seeds pitted, not lying in pulp; flowers dirty violet-brown.

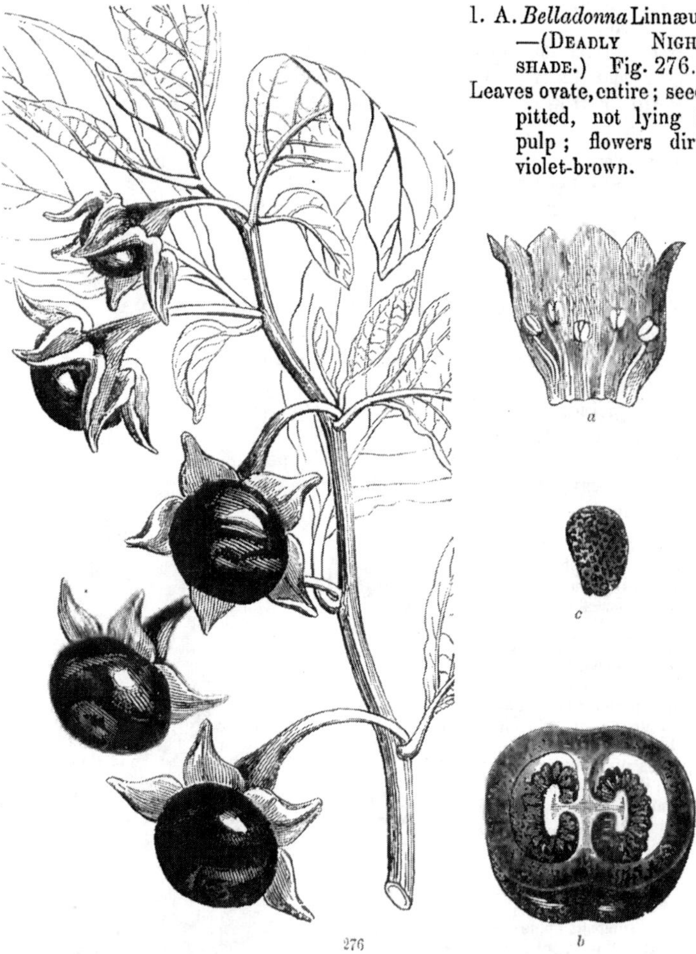

Habitat. Woods.
Quality. Like those of Henbane.
Uses. Nervous irritation, tic-douloureux, rigidity of the os uteri, spasmodic stricture, hooping cough, maladies of the eyes, epilepsy, hysteria, mania, chorea.

Fig. 276.—Atropa Belladonna in fruit; *a*, a corolla laid open; *b*, a cross section of fruit; *c*, seed.

NICOTIANA. *Linnæus.*

Corolla funnel-shaped, with a plaited limb. *Stigma* capitate. *Capsule* 2-celled, 4-valved at the point.

1. N. *Tabacum* Linnæus.—(AMERICAN TOBACCO.) Fig. 277.

Leaves ovate or oblong, lanceolate, acuminate, the lower decurrent; corolla much longer than the calyx, pink, with acute segments.

Habitat. Tropical America.
Quality. Narcotic, purgative, emetic, anodyne; relaxes muscular fibre.
Uses. Colic, hernia, constipation, ischuria and dysuria, tetanus, dropsy. A favourite luxury. Constitutes all the American cigars and Indian cheroots.

2. N. *rustica* Linnæus.—(SYRIAN TOBACCO.)

Leaves stalked, ovate, obtuse, entire; corollas not much longer than the calyx, green, with rounded segments.

Habitat. Levant.
Quality and *Uses.* Like the last, but milder; forms Turkish, Syrian, Latakia Tobaccos.

3. N. *persica* Lindley.—(TOBACCO OF SHIRAZ.)

Stem-leaves amplexicaul, oblong, acuminate; corolla much longer than the calyx, white, with ovate emarginate unequal segments.

Habitat. Persia.
Quality and *Uses.* Like the first, but much more fragrant and agreeable; not used in medicine. Forms the finest Persian Tobacco; but not suited to cigars, from the difficulty of making it burn.

277

CAPSICUM. *Linnæus.*

Corolla rotate. *Stamens* projecting, converging, opening by slits. *Fruit* a dry berry.

1. C. *annuum* Linnæus.—(CHILLI.)

Leaves ovate, acuminate, entire, smooth; fruit long, conical.

Habitat. Mexico, where it was called *Tchilli* according to Hernandez.
Quality. Stimulant, rubefacient, vesicant; in excess dangerous.
Uses. This and other species furnish the well-known condiment called Cayenne Pepper. The fruit and seeds are a powerful stimulant, without any narcotic property. Cayenne pepper consists principally of the ground seeds. It is employed in medicine, in combination with Cinchona, in intermittents and lethargic affections, and also in atonic gout, dyspepsia accompanied by flatulence, tympanitis, paralysis, &c. Its most valuable application appears, however, to be in cynanche maligna and scarlatina maligna, used either as a gargle or administered internally.

2. C. *frutescens* Linnæus.—(GOAT PEPPER;)

3. C. *baccatum* Linnæus.—(BIRD PEPPER;)

Quality. Have similar properties, but are more acrimonious.

Fig. 277.—Flowers of Nicotiana Tabacum.

Natural Order, Asclepiads; *Asclepiadaceæ* (V. K., p. 623.)

Prevailing Quality. Acrid, purgative, emetic.

ASCLEPIAS. *Linnæus.*

278

Coronet consisting of 5 cucullate processes, each bearing a horn in its inside. Fig. 278.

1. A. *tuberosa* Linnæus. A. *decumbens* Linn.—(BUTTERFLY WEED.)

Stem ascending, hispid; leaves linear-oblong, hirsute; umbels rather corymbose; pedicels downy; flowers orange red.

Habitat. United States.
Quality. Root expectorant, diaphoretic; a mild tonic and stimulant.
Uses. Catarrh, pneumony, pleurisy.

2. A. *curassavica* Linnæus.

Leaves lanceolate, acuminate, membranous, smooth; peduncles shorter than the leaves; umbels many-flowered; lobes of corolla acuminate; flowers scarlet.

Habitat. West Indies.
Quality. Emetic and purgative.
Uses. In gleets and fluor albus.

GOMPHOCARPUS. *R. Brown.*

Coronet consisting of 5 cucullate processes, not containing a horn.

1. G. *fruticosus* R. Brown.

Branches minutely hoary; leaves linear, smooth, revolute at the edge; processes truncate; flowers white; follicles prickly.

Habitat. Levant.
Quality and *Uses.* Like those of Solenostemma Argel. Rarely seen as an adulteration of the senna that comes to England.

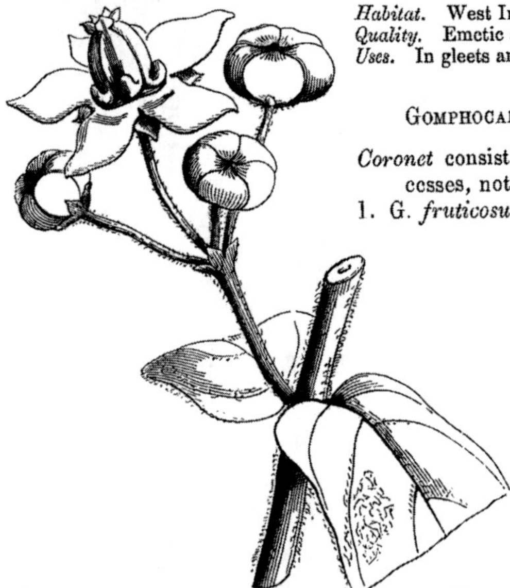

279

CALOTROPIS. *R. Brown.*

Coronet consisting of 5 blunt saccate processes adhering to the base of the filaments.

1. C. *gigantea* R. Br.—(MUDAR PLANT.) Fig. 279.

Processes of the coronet short, incurved, blunt; leaves cordate, ovate, acute.

Fig. 278.—Flower of an Asclepias; 279. Calotropis gigantea.

Habitat. East Indies.
Quality. Alterative and sudorific.
Uses. Elephantiasis, venereal diseases, chronic cutaneous affections, worms.

VINCETOXICUM. *Mœnch.*

Coronet without accessory processes, scutelliform, fleshy, with 5-10 lobes; *tube* very short

1. V. *officinale* Mœnch. *Cynanchum vincetoxicum.*

Branches with two hairy lines; leaves cordate, ovate-lanceolate, ciliated; cymes shorter than the leaves; stem erect; flowers white.

Habitat. South of Europe.
Quality. Emetic, purgative.
Uses. An old antidote to poisons.

SOLENOSTEMMA. *Hayne.*

Coronet raised on a long shaft, with 5 obtuse concave processes at its base, opposite the lobes of the corolla.

1. S. *Argel* Hayne. *Cynanchum Argel* Delile.—(ARGHEL.) Fig. 280.

Stem 2 feet high, erect, branched; leaves lanceolate, leathery, nearly veinless, wrinkled and downy on the under side.

280

281

Habitat. Egypt.
Quality. Acrid, purgative.
Uses. Forms a considerable proportion of many samples of Alexandrian senna, but readily known by the thick leathery texture of the leaves, which are veinless, wrinkled, and downy.

HEMIDESMUS. *R. Brown.*

Coronet 0. *Corolla* rotate. *Filaments* not united upwards; *anthers* free from the stigma, with 20 pollen masses.

1. H. *indicus* R. Brown.—(INDIAN SARSAPARILLA.) Fig. 281.

Smooth, twining; leaves narrow oblong, obtuse at each end; flowers axillary, nearly sessile.

Fig 280.—Solenostemma Argel in flower; 281. Hemidesmus indicus.

Habitat. East Indies. *Quality.* Alterative, sudorific, diuretic. *Uses.* A substitute for Sarsaparilla.

Natural Order, **Sebestens**; *Cordiaceæ* (V. K., p. 628.)

Prevailing Quality. Mucilaginous, emollient.

CORDIA. *Plumier.*

Calyx tubular, not striated, toothed. *Corolla* hypocrateriform or funnel-shaped. *Drupe* pulpy, surrounded by the permanent calyx.

1. C. *Myxa* Linnæus. Fig. 282. Branches smooth, round; leaves stalked, ovate, occasionally repand, smooth above, rather rough beneath; tube of corolla as long as calyx; drupe egg-shaped, mucronate.

Habitat. East Indies.
Quality. Fruit succulent, mucilaginous, emollient.
Uses. Yields the Sebesten Plums, formerly employed in the preparation of lenitive electuary; also as a pectoral medicine.

2. C. *latifolia* Roxburgh. Branches angular; leaves stalked, roundish-ovate, sometimes rather cordate and repand, nearly smooth above; tube of corolla rather shorter than calyx; drupe roundish obovate.

Habitat. East Indies.
Quality and *Uses.* Like the last, but of better quality.

N.B.—Under the name Sebesten Plums, Sebestans, or Sepistans, two sorts of Indian fruit have been employed as pectoral medicines, for which their mucilaginous qualities, combined with some astringency, have recommended them. They are believed to have been the Persea of Dioscorides. According to Mr. Colebrooke this is a larger and more mucilaginous sort than that described by European writers on Materia Medica, which is the produce of the last species. Linnæus applied the name of Sebesten to an American species of this genus, not known in medicine.

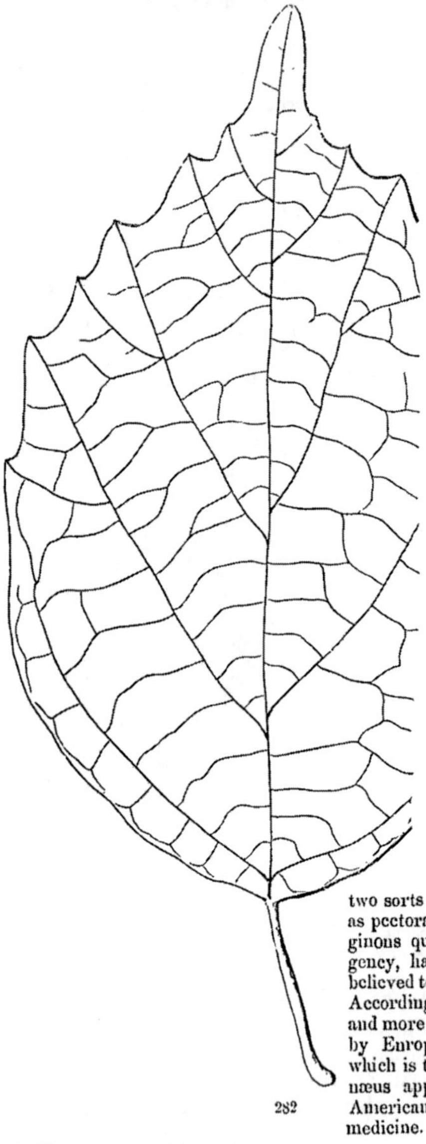

282

Fig. 282.—Leaf of Cordia Myxa.

Natural Order, Bindweeds; *Convolvulaceæ* (V. K., p. 630.)

Prevailing Quality. Purgative.

CONVOLVULUS. *Linnæus.*

Calyx naked. *Stigmas* 2, linear. *Ovary* 2-celled, with 4 ovules.

1. C. *Scammonia* Linnæus.—(SCAMMONY.) Fig. 283.

Perennial; stem smooth; leaves sagittate with truncate or lobed auricles; peduncles very long, many-flowered; sepals truncate, much shorter than the calyx.

Habitat. Levant.
Quality. A powerful drastic purgative.
Uses. Constipation, worm cases, dropsy.

283

2. C. *arvensis* Linnæus.—(SMALLER BINDWEED.) Fig. 284.

Perennial, creeping rooted; leaves sagittate, auricled; sepals roundish-ovate.

Habitat. Hedges.
Quality and *Uses.* Like Scammony, but less active.

284

3. C. *dissectus* Cavanilles.

Perennial; stem downy; leaves deeply 5-parted or sagittate, linear; peduncles 1-2-flowered; calyx downy, nearly as long as the corolla.

Habitat. South America and South Africa.
Quality. Abounds in prussic acid.
Uses. Gives its flavour to the liquor called noyeau imported from Martinique, &c., where the plant is called "the Noyeau Vine."

Fig. 283.—Leaf of Convolvulus Scammonia; 284. Convolvulus arvensis.

CALYSTEGIA. *R. Brown.*

Calyx concealed within 2 large leafy bracts. *Stigmas* 2 ; linear. *Ovary* 2-celled at base, 1-celled at apex.

1. C. *sepium* R. Brown.—(LARGER BINDWEED.) Fig. 285.

Stem twining, smooth, angular ; leaves large, smooth, sagittate, truncate at base ; peduncles angular, 1-flowered ; flowers large.

285

a

Habitat. Hedges.
Quality and *Uses.* Like Scammony, but less active.

IPOMŒA. *Linnæus.*

Calyx naked. *Stigmas* 2, capitate. *Ovary* 2-celled, 4-seeded.

1. I. *pandurata* Meyer.

Stem slender, rather downy ; leaves cordate, entire or pandurate, or even 3-lobed ; peduncles 1-3-flowered, longer than the petioles ; sepals small, ovate ; corolla large, white and purple.

Habitat. West Indies and Southern States of the North American Union.
Quality. Powdered root acts like rhubarb.
Uses. Calculous cases, gravel.

Fig. 285.—Calystegia sepium; *a*, the two bracts forced aside to show the calyx.

2. I. *operculata* Martius.

Stem quadrangular, winged, smooth ; leaves palmate, smooth ; peduncles 1-flowered, longer than the petiole ; sepals orbicular, the outer large ; corolla long, tubular, white ; capsule operculate.

Habitat. Brazil.
Quality and *Uses.* Like those of jalap, but weaker.

3. I. *macrorhiza* Michaux. *Convolvulus Jalapa* Linnæus.

This plant, inhabiting the sandy soil of Georgia and Carolina, with white insipid farinaceous roots weighing from 40 to 50 lbs., is said to possess no purgative properties whatever. Dr. Baldwin administered six drachms of the powdered root without effect ; in fact it contains little or no resin, but like Batatas consists chiefly of saccharine and farinaceous matter.

Exogonium. *Choisy.*

Calyx naked. *Corolla* tubular. *Stamens* prominent. *Stigma* capitate. *Ovary* 2-celled, 4-seeded.

1. E. *Purga* Bentham.—(True Jalap.) Fig. 286.

Leaves cordate, acuminate, smooth ; peduncles 2-flowered; sepals smooth, the outer shortest ; corolla large, crimson, hypocrateriform ; root tuberous.

Habitat. Mexico, near Xalapa.
Quality. A powerful drastic purgative.
Uses. Constipation, worm cases, water in the brain, dropsies, retention of the catamenia.

Batatas. *Rumphius.*

Calyx naked. *Stigmas* 2, capitate. *Ovary* 4-celled.

1. B. *edulis* Choisy. *Convolvulus Batatas* Linnæus. — (Sweet Potato.)

Stem creeping widely ; leaves cordate, acute, angular, stalked ; peduncles longer than the petiole, 3-4-flowered ; sepals mucronate ; corolla large, purple ; root very large, tuberous.

Habitat. East Indies ; cultivated in all tropical countries.
Quality. Sweet, nutritive, laxative.
Uses. The tubers are largely consumed for food in all hot countries, where they occupy the place of potatoes with us.

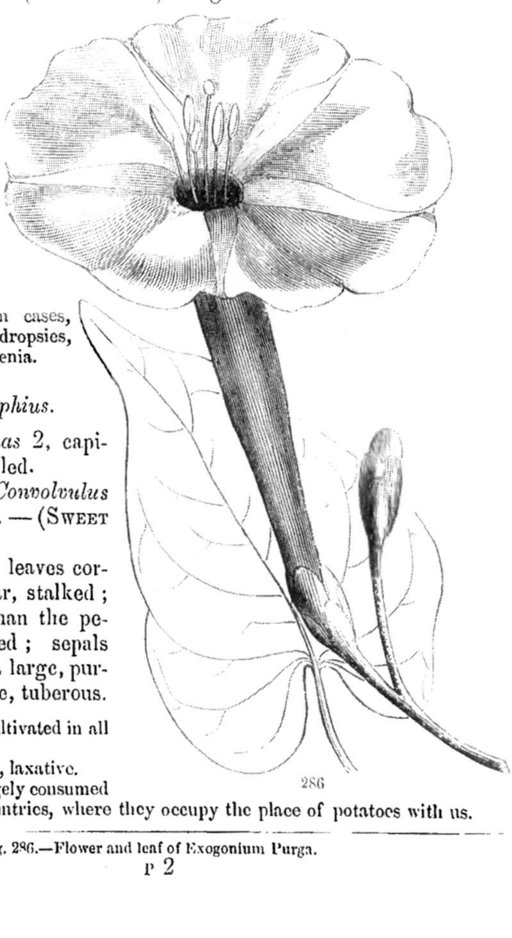

286

Fig. 286.—Flower and leaf of Exogonium Purga.

PHARBITIS. *Choisy.*

Calyx naked. *Stigma* capitate. *Ovary* 3-celled, with 2 seeds in each cell.

1. P. *Nil* Choisy. *Convolvulus Nil* Linnæus.—(CONVOLVULUS MAJOR.)

Stem hairy backwards; leaves cordate, entire or 3-lobed, hairy; peduncles 2-3-flowered, generally longer than the petiole; sepals hispid at the base; corolla large, pallid, showy.

Habitat. Tropical countries.
Quality. Seeds purgative, when roasted.
Uses. A common purgative in India, under the name of Kala Dana. Said to be an effectual quick cathartic. Seeds are roasted like coffee, powdered, and administered in doses of from 30 to 40 grains, in any convenient vehicle.

THE BIGNONIAL ALLIANCE (V. K., p. 668.)

Natural Orders of Bignonials.

Pedaliads (*Pedaliaceæ.*) Placentæ parietal.

Bignoniads (*Bignoniaceæ.*) Placentæ axile. Albumen 0.

Linariads * (*Scrophulariaceæ.*) Placentæ axile. Albumen abundant.

287

Natural Order, Pedaliads; *Pedaliaceæ* (V. K., p. 669.)

Prevailing Quality. Emollient, oily.

SESAMUM. *Linnæus.*

Upper lobe of *calyx* smaller. *Corolla* long-campanulate, with an unequal 5-lobed border. *Capsule* oblong, 4-cornered, 2-valved. *Seeds* thick; apterous.

1. S. *indicum* Linnæus.—(SESAME. TIL or TEEL.) Fig. 287.

Stem erect, downy; leaves lanceolate, downy, entire, or lobed.

Habitat. East Indies.
Quality. Emollient, demulcent.
Uses. Oil of seeds employed instead of Salad-oil, but is apt to become rancid. Meal of seeds for poultices, like Linseed.

Fig. 287.—Sesamum indicum, less than the natural size; 1, a ripe fruit; 2, a valve of the fruit; 3, a seed; 4, a cross section of it.

* The ambiguity of the term "Figworts," employed by me on former occasions, induces me to change it for the more expressive Linariads.

Natural Order, **Bignoniads**; *Bignoniaceæ* (V. K., p. 675.)

Prevailing Quality. Uncertain.

CATALPA. *Scopoli.*

Corolla campanulate. *Stamens* 5, 3 being sterile. *Capsule* long, siliquiform.

1. C. *syringifolia* Sims.—(CATALPA TREE.)

Leaves membranous, ovate, cordate, acuminate, nearly entire, downy beneath; flowers panicled.

Habitat. United States.
Quality. Emollient.
Uses. In Italy a decoction of the fruit for coughs and hoarseness; bark said to be bitter. [According to Kæmpfer a nearly allied species, or perhaps the same, found in Japan, has extremely bitter leaves and bark, and a decoction of the pods is employed in asthmatic complaints; the leaves are also used for fomentations.]

TECOMA. *Jussieu.*

Corolla short-tubed, with a dilated orifice and an irregular 5-lobed limb. *Rudiment* of a 5th stamen. *Capsule* 2-valved, with the partition opposite the valves.

1. T. *stans* Jussieu.

Erect. Leaves unequally pinnated, with 3 pairs of lanceolate, deeply serrated, acuminate leaflets; flowers panicled, yellow.

Habitat. West Indies.
Quality. Roots bitter, diuretic.

Natural Order, **Linariads**; *Scrophulariaceæ* (V. K., p. 681.)

Prevailing Quality. Purgative, emetic, bitter; narcotic.

DIGITALIS. *Linnæus.*

Calyx 5-parted. *Corolla* campanulate, with an oblique 4-cleft limb. *Stamens* 4; *anthers* with divaricating lobes. *Capsule* 2-celled, 2-valved, opening septicidally..

1. D. *purpurea* Linnæus.—(FOX-GLOVE.) Fig. 288.

Leaves downy, crenated; corolla large, purple or white, quite smooth.

Habitat. Roadsides.
Quality. Diuretic, emetic, purgative, narcotic; reduces the pulse.
Uses. In fevers, inflammation, dropsy, hæmorrhages, diseases of the heart, phthisis, insanity, &c.

288

Fig 288.—Flowers of Digitalis purpurea.

289

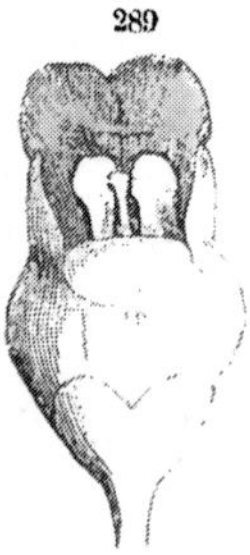

SCROPHULARIA. *Linnæus.*

Corolla nearly globose, with a small 5-lobed limb; the lowest lobe reflexed. *Stamens* 4, with an additional barren spathulate one.

1. S. *nodosa* Linnæus.—(FIGWORT.) Fig. 289.

Leaves smooth, doubly serrated, the lower serratures longer and sharper than the others.

Habitat. Ditches.
Quality. Emetic, purgative, diuretic, narcotic.
Uses. Leaves in fomentations; ointment in skin diseases.

290

a

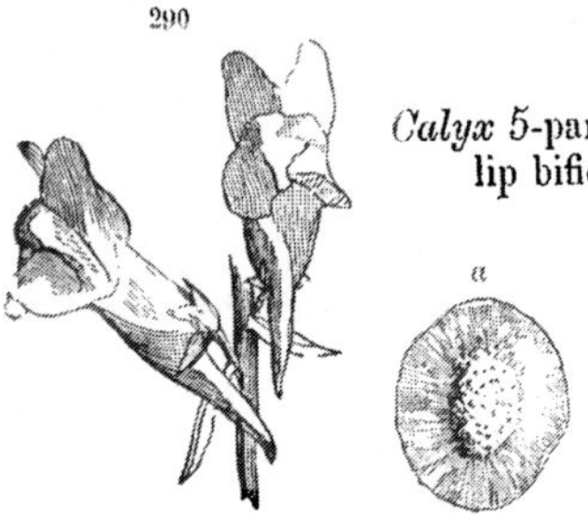

LINARIA. *Tournefort.*

Calyx 5-parted. *Corolla* ringent, spurred, the upper lip bifid.

1. L. *vulgaris* Miller.—(TOAD-FLAX.) Fig. 290.

Leaves numerous, lanceolate-linear, thickly covering the upright stem; rachis and pedicels covered with glandular hairs.

Habitat. Hedgerows and plantations.
Quality. Purgative, diuretic, bitter.
Uses. Chronic diseases of skin; decoction a fly poison.

291

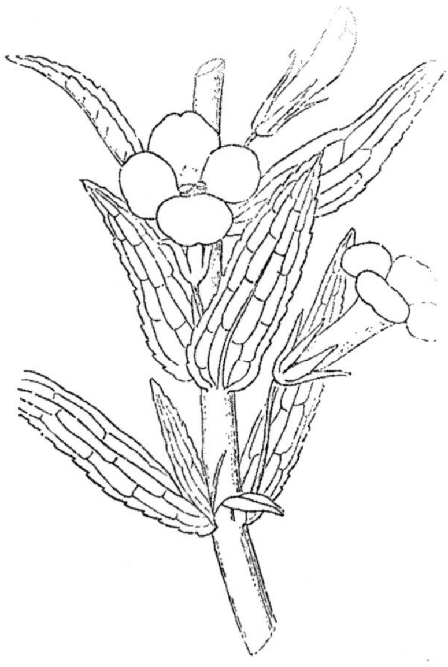

GRATIOLA. *Linnæus.*

Calyx 5-parted. *Upper* lip of *corolla* bifid, *lower* trifid. Only 2 *stamens* fertile; *anthers* pendulous.

1. G. *officinalis* Linnæus.—(HEDGE HYSSOP.) Fig. 291.

Leaves sessile, lanceolate, 3-nerved, serrulated, entire at the base; flowers solitary, axillary.

Habitat. Meadows in Europe.
Quality. A violent cathartic, diuretic, emetic; an acrid bitter poison.
Uses. Visceral obstructions, liver affections, dropsy, scurvy, venereal diseases, hypochondriasis.

A very active plant, formerly called *Gratia Dei.* It is extremely bitter, acts

Fig. 289.—Flower of a Scrophularia, seen in front; 290. Linaria vulgaris flowers; *a*, a seed magnified; 291. Gratiola officinalis.

violently both as a purgative and emetic, and has been said to be the basis of the gout medicine, called *Eau Médicinale*. In over-doses it is a violent poison, and according to Haller, renders, by its abundance, some of the Swiss meadows useless as pastures.

EUPHRASIA. *Linnæus.*

Calyx tubular, 4-toothed. *Corolla* bilabiate, with a flat limb. *Capsule* obtuse, many-seeded. *Seeds* striated, apterous.

1. E. *officinalis* Linnæus.—(EYEBRIGHT.)

Leaves ovate, with about 5 teeth on each side; lobes of the corolla veiny, lobed.

Habitat. Downs, meadows, woods.
Quality. Slightly bitter and aromatic.—"Nearly inert."—*Pereira.*
Uses. Catarrhal inflammations of the eye, cough, hoarseness.

FRANCISCEA. *Pohl.*

Calyx 5-toothed. *Corolla* hypocrateriform, with a very slightly unequal limb. *Stamens* 4. *Capsule* dry. *Seeds* large, immersed in pulp.

1. F. *uniflora* Pohl. F. *Hopeana* Hooker.—(MANACA.) Fig. 292.

A smooth shrub; leaves oblong or obovate, obtuse; flowers solitary, purple changing to white, with a narrow tubular calyx.

292

Habitat. Brazil.
Quality. Purgative, emetic, emmenagogue, alexipharmic; nauseously bitter.
Uses. Root and bark employed largely in Brazil against syphilis, under the name of Mercurio vegetal.

VERBASCUM. *Linnæus.*

Corolla rotate. *Stamens* 5, very unequal, the upper nearly abortive.

1. V. *Thapsus* Linnæus.—(MULLEIN. HIGH TAPER.) Fig. 293.

Leaves woolly, crenated, decurrent from one to the other; filaments covered with white wool, the two longest smooth or nearly so.

Fig. 292.—Franciscea uniflora.

Habitat. Roadsides.
Quality. Seeds and flowers poisonous; foliage acrid and bitterish.

293

a

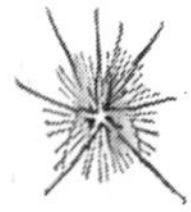

b

THE ECHIAL ALLIANCE (V. K., p. 649.)

Natural Orders of Echials.

Borageworts (*Boraginaceæ*). Flowers symmetrical.

Labiates (*Lamiaceæ*). Flowers unsymmetrical. Nuts 4.

Verbenes (*Verbenaceæ*). Ditto. Nuts confluent.

Natural Order, Borageworts; *Boraginaceæ* (V. K., p. 655.)

Prevailing Quality. Mucilaginous, inert.

BORAGO. *Linnæus.*

Corolla rotate. *Filaments* bifid, with their inner leg antheriferous.

1. B. *officinalis* Linnæus.—(BORAGE.)

Lower leaves elliptical obtuse, tapering to the base; lobes of the corolla ovate, acuminate, flat.

Habitat. Waste places.
Quality and *Uses.* Commonly employed to cool beverages in which its leaves are steeped.

ANCHUSA. *Linnæus.*

Corolla hypocrateriform, with 5 inflexed scales in the orifice. *Nuts* surrounded at the base by a tumid edge.

1. A. *tinctoria* Linnæus.—(ALKANET.)

Stem herbaceous, procumbent, rough with hairs; leaves lanceolate, obtuse, hoary; calyx hairy, rather shorter than the tube of the corolla; nuts warted.

Habitat. South of Europe and Levant.
Quality. Roots yield a reddish colouring matter.
Used to colour fatty substances; alkalies render it blue.

Fig. 293.—Part of the flower-spike of Verbascum Thapsus; *a*, one of the stellate hairs; *b*, a capsule magnified.

Natural Order, **Labiates**; *Lamiaceæ* (V. K., p. 659.)

Prevailing Quality. Aromatic, tonic.

SALVIA. *Linnæus.*

Stamens ascending, 2, with half an anther borne on one end of a divaricating connective.

1. S. *officinalis* Linnæus.—(GARDEN SAGE.)

A low shrub; leaves hoary, crenulate, rugose; bracts deciduous; calyx-teeth spiny.

Habitat. South of Europe.
Quality. Stomachic, aromatic, bitter.
Uses. As an ingredient in culinary seasoning.

2. S. *Sclarea* Linnæus.—(CLARY.)

Stem herbaceous, villous; leaves green, hairy, cordate, rugose; bracts large, coloured, deciduous; calyxes spiny.

Habitat. Middle of Europe.
Quality. Stimulant, aromatic, bitter.
Uses. In the preparation of a domestic wine.

ROSMARINUS. *Linnæus.*

Stamens ascending, 2, with 2-celled anthers, and toothed filaments. *Calyx* bilabiate $\frac{1}{2}$.

1. R. *officinalis* Linnæus.—(ROSEMARY.) Fig. 294.

A bush; leaves narrow, sessile, undivided.

Habitat. Hills in the South of Europe.
Quality. Carminative, stimulant.
Uses. Hypochondriasis; oil in the preparation of unguents for the hair.

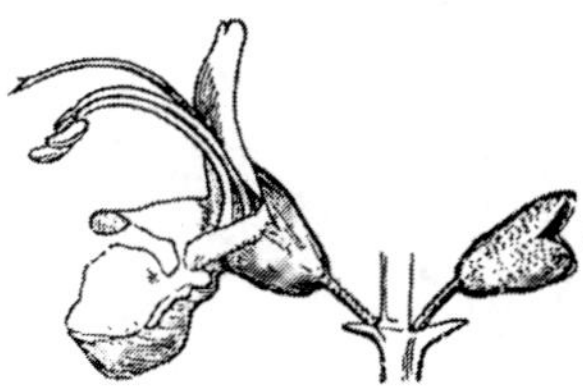

a

294

LAVANDULA. *Linnæus.*

Stamens declinate, 4, included. *Corolla* bilabiate $\frac{2}{3}$.

1. L. *vera* De Candolle.—(COMMON LAVENDER.)

Floral leaves rhomboid-ovate.

Habitat. Basin of the Mediterranean.
Quality. Extremely fragrant; stimulant, tonic, stomachic, cordial.
Uses. Hysteria, headache; and as a perfume; the tincture for faintness, flatulence, &c.

Fig. 294.—Rosmarinus officinalis; *a*, flowers magnified.

2. L. *Spica* De Candolle.—(FRENCH LAVENDER.)
Floral leaves lanceolate-linear, or subulate.

Habitat. Basin of the Mediterranean.
Quality. Yields oil of Spike.
Uses. By painters on porcelain, and by artists in the preparation of varnish.

MENTHA. *Linnæus.*

Stamens straight, 4. *Corolla* nearly equal, little longer than the calyx. *Anthers* each with 2 parallel cells. *Calyx* 5-toothed.

1. M. *viridis* Linnæus.—(MINT. SPEARMINT.) Fig. 295.
Leaves glabrous, sessile, lanceolate, acute, serrate ; spikes lax, cylindrical; bracts subulate ; throat of calyx naked.

Habitat. Marshy places.
Quality. Aromatic, carminative, stimulant, tonic.
Uses. Leaves as sauce and salad ; against colic.

The herb has a strong, peculiar and pleasant odour, with an aromatic bitter taste, followed by a sense of coldness when air is drawn into the mouth. It is an aromatic and carminative, employed in flatulence and to relieve the pain of colic. Various preparations are ordered by Pharmacopœias, of which *oil of spearmint*, and *spearmint* water are the most common.

295

2. M. *piperita* Linnæus.—(PEPPERMINT.)
Leaves stalked, oblong, acute, serrated; spikes oblong-cylindrical, interrupted below ; calyx-teeth straight in fruit ; throat naked.

Fig. 295.—Mentha viridis.

Habitat. Meadows, rare.
Quality. Aromatic, carminative, stomachic, stimulant.
Uses. Against flatulence, nausea, colic, &c.—Peppermint is an aromatic stimulant, and the most pleasant of all the mints. It is employed principally to expel flatus, to cover the unpleasant taste of other medicines, and to relieve nausea and griping pains of the alimentary canal. The volatile oil is sometimes taken as an anti-spasmodic; it is what gives their flavour to peppermint lozenges.—*Pereira.*

3. M. *Pulegium* Linnæus.—(PENNYROYAL.) Fig. 296.
Stems prostrate; leaves elliptical, obtuse, nearly entire; verticillasters remote, axillary; calyx, when in fruit, closed by a ring of hairs.

Habitat. Moist heaths and downs.
Quality. Aromatic, carminative, emmenagogue, antispasmodic.
Uses. Obstructed menstruation, hysteria, hooping cough.

HYSSOPUS. *Linnæus.*

Stamens straight, diverging, 4. *Calyx* with 15 ribs.

1. H. *officinalis* Linnæus.—(HYSSOP.)
Leaves lanceolate, entire; flowers in one-sided verticillate racemes.

Habitat. South of Europe.
Quality. Stimulating, stomachic, carminative.
Uses. Against flatulence and in hysterical complaints.

296

MELISSA. *Linnæus.*

Stamens ascending, 4. *Calyx* ribbed, bilabiate. Tube of *corolla* longer than calyx, without any ring.

1. M. *officinalis* Linnæus.—(BALM.)
Leaves ovate, crenate-serrate, the lower cordate; verticillasters secund: bracts ovate.

Habitat. South of Europe.
Quality. Stimulant, tonic.
Uses. Balm-tea as a diaphoretic in fevers; an emmenagogue; in amenorrhœa and chlorosis; as an exhilarating drink; in hypochondriasis.

Fig. 296.—Mentha Pulegium.

OCYMUM. *Linnæus.*

Stamens declinate, 4. *Calyx* of the fruit deflexed, with a large ovate upper lobe.

1. O. *Basilicum* Linnæus.—(BASIL.)

Leaves narrowed at the base, ovate or oblong, remotely serrated ; verticillasters 6-flowered ; calyxes in fruit completely bent down, more than 2 lines long, the upper lip round ; stamens about twice as long as corolla.

Habitat. East Indies.
Quality. Stimulant, aromatic.
Uses. An ingredient in culinary seasoning ; a palliative in the pains of childbirth.

ORIGANUM. *Linnæus.*

297

Stamens straight, diverging, 4. *Calyx* 10-ribbed, equally 5-toothed, with a villous orifice. *Spikes* loose, with broad bracts.

1. O. *vulgare* Linnæus.—(WILD MARJORAM.) Fig. 297.

Leaves ovate, acute ; bracts without glands on the upper side.

Habitat. Chalky pastures.
Quality. Tonic, stimulant, fragrant.
Uses. As a seasoning in cookery ; powder an errhine ; tea for nervousness. This plant yields what is called *oil of thyme* in the shops, a common remedy for the pain of carious teeth. It is frequently used, mixed with olive oil, as a stimulating liniment against baldness, in rheumatic complaints, and against sprains and bruises.

THYMUS.

Stamens straight, diverging, 4. *Calyx* 10-ribbed, 2-lipped, $\frac{3}{2}$, with a villous orifice.

1. T. *Serpyllum* Linnæus.—(GARDEN THYME.)

Leaves flat ; upper lip of corolla ovate, nearly quadrangular.

Habitat. Hills, heaths, and exposed places.
Quality. Fragrant, stimulating, carminative.
Uses. Chiefly as an ingredient in culinary seasoning.

Fig. 297.—Origanum vulgare.

TEUCRIUM.

Stamens 4, ascending, prominent. *Corolla* with upper lip 2-parted, and declinate, as long or longer than the lower, which is large and concave.

1. T. *Marum* Linnæus.—(CAT THYME.)

Verticillasters 2-6-flowered, racemose; leaves hoary, entire, with a few teeth near the point; a small shrub.

Habitat. Basin of the Mediterranean.
Quality. Stimulating, aromatic.
Uses. A very remarkable feline aphrodisiac.

NEPETA. *Linnæus.*

Stamens 4, ascending, the uppermost longest. *Calyx* equal, 15-nerved, with the upper lobe rather broader than the others. Cells of the *anthers* diverging.

1. N. *Cataria* Linnæus.—(CAT MINT.)

Leaves ovate, acute, serrato-crenate, cordate, hoary beneath; verticillasters panicled; flowers white; nuts smooth.

Habitat. Hedges.
Quality. Stimulating.
Uses. In amenorrhœa; a remarkable feline aphrodisiac.

2. N. *Glechoma* Bentham. *Glechoma hederacea* Linn.—(GROUND IVY.)

Stem creeping; leaves crenate, reniform.

Habitat. Dry ditches, plantations, &c.
Quality. Stimulant, pectoral, aromatic.
Uses. A tea prepared from the leaves in great repute among the poor.

SATUREIA. *Linnæus.*

Stamens straight, diverging, 4. *Calyx* 10-ribbed, 5-toothed, nearly equal, with a nearly hairless orifice. Upper lip of *corolla* ovate.

1. S. *hortensis* Linnæus.—(SUMMER SAVORY.)

An annual; leaves linear-lanceolate, pointless.

Habitat. South of Europe.
Quality. Aromatic, carminative.
Uses. An ingredient in culinary seasoning.

2. S. *montana* Linnæus.—(WINTER SAVORY.)

A small shrub. Leaves lanceolate, mucronate, marked with glandular dots on each side.

Habitat. South of Europe.
Quality and *Uses.* As the last.

MARRUBIUM. *Linnæus.*

Calyx tubular, 10-ribbed, with 5-10 spiny equal teeth. *Stamens* 4, included. *Corolla* $\frac{1 \text{ or } 2}{3}$.

1. M. *vulgare* Linnæus.—(HOREHOUND.) Fig. 298.

Leaves ovate, crenate, rugose, downy; verticillasters almost globose; calyx-teeth hooked, smooth above the middle.

Habitat. Waste places, hedgerows, &c.
Quality. Tonic, stimulant, laxative.
Uses. Coughs ; uterine and hepatic affections.

SCUTELLARIA. *Linnæus.*

Calyx 2-lipped, the upper lip truncate, external to the lower lip, which is truncate and entire. *Gynophore* long and curved.

1. S. *galericulata* Linnæus.—(COMMON SCULLCAP.) Fig. 299.

Leaves oblong-lanceolate, cordate, remotely crenate ; flowers axillary, opposite, racemose.

298

299

Habitat. Wet ditches.
Quality and *Uses.* An old exploded remedy for hydrophobia ; also formerly used against intermittents ; has little or no useful effect upon the constitution, unless as a weak tonic.

Natural Order, Verbenes; *Verbenaceæ* (V. K., p. 663.)

Prevailing Quality. Aromatic, subacrid.

STACHYTARPHETA. *Vahl.*

Ovules erect, solitary. *Fruit* a dicoccous capsule. *Flowers* spiked. *Stamens* 2.

1. S. *jamaicensis* Vahl.

Nearly smooth ; leaves oval or roundish ovate, coarsely serrated ; spikes very long ; bracts aristate.

Habitat. West Indies.
Quality. Reported to be purgative and anthelmintic.
Uses. Juice as a purgative for children ; fresh leaves applied to ulcers ; when dried, form a bad kind of tea, sometimes sold in Austria as Brazilian Tea.

Fig. 298.—Marrubium vulgare ; 299.—Scutellaria galericulata ; *a*, its calyx.

Lippia. *Linnæus.*

Ovules erect, solitary. *Fruit* a dicoccous capsule. *Flowers* capitate. *Stamens* 4. *Calyx* membranous.

1. L. *Pseudo-thea* Schauer. *Lantana Pseudo-thea* Aug. de St. Hilaire.

An erect viscid shrub. Leaves erect, stiff, lanceolate, serrate, entire at the base; heads of flowers hemispherical; bracts acute, as long as the tube of the corolla.

Habitat. Brazil.
Quality. Aromatic, fragrant.
Uses. In coughs, rheumatism, medicated baths; when dried, forms an agreeable tea.

Vitex. *Linnæus.*

Ovules pendulous. *Fruit* a drupe, with a 4-celled stone. *Corolla* $\frac{2}{3}$.

1. V. *Agnus castus* Linnæus.—(Chaste Tree.) Fig. 300.

Downy; leaves white at the back, digitate; leaflets lanceolate, acuminate, entire or nearly so; calyx campanulate; corolla three times as long as the calyx; inflated in the throat.

Habitat. Basin of the Mediterranean.
Quality. Fruit acrid.
Uses. Seeds used in Smyrna as an external application against colic; taken internally, act as powerful aphrodisiacs.

2. V. *Negundo* Linnæus.

In India a decoction of the aromatic leaves of this plant helps to form the warm bath for women after delivery; bruised they are applied to the temples for headache; pillows stuffed with them are put under the head to remove catarrh and the headache attending it.

3. V. *trifolia* Linnæus.

The leaves of this Indian species are a powerful discutient and are employed by the Malays to remove the boss. The leaves are given in decoction and infusion, and formed into a cataplasm which is applied to the enlarged spleen.

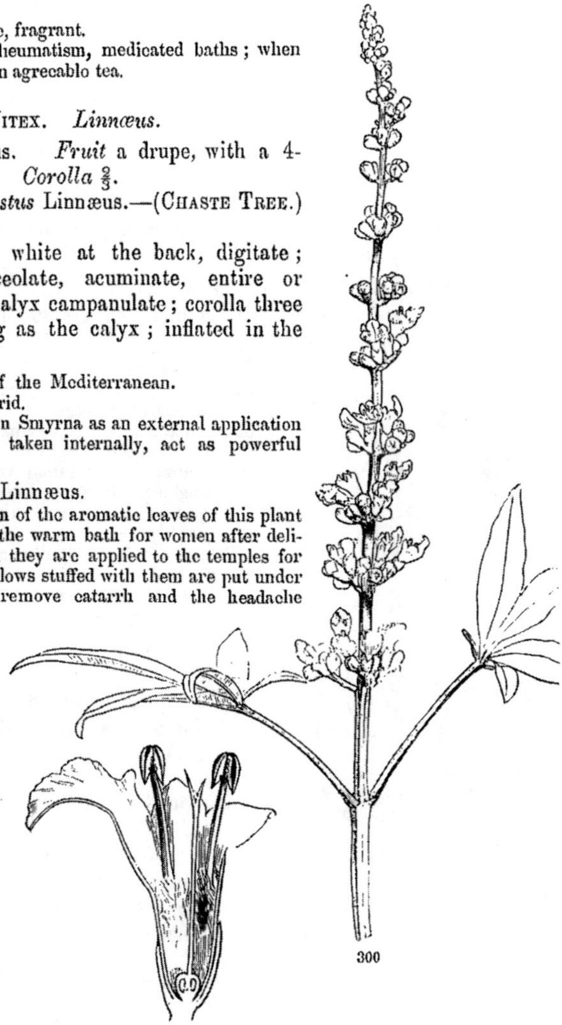

Fig. 300.—Vitex Agnus castus; *a*, flower divided perpendicularly and magnified.

THE CAMPANAL ALLIANCE.

Natural Orders of Campanals.

Lobeliads (*Lobeliaceæ.*) Anthers syngenesious. Ovules 00.

Valerianworts (*Valerianaceæ.*) Anthers free. Ovule 1, pendulous.

Composites (*Asteraceæ.*) Anthers syngenesious. Ovule 1, erect.

Natural Order, **Lobeliads**; *Lobeliaceæ* (V. K., p. 692.)

Prevailing Quality. Narcotico-acrid.

LOBELIA. *Linnæus.*

Corollas $\frac{2}{3}$, the upper lip split to the base. *Stigma* surrounded by a fringe. *Capsule* 2-3-celled.

1. L. *inflata* Linnæus.—(INDIAN TOBACCO.) Fig. 301.

Smooth, erect; leaves irregularly toothed; flowers small, blue, racemose; capsule ovate, inflated.

301

Habitat. United States.

Quality. Diaphoretic, expectorant; emetic; an acro-narcotic poison.

Uses. Asthma, hernia, croup, hooping-cough. In small doses it is expectorant and diaphoretic, exciting expectoration without the pain of coughing. In such doses as a common tea spoonful of the seeds and leaves, in which quantity irregular practitioners have ventured to give it, it frequently proves fatal in five or six hours. It has been used instead of tobacco, in the form of enema, in strangulated hernia.

2. L. *cardinalis* Linnæus.—(CARDINAL FLOWER.)

Downy, erect; leaves oblong-lanceolate, irregularly toothed; bracts glandular; flowers scarlet, racemose; calyx with a smooth hemispherical tube.

Habitat. United States.

Quality and *Uses.* Like the next.

3. L. *siphilitica* Linnæus.

Hairy, erect; leaves ovate, irregularly toothed; flowers racemose, blue; calyx hairy, with a hemispherical tube and lanceolate auriculate lobes.

Habitat. United States.

Quality. Emetic, cathartic, diuretic.

Uses. Root supposed to be useful in syphilis, but now disused.

Fig. 301.—Lobelia inflata.

Natural Order, Valerianworts; *Valerianaceæ* (V. K., p. 697.)

Prevailing Quality. Stimulating, aromatic.

VALERIANA. *Linnæus.*

Calyx pappose. *Corolla* spurless. *Stamens* 3.

1. V. *officinalis* Linnæus.—(TRUE VALERIAN.) Fig. 302.

Leaves all pinnatifid, with 7-8 pairs of lanceolate segments; corymb becoming panicled; fruit smooth.

Habitat. Damp meadows in Europe.

Quality. Nervine, tonic, antispasmodic; a remarkable feline stimulant.

Uses. Epilepsy, chorea, spasms, hypochondriasis, hysteria.

2. V. *Phu* Linnæus.

Radical leaves oblong: cauline pinnatifid with oblong lobes; corymb panicled; stigmas slender; fruit smooth.

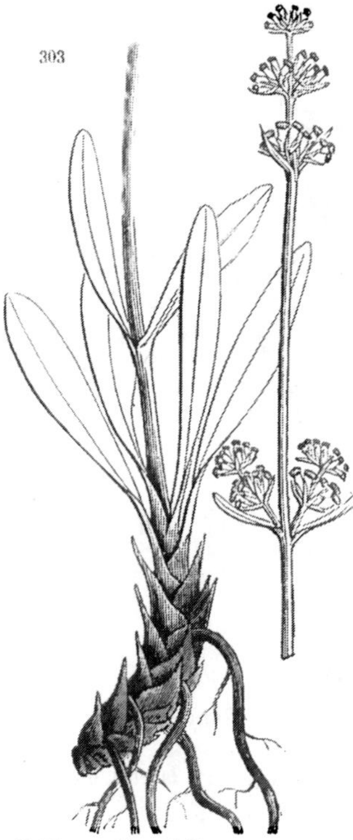

303

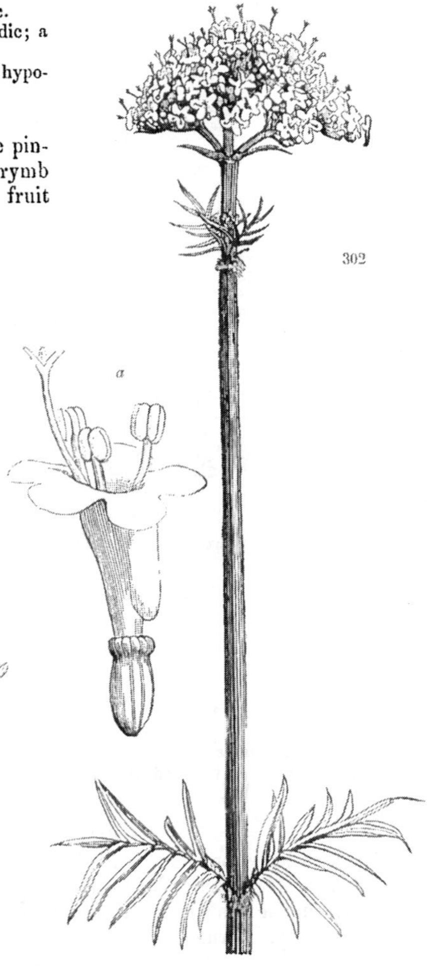

302

Habitat. Alps of Europe.

Quality and *Uses.* As in the last. This, or V. Dioscoridis, supposed to be the Φοῦ of the Greeks, and to be strongest of the European Valerians.

Fig. 302.—Valeriana officinalis; *a*, flower magnified; 303. Valeriana celtica.

3. V. *celtica* Linnæus. Fig. 303.

Smooth; leaves entire, obtuse, those next the root obovate; stem simple; flowers in interrupted spikes; fruit hairy.

Habitat. European Alps.

Quality and *Uses.* Like those of Nardostachys, as a substitute for which this is largely employed by Eastern nations.

4. V. *dioica* Linnæus. Fig. 304.

Radical leaves ovate: cauline pinnatifid with linear lobes; flowers diœcious; stigmas connate; fruit smooth.

304

b c a

Habitat. Boggy places all over Europe.

Quality and *Uses.* As in V. officinalis, but more feeble.

NARDOSTACHYS. *De Candolle.*

Calyx with leafy toothed lobes. *Corolla* spurless. *Stamens* 4.

1. N. *Jatamansi* De Candolle.—(SPIKENARD.)

Stem villous; leaves lanceolate, downy; fascicles of flowers lateral, opposite, stalked and terminal.

Habitat. Himalayas.

Quality and *Uses.* Stimulant, bitter; used as a perfume by eastern nations, and against hysteria and epilepsy.

Fig. 304.—Valeriana dioica; *a*, a male flower; *b*, a female; *c*, a ripe fruit.

VALERIANELLA. *Mœnch.*

Calyx with an imperfect toothed limb. *Corolla* spurless. *Stamens* 3.

1. V. *olitoria* Mœnch.—(LAMB'S LETTUCE.)

Fruit globose, smooth, compressed, oblique, with scarcely any limb ; leaves linear-oblong, nearly entire ; stem with rough angles.

Habitat. Cornfields and gardens.
Quality. Insipid.
Uses. Occasionally grown as a small salad.

Natural Order, **Composites** ; *Compositæ* (V. K., p. 702.)

Prevailing Quality. Bitter, and tonic.

* CORYMBIFEROUS COMPOSITES.

ARNICA. *Linnæus.*

Pappus hairy. *Florets of the ray* ♀, *of the disk* ⚥. *Stigmas* clavate, terminated by a hairy cone. *Bracts* forming a cylindrical involucre. *Receptacle* naked. *Achænia* wingless, striated.

1. A. *montana* Linnæus.—(MOUNTAIN TOBACCO.)

Radical leaves obovate, 5-nerved ; stem few-flowered ; bracts glandular.

Habitat. Alpine meadows.
Quality. Acrid, nauseous, emetic, causes constipation.
Uses. Typhoid fevers ; amaurosis, paralysis, dropsy, chlorosis, amenorrhœa, dysentery, &c.

EUPATORIUM. *Linnæus.*

Pappus hairy. *Florets* all tubular, ⚥. *Stigmas* clavate. *Bracts* imbricate, oblong. *Receptacle* naked.

1. E. *cannabinum* Linnæus.

Leaves stalked, 3-5-parted, with lanceolate-serrated segments, the middle of which is longest.

Habitat. Common by the side of ponds and ditches.
Quality. Root bitter, aromatic, pungent.
Uses. Root purgative ; not now employed.

2. E. *glutinosum* Lamarck.

A shrub. Branches smooth, glutinous ; leaves stalked, cordate, lanceolate, acuminate, serrated, extremely wrinkled, smooth on the upper side, downy on the lower.

Habitat. Peru.
Quality and *Uses.* Said to be the true Matico : but this is questioned. *See* PIPER.

INULA. *Linnæus.*

Pappus pilose, uniform, in one row. *Florets of the ray* ♀ ; *of the disk* ⚥. *Anthers* with 2 bristles. *Bracts* imbricated, in many rows. *Receptacle* naked.

1. I. *Helenium* Linnæus.—(ELECAMPANE.)

Leaves unequally dentato-serrate, downy beneath, those of the stem cordate-ovate, acuminate, amplexicaul.

Habitat. Europe, in damp meadows.
Quality. An aromatic tonic. Diaphoretic, diuretic, expectorant, emetic.
Uses. Root in coughs, dyspepsia, exanthemata.

TUSSILAGO. *Linnæus.*

Pappus hairy. *Florets of the ray* ♀, in many rows, ligulate ; *of the disk* ♂. *Receptacle* naked. *Bracts* with a membranous edge.

1. T. *Farfara* Linnæus.—(COLTSFOOT.) Fig. 305.

Leaves angular, cordate, appearing after the scapes, which are covered with scales, and carry each a cylindrical yellow flower-head.

305

Habitat. Waste places.
Quality. Emollient, demulcent, tonic.
Uses. A common remedy for troublesome coughs ; formerly smoked, now used in decoction.

HELIANTHUS. *Linnæus.*

Pappus paleaceous, deciduous. *Florets of the ray* 0; *of the disk* ⚥. *Anthers* without tails. *Bracts* imbricated. *Achænia* flat, all of the same form.

1. H. *tuberosus.*—(JERUSALEM ARTICHOKE.)

A tall tuberous perennial. Leaves triple-nerved, serrated, scabrous, the lower cordate, the upper ovate-lanceolate ; heads small.

Habitat. Brazil.
Quality. Tubers nutritious, slightly aromatic.
Uses. A common esculent.

PYRETHRUM. *Linnæus.*

Pappus an elevated membranous border. *Florets of the ray* ♀, in one row ; *of the disk* ⚥. *Bracts* forming a hemispherical flower-head. *Receptacle* rather convex, naked. *Achænia* angular, not winged.

1. P. *Parthenium* Smith.—(FEVERFEW.)

Leaves downy, pinnated ; leaflets elliptical, obtuse, pinnatifid, the segments rather toothed.

Habitat. Woods and gardens. *Quality.* Bitter, tonic.
Uses. A decoction a favourite popular remedy for slight fevers.

Fig. 305.—Tussilago Farfara.

Tanacetum. *Linnæus.*

Pappus a slight membranous border. *Florets* all tubular, of the ray ♀, *of the disk* ⚥. *Bracts* forming a hemispherical flower-head. *Receptacle* naked. *Achænia* oblong, angular, with a large epigynous disk.

1. T. *vulgare* Linnæus.—(Tansy.)

Leaves bipinnatifid, with serrated segments.

Habitat. Road-sides.
Quality. Aromatic, bitter, tonic, anthelmintic.
Uses. Dyspepsia, intermittents, gout; as an ingredient in puddings and cakes; in worm cases.

Achillea. *Linnæus.*

Pappus 0. *Florets of the ray* short, ♀; *of the disk* ⚥, with a flattened winged tube. *Bracts* forming an ovate or oblong imbricated flower-head. *Receptacle* scaly, subconvex. *Achænia* compressed.

306

1. A. *Millefolium* Linnæus.—(Milfoil.) Fig. 306.

Leaves woolly, those of the stem lanceolate or nearly linear, bipinnatifid, with deeply divided pinnæ; the rachis scarcely at all toothed.

Habitat. Road-sides.
Quality. Strong-scented, bitter, stimulating, tonic.
Uses. Its astringent leaves have been used to staunch wounds.

2. A. *nobilis* Linnæus.

Leaves woolly, oval, bipinnatifid, with deeply divided pinnæ; the rachis toothed from the point to the middle.

Habitat. Fields in Europe.
Quality and *Uses.* As in the last.

307

3. A. *Ptarmica* Linnæus. — (Sneezewort.) Fig. 307.

Leaves lanceolate, acute, sharply and finely serrated.

Habitat. Europe, in damp places.
Quality. Acrid, burning.
Uses. Root a substitute for Anacyclus Pyrethrum; powdered leaves produce sneezing.

Artemisia. *Linnæus.*

Pappus 0. *Florets* few, all tubular; *of the disk* ⚥; *of the ray* in one row. *Bracts* forming a roundish imbricated head. *Receptacle* naked or hairy. *Achænia* obovate, with a small epigynous disk.

1. A. *Abrotanum* Linnæus.—(Southernwood.)

Leaves downy beneath, not auricled, bipinnate, with extremely narrow segments; flower-heads hoary, nearly round; receptacle naked.

Habitat. Common in gardens.
Quality. Fragrant, bitter, acrid.
Uses. Leaves dried to drive away moths from linen; an ingredient in some Continental beer.

Fig. 306.—Portion of the inflorescence of Achillea Millefolium; 307. Leaf of Achillea Ptarmica.

2. A. *Absinthium* Linnæus.—(WORMWOOD.) Fig. 308.

Leaves hoary, 2-3-pinnatifid, with lanceolate obtuse segments; receptacle hairy.

308

Habitat. Thickets and mountainous places in Europe.

Quality. Aromatic, bitter, tonic; produces headache and giddiness.

Uses. Dyspepsia, intermittents, worm cases.

3. A. *Sieberi* Besser. *Art. Contra* Linnæus.—(SEMEN CONTRA. SEMENCINE.) Fig. 309.

Leaves very small, stiff, hairless, glandular, 3-5-parted; the middle lobe pinnatifid, the latter trifid and linear; flower-heads very small, panicled, few-flowered, glandular.

Habitat. Palestine.

Quality. Odour nauseous, balsamic, bitter; sialagogue.

Uses. The leaves and flower-heads a celebrated remedy for worms. Several allied species furnish the Semen Cinæ or Semencine of the Continental druggists.

309

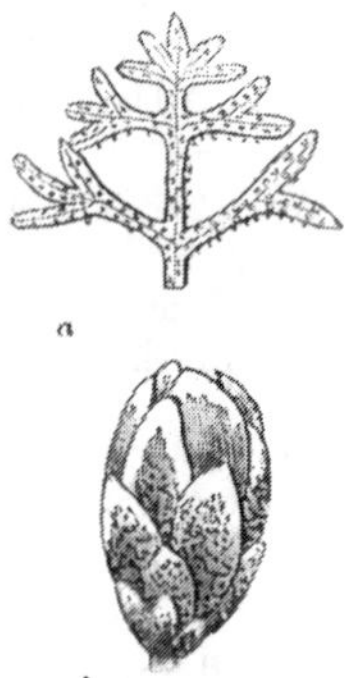

4. A. *Moxa* De Candolle.—(MOXA WEED.)

A shrub. Leaves hoary, with a loose separable down, bipinnatifid, with linear-lanceolate obtuse segments; heads middle-sized, globose, in racemose panicles.

Habitat. China.

Quality. The loose wool, or the beaten tops, form an inflammable substance, called Moxa, employed to produce eschars, instead of the actual cautery.

Uses. Paralysis of the nerves; sciatica, lumbago, neuralgia, spasmodic asthma, white swelling, stiff joints, visceral diseases.

Fig. 308.—Artemisia Absinthium; 309. Artemisia Sieberi; *a*, a leaf; *b*, a flower-head; both magnified.

5. A. *Dracunculus* Linnæus.—(Tarragon. Estragon *Fr.*)
Leaves green, smooth, linear-lanceolate, undivided; flower-heads nearly round.

Habitat. Siberia. Common in gardens.
Quality. Warm, aromatic, slightly sialagogue.
Uses. Chiefly employed as a pickle, and to flavour vinegar. The names, Dracunculus, Estragon, Tragon, Tarragon, are said to allude to the convolutions of the root, which have been compared to a dragon's tail.

Anthemis. *Linnæus.*

Pappus 0. *Florets of the ray* ♀ or 0, in one row; *of the disk* ⚥. *Bracts* imbricated. *Receptacle* conical, scaly. *Achænia* obscurely 4-cornered.

1. A. *nobilis* Linnæus.—(Chamomile.) Fig. 310.
Perennial, prostrate; leaves pinnate, downy, the lobes pinnatifid; receptacle long, conical.

Habitat. Pastures on gravel.
Quality. Aromatic, bitter, tonic, emetic.
Uses. Intermittents, dyspepsia, flatulence, colic, eructation.

310

a

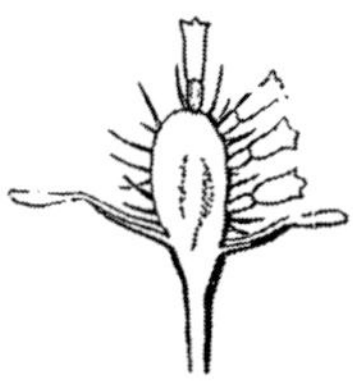

Anacyclus. *Linnæus.*

Like Anthemis, except that the *Achænia* are winged and obcordate.

1. A. *Pyrethrum* De Candolle. *Anthemis Pyrethrum* Linnæus.—(Pellitory of Spain.)
Stems procumbent, downy; radical leaves nearly smooth, pinnate, with pinnatifid segments and linear subulate lobes; branches monocephalous.

Habitat. Barbary, Spain, Levant.
Quality. Rubefacient; a very powerful local irritant; sialagogue.
Uses. As tincture for toothache; chewed for palsy of the tongue, and neuralgia of the head and face; as a gargle in relaxed uvula.

Fig. 310.—Anthemis nobilis; *a*, the receptacle with a few florets adhering to it.

GUIZOTIA. *Cassini.*

Pappus 0. *Florets of the ray* ♀, in one row; *of the disk* ⚥, covered with thick jointed hairs below the middle. *Bracts* in two rows, leafy. *Receptacle* paleaceous. *Achænia* angular.

1. G. *oleifera* De Candolle. *Verbesina sativa* Roxburgh.—(RAM-TIL.)

Stem downy at the upper end; leaves half amplexicaul, ovate-lanceolate or cordate, remotely serrated, rather rough; outer bracts more leafy and spreading than the others.

Habitat. East Indies, Abyssinia.
Uses. Seeds yield a large quantity of bland oil; employed like Olive oil.

311

CALENDULA. *Linnæus.*

Florets of the ray ⚥; *of the disk* 0. *Achænia* deformed, curved, variously toothed, and muricated at the back.

1. C. *officinalis* Linnæus. — (POT MARIGOLD.) Fig. 311.

An annual. Stem erect; lower leaves obovate, on long stalks; florets large, deep yellow; achænia boat-shaped.

a b

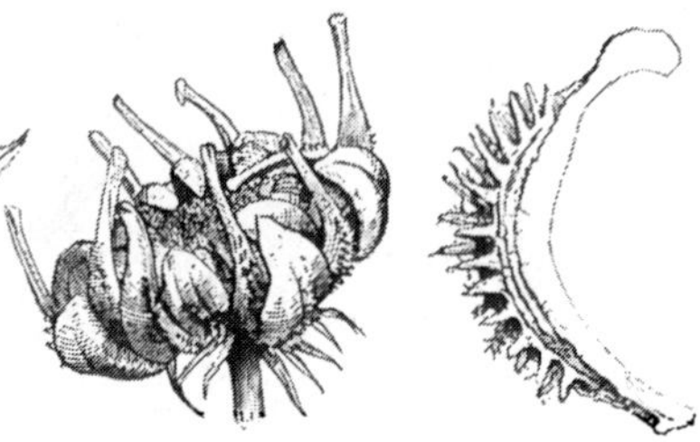

Habitat. Common in gardens.
Uses. It was formerly much employed as a carminative; it is chiefly used now to adulterate saffron.

---

* * CYNARACEOUS COMPOSITES.

CENTAUREA. *Linnæus.*

Pappus pilose, in many rows; second row longest. *Bracts* imbricated, spiny or lacerated. *Filaments* papillose. *Receptacle* chaffy. *Achænia* attached obliquely.

1. C. *Calcitrapa* Linnæus.—(STAR THISTLE.)

Fig. 311.—Calendula officinalis; *a*, a capitulum in fruit; *b*, an achænium.

Bracts smooth, palmate, spiny, with a strong channelled central spine; heads lateral; leaves deeply pinnatifid.

Habitat. Gravelly and sandy places.
Quality. Roots bitter.
Uses. Employed as a substitute for the Cnicus Benedictus.

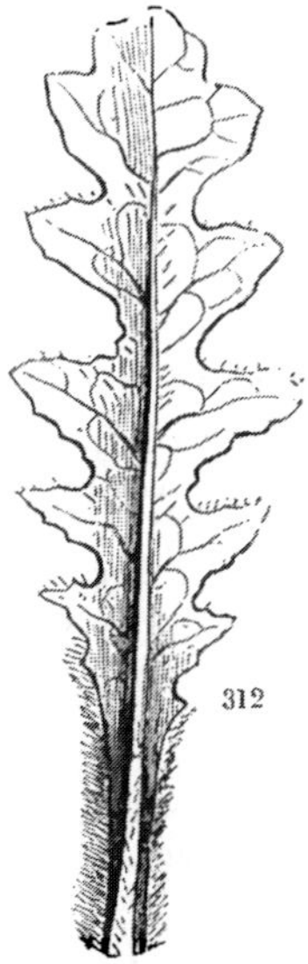

312

CNICUS. *Linnæus.*

Bracts of the involucre coriaceous, extended into a long hard pinnated spine. *Achænia* regularly furrowed, smooth, with a broad lateral scar. *Pappus* triple; the outer horny, short; the next composed of 10 long bristles; the third of 10 short bristles.

1. C. *Benedictus* Linnæus. Fig. 312.

A branched rather shaggy annual; leaves amplexicaul, rather decurrent, half pinnatifid; heads terminal, enveloped in leaves; florets yellow.

Habitat. Levant, Persia; now in South America.
Quality. Tonic, diaphoretic, emetic.
Uses. Weak digestion; warm infusion in chronic diseases.

CYNARA. *Linnæus.*

Bracts of the involucre fleshy at the base, emarginate, with a hard point. *Receptacle* fringed.

1 C. *Scolymus* Linnæus.—(The ARTICHOKE.)

Leaves rather spiny, pinnatifid and undivided; scales of the involucre ovate.

Habitat. South of Europe.
Uses. The receptacle or "bottom" of the flower-head is largely employed as a delicate esculent. The pappus forms the "choke."

313

ARCTIUM. *Linnæus.*

Pappus short, pilose, distinct. *Bracts* forming a globular head, armed with hard hooks, and constituting a bur.

1. A. *majus* Schkuhr. *Lappa major.*—(BURDOCK.) Fig. 313.

Involucre nearly smooth; bracts all subulate and hooked, longer than the florets; heads rather corymbose.

Habitat. Road-sides, waste places, old gravel pits.
Quality. Resolvent, diaphoretic, diuretic; achænia diuretic and purgative.
Uses. Roots, leaves, and fruit, as an alterative and resolvent in gouty, rheumatic, calculous, and venereal complaints. N.B.—The smaller Burdock, Arctium minus, known by its cobwebby heads, placed in racemes, has the same property.

Fig. 312.—Leaf of Cnicus Benedictus: 313. Arctium majus.

* * * CICHORACEOUS COMPOSITES.

TARAXACUM.

Pappus filiform, very soft, deciduous. *Bracts* in 2 rows; the outer short and lax. *Achænia* compressed, beaked, muricated. *Receptacle* naked.

1. T. *Dens Leonis* Desfontaines.—(DANDELION. DENT DE LION.)

Leaves runcinate, toothed; achænia linear, obovate, blunt, scaly, muricated, with a long beak.

Habitat. Pastures and waste places, everywhere.
Quality. Stomachic, tonic, aperient, diuretic.
Uses. Weak digestion, hepatic affections, dropsy, dyspepsia, cutaneous diseases, uterine obstructions.

LACTUCA. *Linnæus.*

Pappus filiform, soft, deciduous. *Heads* few-flowered. *Bracts* in 2-4 rows, outer shorter, with a membranous edge. *Receptacle* naked. *Achænia* compressed, contracted into a filiform beak, which is not muricated.

1. L. *virosa* Linnæus.—(ACRID LETTUCE.)

Leaves with a prickly keel, auricled, toothed or sinuate, mucronate; achænia black, as long as the white beak.

Habitat. Dry banks.
Quality. Narcotic, subacrid.
Uses. Yields Lettuce-opium or Lactucarium.

2. L. *Scariola* Linnæus.—(PRICKLY LETTUCE.)

Leaves with a prickly keel, sagittate, sinuate, amplexicaul; achænia pale, as long as the white beak.

Habitat. Waste places.
Quality and *Uses.* As in the last.

3. L. *sativa* Linnæus.—(GARDEN LETTUCE.)

Leaves oblong, erect, narrower at the base, smooth on the keel; with a long leafy flowering stem.

Habitat. East Indies? Common in gardens.
Quality. Sedative, hypnotic, antiscorbutic, (anaphrodisiac?).
Uses. Leaves largely as salad. Lettuce opium in troublesome coughs, inflammation, nervous disorders, as a substitute for opium, but more uncertain.

SCORZONERA. *Linnæus.*

Pappus feathery, in several rows. *Bracts* imbricated. *Receptacle* naked. *Achænia* neither stalked nor beaked, with a lateral scar.

1. S. *hispanica* Linnæus.—(VIPER'S GRASS.)

Root cylindrical, succulent; branches monocephalous; leaves amplexicaul, lanceolate, wavy; involucres smooth; flowers yellow.

Habitat. Spain and the south of Europe. Common in gardens.
Quality. Root said to be sudorific; nutritious, subaromatic.
Uses. A delicate eatable root; fancied by the Spaniards to be a specific against viper bites.

CICHORIUM. *Linnæus.*

Pappus two rows of minute paleæ. *Bracts* in 2 unequal rows, the outer reflexed from the first, the inner afterwards. *Receptacle* nearly naked. *Achænia* obovate, striated.

1. C. *Intybus* Linnæus.—(Succory. Chicory.) Fig. 314.

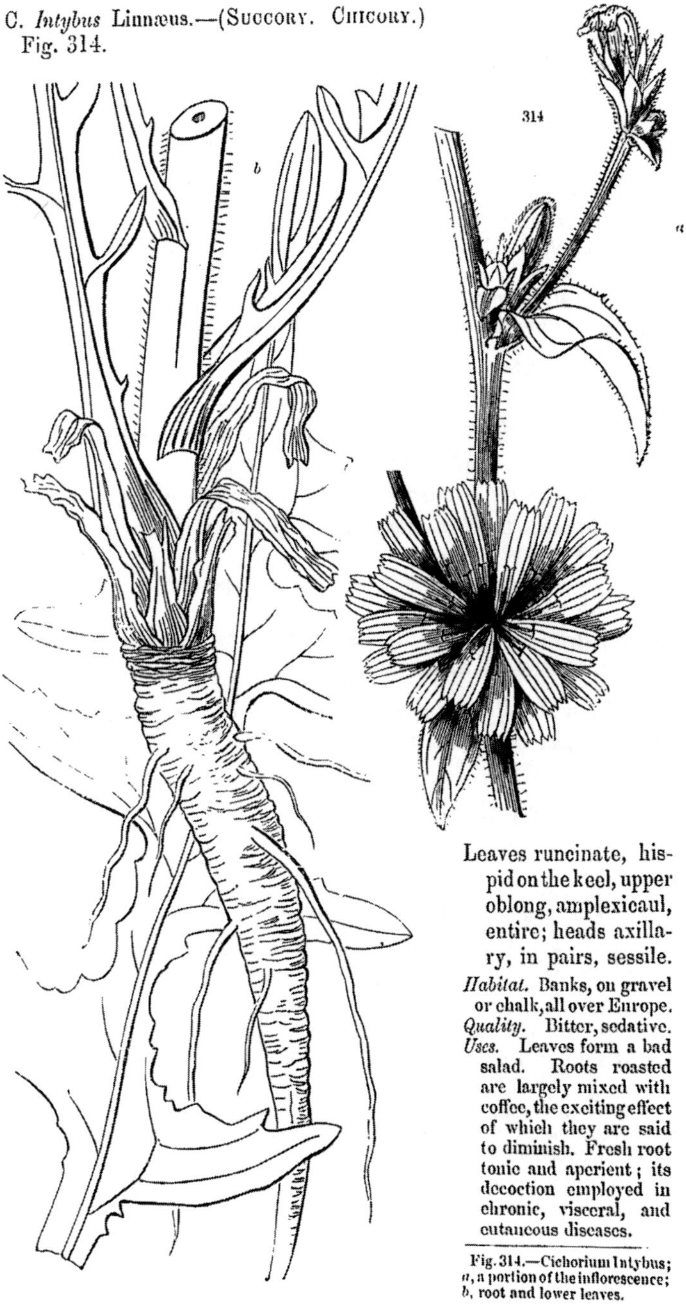

Leaves runcinate, hispid on the keel, upper oblong, amplexicaul, entire; heads axillary, in pairs, sessile.

Habitat. Banks, on gravel or chalk, all over Europe.
Quality. Bitter, sedative.
Uses. Leaves form a bad salad. Roots roasted are largely mixed with coffee, the exciting effect of which they are said to diminish. Fresh root tonic and aperient; its decoction employed in chronic, visceral, and cutaneous diseases.

Fig. 314.—Cichorium Intybus; *a*, a portion of the inflorescence; *b*, root and lower leaves.

2. C. *Endivia* Linnæus.—(ENDIVE.)

Stem smooth, or rather hairy; lower leaves oblong, sinuated, nearly smooth, those next the flower-heads broadly auricled; florets blue or white.

Habitat. East Indies. Common in gardens.
Quality. Bitter.
Uses. Employed largely as a winter salad.

TRAGOPOGON. *Linnæus.*

Pappus feathery, in many rows. *Bracts* in one row, 8-10, united at the base. *Receptacle* punctured. *Achænia* striated, with a long beak.

1. T. *porrifolium* Linnæus.—(SALSAFY. CERCIFIS or SALSIFIS *Fr.*)

Smooth; leaves erect, linear-lanceolate, acuminate, entire; peduncles obconical, fistular; flowers purple; roots cylindrical, succulent.

Habitat. Meadows all over Europe. Common in gardens.
Quality. Said to be sudorific. Root nutritious and subaromatic.
Uses. Roots sweet, tender; much esteemed as an esculent.

THE CINCHONAL ALLIANCE (V. K., p. 757.)

Natural Orders of Cinchonals.

Cranberries (*Vacciniaceæ.*) Stamens epigynous, porandrous.

Cinchonads (*Cinchonaceæ.*) Stamens epipetalous. Leaves opposite, with interpetiolar stipules.

Caprifoils (*Caprifoliaceæ.*) Stamens epipetalous. Leaves opposite, with no stipules.

Stellates (*Galiaceæ.*) Stamens epipetalous. Leaves verticillate, with no stipules, Fruit didymous.

Natural Order, **Cranberries**; *Vacciniaceæ* (V. K., p. 757.)

Prevailing Quality. Uncertain.

OXYCOCCUS. *Persoon.*

Corolla rotate, 4-parted, revolute. *Stamens* 8. *Berry* 4-celled.

1. O. *palustris* Persoon.—(CRANBERRY.)

Branches creeping, filiform; leaves oval, entire, rolled back at the edge.

Habitat. Bogs over all the north of Europe.
Quality. Fruit largely used in tarts and puddings.

VACCINIUM. *Linnæus.*

Corolla 4-5-cleft, campanulate or urceolate. *Stamens* 8-10. *Berry* 4-5-celled, many-seeded.

1. V. *Vitis Idæa* Linnæus.—(RED WHORTLEBERRY.)

A dwarf shrub; leaves evergreen, obovate, emarginate, rolled back at the edge, marked beneath with black dots.

Habitat. North of Europe and America.
Uses. The berries form one of the most agreeable of marmalades. Leaves sometimes used to adulterate samples of Arctostaphylos uva ursi.

2. V. *uliginosum* Linnæus.—(WHORTLEBERRY.)

Stem much branched, ferruginous; leaves small, obovate, entire, downy and glaucous on the under side.

Habitat. All Europe, from the polar region to the Mediterranean.
Quality. Fruit succulent; when fermented, producing a heady liquor.

Natural Order, Cinchonads; *Cinchonaceæ* (V. K., p. 761.)

Prevailing Quality. Tonic, emetic.

CINCHONA. *Linnæus.*

Corolla tubular, valvate. *Anthers* 5, included. *Capsule* ovate, 2-celled, 2-valved, septicidal, with winged seeds.

1. C. *micrantha* Ruiz and Pavon. C. *scrobiculata* Humboldt.—(SILVER OR HUANUCO BARK TREE.)

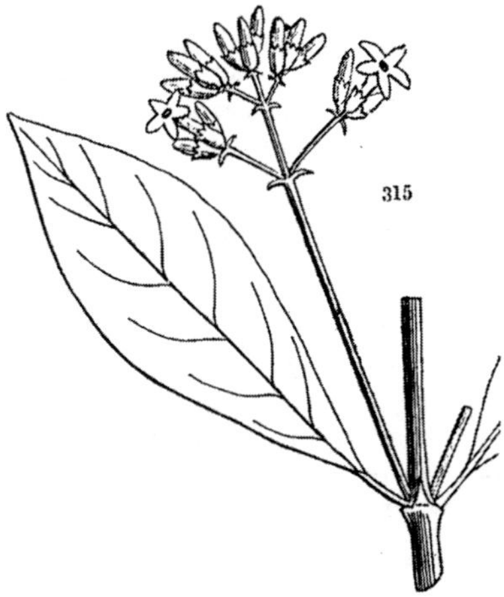

315

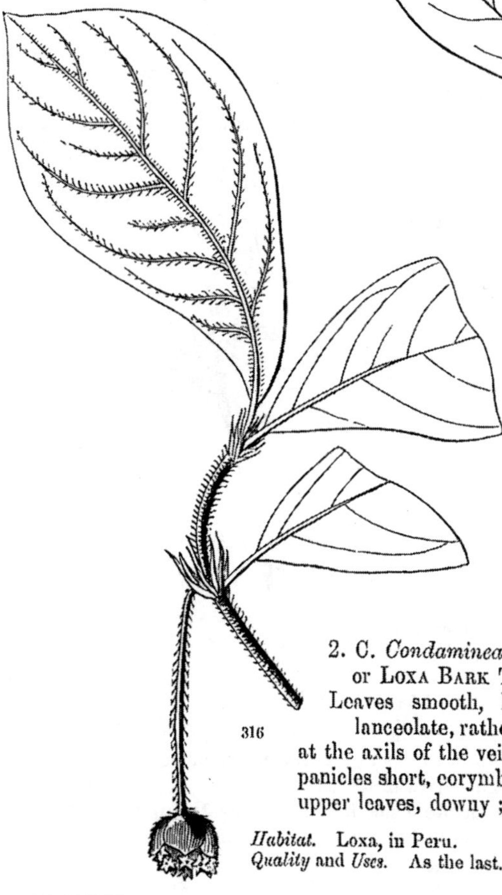

Branches quadrangular, smooth; leaves oblong, very large, membranous, pitted at the axils of the veins; flowers in a loose leafless downy panicle.

Habitat. Peru.
Quality. Stimulant, tonic, corroborant, irritant, astringent.
Uses. Intermittents, inflammatory diseases, rheumatism, atony, debility, &c.

2. C. *Condaminea* Humboldt.—(CROWN OR LOXA BARK TREE.) Fig. 315. Leaves smooth, lanceolate or ovate-lanceolate, rather thin, with a deep pit at the axils of the veins on the under side; panicles short, corymbose, in the axils of the upper leaves, downy; flowers small.

316

Habitat. Loxa, in Peru.
Quality and *Uses.* As the last.

Fig. 315.—Cinchona Condaminea; 316. Cephaëlis Ipecacuanha in flower.

CEPHAËLIS. *Swartz.*

Flowers in heads, surrounded by a leafy involucre. *Calyx* with a short-lobed limb. Lobes of *corolla* small, obtuse. *Anthers* included. *Fruit* succulent, 2-celled, with the stones striated on the external side.

1. C. *Ipecacuanha* A. Richard.—(TRUE IPECACUANHA.) Fig. 316.

A creeping herbaceous plant; leaves oblong-lanceolate, rough above, downy beneath; stipules multifid; heads long-stalked, pendulous.

Habitat. Woods of Brazil.
Quality. Emetic, narcotic.
Uses. Hooping-cough, croup, asthma, cases of poisoning, mucous catarrh, bronchial hæmorrhage, indigestion, dysentery, &c.

RICHARDSONIA. *Kunth.*

Calyx with a globose tube and a 4-7-parted limb. *Corolla* obconical, valvate. *Stamens* 3-5, projecting. *Style* 3-4-cleft, with capitate stigmas. *Capsules* composed of 3-4 indehiscent 1-seeded shells.

1. R. *scabra* Aug. de St. Hilaire.—(WHITE IPECACUANHA.) Fig. 317.

Leaves ovate-lanceolate, rough at the edge; teeth of the stipules shorter than their tube; heads many-flowered; calyx-lobes triangular, ciliated.

317

Habitat. Tropical America in many places.
Quality and *Uses.* As in Cephaëlis Ipecacuanha.

UNCARIA. *Schreber.*

Flowers in globular heads. *Calyx* 5-cleft. *Corolla* funnel-shaped, with a naked mouth. *Capsules* 2-celled, clavate, with numerous winged seeds.

1. U. *Gambir* Roxburgh.—(GAMBIR PLANT.)

Leaves ovate-lanceolate, acute, smooth; stipules ovate; peduncles axillary, solitary, with 2 bracts in the middle, the lower barren and hooked.

Habitat. East Indies, Malay Archipelago.
Quality. One of the most powerful of pure astringents.
Uses. Yields Gambir, a substitute for Catechu.

Fig. 317.—Richardsonia scabra.

MANETTIA. *Mutis.*

Calyx permanent. *Corolla* funnel-shaped, 4-cleft. *Anthers* 4, sessile in the hairy mouth. *Capsules* 2-valved, septicidal, with numerous winged seeds.

1. M. *cordifolia* Martius.

Stem twining, terete, rough; leaves ovate, cordate, acute, downy on each side; peduncles axillary, 1-flowered.

Habitat. Woods of Brazil. *Quality.* Bark of root emetic.
Uses. Regarded in Brazil as a valuable remedy in dropsy and dysentery.

CHIOCOCCA. *P. Browne.*

Calyx 5-toothed. *Corolla* short, funnel-shaped, smooth inside. *Stamens* 5, in the bottom of the corolla. *Fruit* succulent, crowned by the calyx, with 2 papery stones.

1. C. *densifolia* Martius. Fig. 318.

Leaves ovate, rather cordate; stipules broad at the base; racemes many-flowered; corolla much longer than the calyx.

Habitat. Tropical America.
Quality. Bark of root violently emetic and drastic; diuretic.
Uses. Dropsy, visceral obstructions, snake-bites? hydrophobia?

COFFEA. *Linnæus.*

Calyx 4-5-cleft. *Corollas* funnel-shaped, with 4-5 oblong spreading twisted lobes. *Fruit* a compressed drupe, furrowed along the side, crowned by the calyx. *Seeds* solitary, plano-convex, with a deep furrow along the flat side. *Putamen* like parchment.

1. C. *arabica* Linnæus.—(COFFEE TREE.) Fig. 319.

Leaves oblong, ovate, acuminate, smooth; peduncles axillary, short, clustered; corolla 5-cleft; stamens projecting.

Habitat. Arabia Felix and Nubia.
Quality. Antisoporific, stimulating; apt to produce constipation.
Uses. Counteracts opium, relieves intoxication, removes headache; in dyspepsia, diarrhœa, intermittents; some nervous disorders; spasmodic asthma.

Fig. 318.—Chiococca densifolia; 319. Coffea arabica.

Natural Order, Caprifoils; *Caprifoliaceæ* (V. K., p. 767.)

Prevailing Quality. Subacrid, emetic.

SAMBUCUS. *Linnæus.*

Calyx 5-toothed. *Corolla* rotate, 5-cleft. *Stamens* 5. *Stigmas* 3, sessile. *Berry* with 3-5 seeds.

1. S. *nigra* Linnæus.—(ELDER BUSH.) Fig. 320.

Stem almost arborescent; leaves pinnated; leaflets ovate-lanceolate, serrated; corymb 5-rayed at the first division.

Habitat. Woods and hedgerows all over Europe.

Quality. Flowers stimulant and sudorific; fruit aperient and diuretic; inner bark hydragogue, cathartic, emetic.

Uses. Flowers form an ointment and a distilled water; berries make a grateful wine; bark in dropsies.

320

321 b

a

VIBURNUM. *Linnæus.*

Calyx 5-toothed. *Corolla* rotate, campanulate, or tubular, 5-lobed. *Stamens* 5. *Stigmas* 3, sessile. *Berry* 1-seeded.

1. V. *Opulus* Linnæus. Fig. 321.

Leaves 3-5-lobed, with acute toothed lobes; flowers neuter and radical in the circumference of the corymb.

Fig. 320.—Sambucus nigra in fruit; 321. Viburnum Opulus; *a*, a leaf; *b*, a flower, magnified.

Habitat. Marshes; common in gardens, with all the flowers neuter and radiant, forming the "Gueldres Rose."
Quality. Leaves emetic, drastic; fruit austere.
Uses. Leaves as those of Sambucus; fruit a miserable food for savage northern nations.

Natural Order, **Stellates**; *Galiaceæ* (V. K., p. 768.)

Prevailing Quality. Astringent.

RUBIA. *Linnæus.*

Fruit succulent, didymous. *Corolla* rotate.

1. R. *tinctorum* Linnæus.—(MADDER.) Fig. 322.

Leaves in fours, netted, lanceolate, with reversed hooks at the edge; lobes of the corolla taper-pointed.

Habitat. Levant. Commonly cultivated in fields in the south of Europe.
Quality. A mild astringent and tonic. Colours red the bones of animals that feed on it.
Uses. A valuable red dye.

322

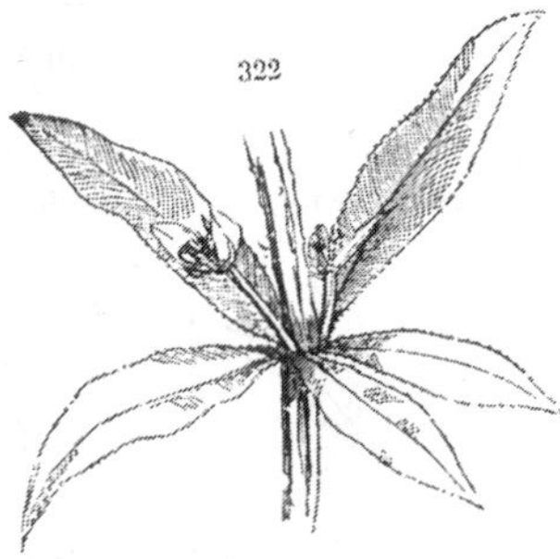

ASPERULA. *Linnæus.*

Fruit dry, didymous. *Corolla* funnel-shaped, or campanulate. *Style* bifid, with capitate stigmas.

1. A. *odorata* Linnæus.—(WOODRUFF.)

Leaves in whorls of 6 and 8, lanceolate, smooth, rough at the edge and keel; fruit covered with hooked bristles.

Habitat. In woods in most parts of Europe.
Quality. Possesses a very agreeable fragrance when dried.
Uses. Said to be diuretic; forms an agreeable herb-tea; dried leaves said to drive away moths from clothes.

THE MYRTAL ALLIANCE (V. K., p. 717.)

Natural Orders of Myrtals.

Myrobalans (*Combretaceæ.*) Leaves dotless. Stamens definite. Ovules 1 or 2, pendulous. Ovary 1-celled.

Myrtleblooms (*Myrtaceæ.*) Leaves dotted. Stamens 00. Ovules axile. Ovary 2- or more- celled.

Natural Order, **Myrobalans**; *Combretaceæ* (V. K., p. 717.)

Prevailing Quality. Astringent.

TERMINALIA. *Linnæus.*

Calyx campanulate, deciduous. *Petals* 0. *Fruit* a wingless juiceless drupe.

1. T. *Bellerica* Roxburgh.—(BELERIC MYROBALAN.)

Leaves alternate, elliptical, entire, acute at each end, smooth, on long stalks, with 2 small glands at the end of the petiole.

Habitat. Mountains of India. *Quality.* Fruit astringent, tonic, attenuant.
Uses. Kernels eaten in India; said to intoxicate; gum, though soluble in water, is inflammable, and burns like a candle.

Fig. 322.—A whorl of leaves of Rubia tinctorum.

2. T. *Chebula* Retzius.

Leaves rather opposite, ovate, acute, stalked, when old smooth on the upper side, but when young coarsely silky; with 2 glands at the end of the petiole, and a few along its sides.

Habitat. The mountains of India.
Quality. Fruit and galls extremely astringent.
Uses. Employed by dyers; dyes yellow with alum, black with salts of iron.

Natural Order, **Myrtleblooms**; *Myrtaceæ* (V. K., p. 734.)

Prevailing Quality. Aromatic, astringent.

CARYOPHYLLUS. *Linnæus.*

Calyx cylindrical, 4-cleft, much longer than broad. *Petals* 4. *Stamens* distinct, in 4 clusters. *Berry* oblong, with 1-2 cells and as many seeds.

1. C. *aromaticus* Linnæus.—(The CLOVE TREE.) Fig. 323.

Leaves ovate-oblong, acuminate at each end; cymes many-flowered.

324 323

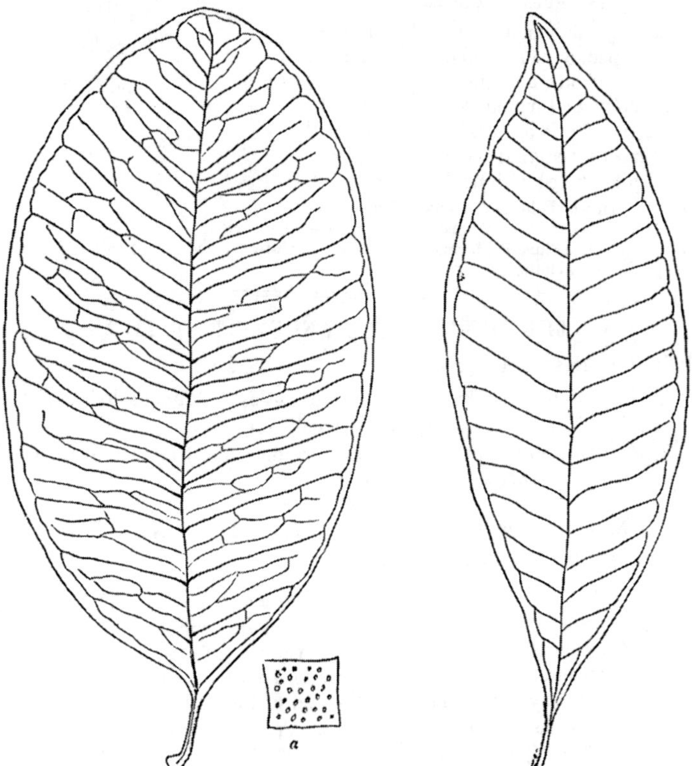

Fig. 323.—Leaf of Caryophyllus aromaticus; 324. Leaf of Eugenia Pimenta; *a*, a portion of it magnified to show the oil-cysts.

Habitat. East Indian Islands.
Quality. Fragrant, sweetish, very agreeable; stomachic, carminative.
Uses. Dried flower-buds much used as flavouring ingredients; in dyspepsia, nausea, flatulence.

EUGENIA. *Linnæus.*

Calyx roundish, 4-parted. *Petals* 4. *Stamens* distinct. *Berry* roundish, 1-2-celled, with as many seeds.

1. E. *Pimenta* De Candolle. *Myrtus Pimenta* Linnæus.—(PIMENTO. ALLSPICE.) Fig. 324.

Leaves oblong or oval, obtuse, smooth; peduncles axillary and terminal, in trichotomous panicles; berry globose, 1-seeded.

Habitat. West India Islands.
Quality. "Intermediate between pepper and cloves."—*Pereira.*
Uses. As a spice in cookery; in weak digestion; to relieve flatulency, &c.

325

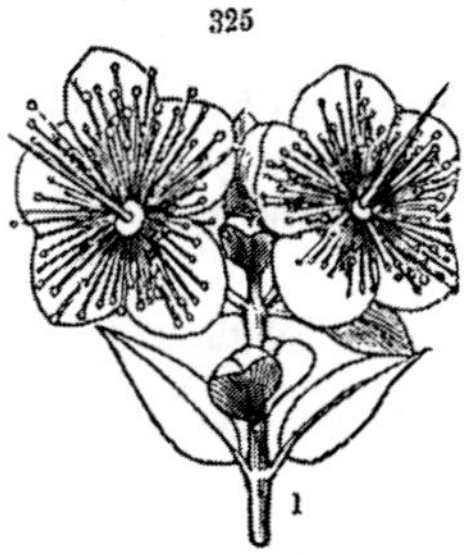

MYRTUS. *Linnæus.*

Calyx roundish, 5-cleft. *Petals* 5. *Stamens* distinct. *Berry* 2-3-celled, many-seeded.

1. M. *communis* Linnæus.—(The MYRTLE.) Fig. 325.

Leaves ovate and lanceolate, acute; pedicels solitary, 1-flowered, about as long as the leaf, with a pair of linear deciduous bracts beneath each flower.

Habitat. Persia. Common in gardens.
Quality. Fragrant, aromatic.
Uses. Dried fruit and flower-buds formerly used as a spice; and are said to be so still in Tuscany; form a kind of wine; flowers yield a distilled water called Eau d'Ange.

326

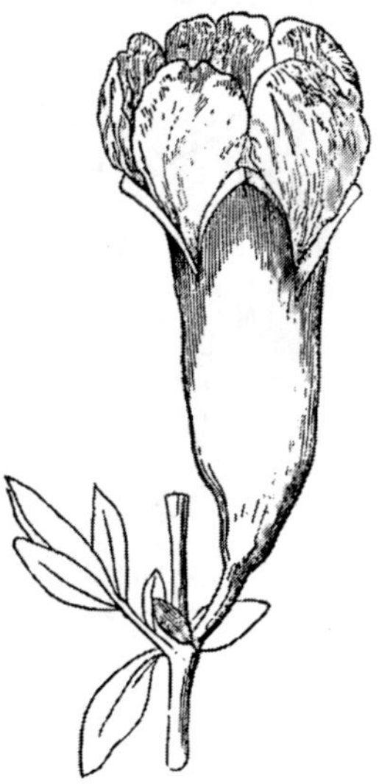

PUNICA. *Linnæus.*

Calyx coriaceous, tubular, 5-7-cleft, valvate. *Petals* 5-7, crumpled. *Fruit* a leathery indehiscent case, with numerous irregular cells, and 00 seeds, covered with pulp.

1. P. *Granatum* Linnæus.—(The POMEGRANATE.) Fig. 326.

A tree; leaves lanceolate.

Habitat. Barbary and the south of Europe.
Quality. Astringent; bark of root emetic and purgative.
Uses. Bark of root in worm cases; rind of fruit for gargles; pulp of seeds refrigerant, in fevers, especially bilious.

MELALEUCA. *Linnæus.*

Calyx 5-parted. *Petals* 5. *Stamens* polyadelphous, in 5 polyandrous phalanges. *Capsule* inclosed in the calyx, adnate to the very branch, with 00 angular seeds.

Fig. 325.—Myrtus communis; 326. Punica Granatum.

1. M. *Cajeputi* Roxburgh. M. *minor* Smith.—(CAJEPUT TREE.)
Leaves alternate, elliptical-lanceolate, 3-5-nerved ; flowers spiked, rather wide apart, with a woolly rachis.

Habitat. Amboyna and other Indian islands.
Quality. Oil a powerful antispasmodic, stimulant and sudorific.
Uses. Low fevers, paralysis, cholera, spasms, colic, chronic rheumatism, stimulating liniments.

THE CACTAL ALLIANCE (V. K., p. 741.)

Natural Order of Cactals.

Indian Figs (*Cactaceæ.*) Sepals and petals 00, undistinguishable. Fruit succulent.

Natural Order, Indian Figs; *Cactaceæ* (V. K., p. 746.)

Prevailing Quality. Uncertain.

OPUNTIA. *Tournefort.*

Stems flat, jointed, obovate or oblong, or ovate, at length confluent in a terete trunk.

1. O. *vulgaris* Miller.—(INDIAN FIG.) Fig. 327.
Spreading ; joints of stem ovate ; prickles all of the same form, very short and numerous.

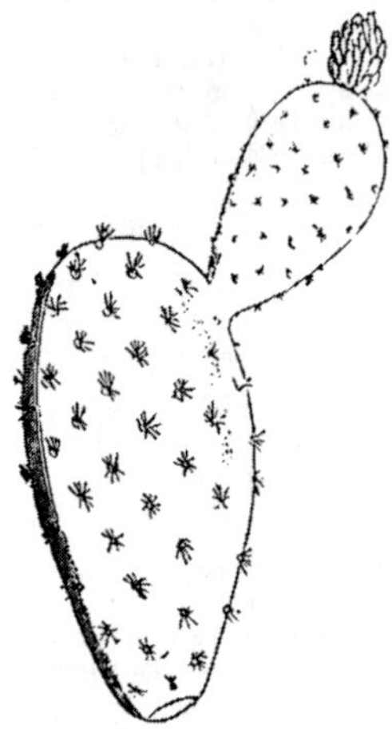

327

b a

Habitat. Southern states of North America. Commonly cultivated in the south of Europe.
Quality and *Uses.* Ripe fruit agreeable, and sold for the table in southern countries ; stains the urine red.

Fig. 327.—Opuntia vulgaris ; *a*, its branches ; *b*, a flower.

THE GROSSAL ALLIANCE (V. K., p. 749.)

Natural Order of Grossals.

Currantworts (*Grossulariaceæ.*) Fruit pulpy. Seeds parietal.

Natural Order, Currantworts; *Grossulariaceæ* (V. K., p. 750.)

Prevailing Quality. Subaromatic.

RIBES. *Linnæus.*

Calyx 5-lobed. *Petals* 5, scale-shaped. *Fruit* a succulent berry.

1. R. *rubrum* Linnæus.—(COMMON CURRANT.)

Leaves angular, bluntly 3-5-lobed, downy beneath, smooth above; racemes pendulous. Unarmed.

Habitat. Hedge rows and woods.
Quality. Juice of the fruit refrigerant and grateful to persons suffering from fever. Largely cultivated for dessert and for cooking.

2. R. *nigrum* Linnæus.—(BLACK CURRANT.)

Leaves angular, 3-5-lobed, with glandular dots on the underside. Unarmed.

Habitat. Woods of Europe and Siberia.
Quality. Tonic, stimulant, aromatic, subacid.
Uses. Fruit much cultivated for dessert and for cooking; forms a very useful domestic conserve employed for sore throat. N.B.—The Gooseberry is Ribes Grossularia.

THE UMBELLAL ALLIANCE.

Natural Orders of Umbellals.

Umbellifers (*Apiaceæ.*) Fruit didymous.

Ivyworts (*Araliaceæ.*) Fruit not didymous. Flowers pentamerous. Leaves alternate.

Cornels (*Cornaceæ.*) Fruit not didymous. Flowers tetramerous. Leaves opposite.

Natural Order, Umbellifers; *Apiaceæ* (V. K., p. 773.)

Prevailing Qualities. Aromatic; stimulating; poisonous.

APIUM. *Linnæus.*

Umbels compound. *Involucre* 0. *Calyx* obsolete. *Fruit* roundish, contracted at the side. *Ridges* 5, narrow, equal, the lateral on the edge. *Vittæ* 1 to each furrow. *Albumen* terete.

1. A. *graveolens* Linnæus.—(CELERY.) Fig. 328.

Smooth; leaves pinnated; leaflets cuneate, cut and toothed at the point.

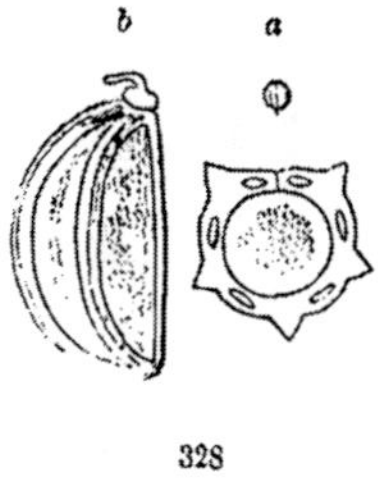

328

Fig. 328.—*a*, Fruit of Apium graveolens, natural size; *b*, half of it magnified; *c*, transverse section of ditto.

Habitat. Ditches, especially in salt marshes.
Quality. Acrid, poisonous when growing in wet places and unblanched.
Uses. When cultivated forms a favourite salad and ingredient in soups.

CICUTA. *Linnæus.*

Umbels compound. *Involucre:* general, obsolete; partial of many subulate bracts. *Calyx* leafy. *Fruit* roundish, contracted at the side. *Ridges* 5, flattish, equal, the lateral at the edge. *Vittæ:* 1 large to each furrow. *Albumen* terete.

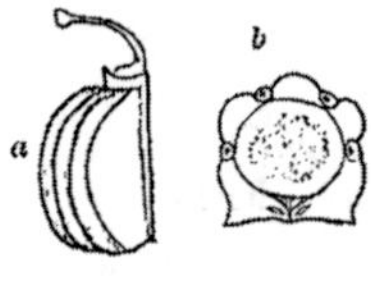

329

1. C. *virosa* Linnæus. Fig. 329.

Leaves tripinnate; leaves linear-lanceolate, acute, serrated.

Habitat. Ditches and river sides.
Quality. A dangerous poison, acting like Conium.

PETROSELINUM. *Hoffmann.*

Umbels compound. *Involucres:* partial of many, general of few bracts. *Calyx* obsolete. *Fruit* ovate, contracted at the side. *Ridges* 5, narrow, equal, the lateral on the edge. *Vittæ* 1 to each furrow. *Albumen* plano-convex.

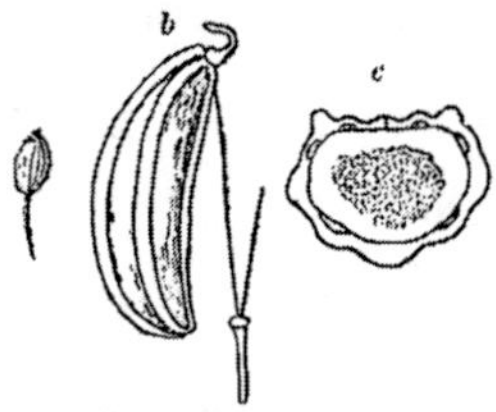

330

1. P. *sativum* Hoffmann.—(PARSLEY.) Fig. 329.

Stem angular; leaves shining, 3-pinnate; leaflets toothed.

Habitat. Common in gardens.
Quality. Pleasant, stimulating, aromatic, diuretic.
Uses. Leaves a common garnish to meat, &c. A favourite pot-herb.

PIMPINELLA. *Linnæus.*

Umbels compound. *Involucres* usually 0. *Calyx* obsolete. *Fruit* contracted at the side, ovate. *Ridges* 5, filiform, equal; the lateral on the edge. *Vittæ* 00. *Albumen* concavo-convex.

331

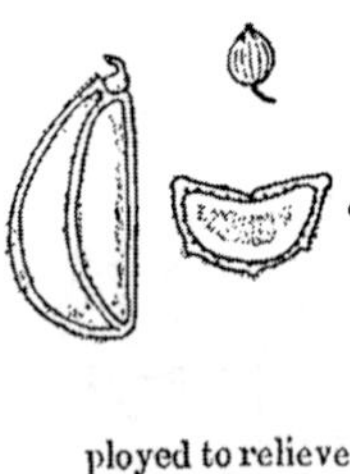

1. P. *Anisum* Linnæus.—(ANISE.) Fig. 330.

Lower leaves roundish-cordate, cut, those of the stem pinnate, with wedge-shaped leaflets; fruit downy.

Habitat. Egypt and Syria. Commonly cultivated.
Quality. Aromatic, stimulant.
Uses. As a flavouring substance for liqueurs, sweetmeats, &c. The officinal preparations, especially the *aqua anisi*, are employed to relieve flatulence, and colicky pains, especially of children. Nurses sometimes take it to promote the secretion of milk. It has also been used in pulmonary affections. Its effects are condimentary, stimulant, and carminative.—*Pereira.*

Fig. 329.—*a*, half a fruit of Cicuta virosa, magnified; *b*, its transverse section; 330. *a*, Fruit of Petroselinum sativum, natural size; *b*, half of it magnified; *c*, its transverse section; 331. *a*, Fruit of Pimpinella Anisum, natural size; *b*, half of it magnified; *c*, its transverse section.

CONIUM. *Linnæus.*

Umbels compound. *Involucre* both general and partial, small. *Calyx* obsolete. *Fruit* ovate, compressed. *Ridges* 5, prominent, wavy. *Vittæ* 0. *Albumen* with a deep furrow on the side next the commissure.

1. C. *maculatum* Linnæus.—(HEMLOCK.) Fig. 332.

Stem and all the parts perfectly hairless; bracts lanceolate, shorter than the partial umbel.

332

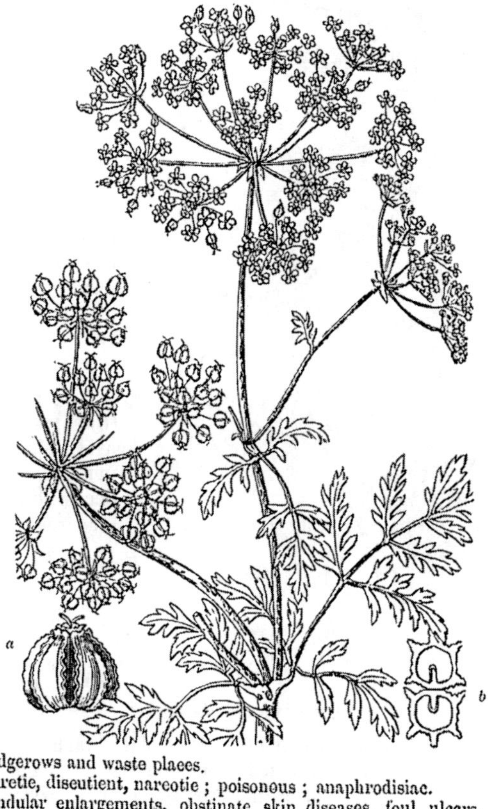

Habitat. Hedgerows and waste places.
Quality. Diuretic, discutient, narcotic; poisonous; anaphrodisiac.
Uses. In glandular enlargements, obstinate skin diseases, foul ulcers, bronchocele, syphilis, hooping cough, tetanus, rheumatism, neuralgia.

ÆTHUSA. *Linnæus.*

Umbels compound. *Involucre:* general 0; partial long, pendulous, halved. *Calyx* obsolete. *Fruit* ovate. *Ridges* 5, raised, thick, acute, the lateral on the edge and broader. *Vittæ* 1 to each furrow. *Albumen* terete.

1. Æ. *Cynapium* Linnæus.—(FOOL'S PARSLEY.) Fig. 333.

Fig. 332.—Conium maculatum; *a*, its fruit magnified; *b*, a transverse section of it.

Bracts longer than the umbel; stalks of the circumference twice as long as the fruit.

Habitat. Hedgerows and waste places.
Quality. Poisonous; narcotic, acrid, emetic.
Uses. Leaves a frequent cause of dangerous accidents, on account of their resemblance to Parsley.

333

CARUM. *Linnæus.*

Umbels compound. *Involucre* small or obsolete. *Calyx* obsolete. *Fruit* oblong, compressed. *Ridges* 5, narrow, equal, the lateral on the edge. *Vittæ* 1 to each furrow. *Albumen* terete.

q v

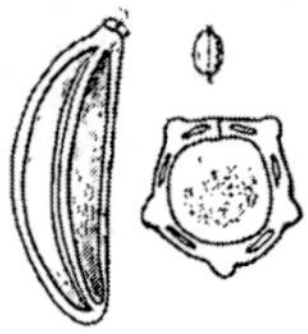

334

1. C. *Carui* Linnæus.—(CARAWAY.) Fig. 334.

Leaves bipinnate; leaflets multifid, the lower pair decussating; stem angular; root fusiform.

Habitat. Europe. Cultivated in gardens.
Quality. An aromatic stimulant and condiment.
Uses. In flatulent colic; chiefly as a flavouring material, for liqueurs and cakes.

CRITHMUM. *Linnæus.*

Umbels compound. *Involucre* of many lanceolate bracts. *Calyx* obsolete. *Fruit* oblong, rather flattened from the back. *Ridges* 5, winged, sharp, the lateral rather the widest. *Vittæ* 00, spread over all the seed. *Albumen* terete.

b a

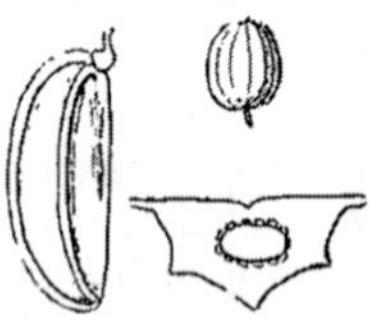

c 335

1. C. *maritimum* Linnæus.—(SAMPHIRE.) Fig. 335.

Leaves fleshy, 2-3 pinnate; leaflets lanceolate, few.

Habitat. Rocky cliffs of the sea coast.
Quality. Aromatic, saline.
Uses. A favourite ingredient in pickles.

Fig. 333.—Æthusa Cynapium; *a* its fruit magnified; 334. *a*, Fruit of Carum Carui, natural size; *b*, half of it magnified; *c*, its transverse section; 335. *a*, Fruit of Crithmum maritimum, natural size; *b*, one of the halves, magnified; *c*, its transverse section.

SIUM. *Linnæus.*

Umbels compound. *Involucres* both partial and compound. *Calyx* minute. *Fruit* compressed from the side. *Ridges* 5, equal, narrow, the lateral on the edge. *Vittæ* 3 or thereabouts to each furrow. *Albumen* subterete.

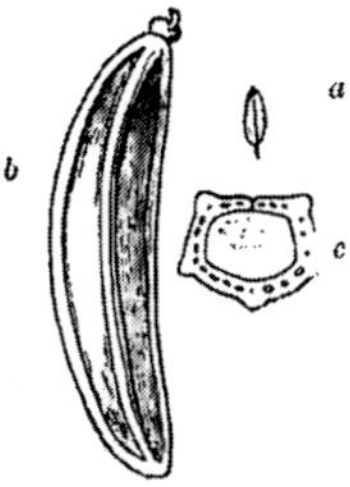

336

1. S. *Sisarum* Linnæus.—(SKIRRET.) Fig. 336.

Root tuberous, fascicled; lower leaves pinnate, with oblong serrated leaflets, the terminal being cordate; the upper ternate with lanceolate leaflets.

Habitat. Japan, China. Common in gardens.
Quality. Roots sweet, succulent, nutritious, subaromatic, employed in cookery in the same way as Scorzonera.

SMYRNIUM. *Linnæus.*

Umbels compound. *Involucres* variable. *Calyx* obsolete. *Fruit* roundish-ovate, compressed, didymous. *Ridges* sharp, thick, the 3 dorsal prominent, the lateral distant, on the edge. *Vittæ* 00. *Albumen* involute.

337

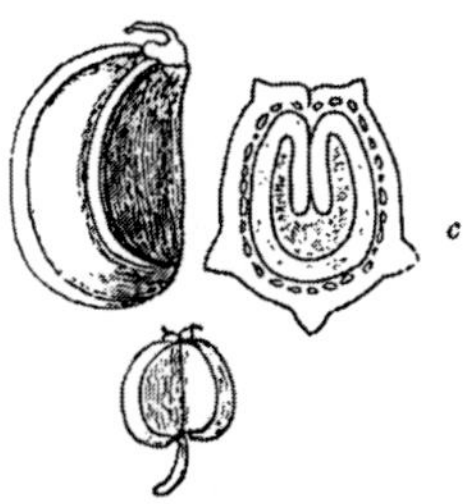

1. S. *Olusatrum* Linnæus. — (ALEXANDERS.) Fig. 337.

Stem terete; leaves ternate, stalked, serrate; fruit black.

Habitat. Waste ground, near ruins.
Quality. Aromatic, rather pleasant when blanched; fruit carminative.
Uses. Formerly instead of Celery; rarely cultivated now.

CORIANDRUM. *Linnæus.*

Umbels compound. *Involucres:* general 0, partial 3-leaved, halved. *Calyx* 5-toothed. *Fruit* spherical. *Ridges:* primary narrow, round, zigzag, obsolete; secondary prominent, filiform. *Vittæ* 0, except on the commissure. *Albumen* hemispherical, concave.

338

1. C. *sativum* Linnæus.—(CORIANDER.) Fig. 338.

Upper leaves multifid; flowers white.

Habitat. Levant; cultivated in gardens.
Quality. Aromatic, stimulant, carminative.
Uses. Fruit in confectionary; and as an adjuvant to other medicine.

DAUCUS. *Linnæus.*

Fruit with both primary and secondary ridges, compressed from the back. *Ridges* prickly, or broken into irregular segments. *Vittæ* 1 beneath each secondary ridge. *Albumen* plano-convex.

Fig. 336.—*a*, Fruit of Sium Sisarum, natural size; *b*, one of the halves, magnified; *c*, its transverse section; 337. *a*, Fruit of Smyrnium Olusatrum, natural size; *b*, one of the halves, magnified; *c*, its transverse section; 338. Fruit of Coriandrum, magnified; *b*, transverse section of one of its halves.

1. D. *Carota* Linnæus.—(Carrot.) Fig. 339.

339

Stem hispid; leaflets pinnatifid, opaque, with cuspidate segments; bracts of the involucre trifid and pinnatifid.

Habitat. Roadsides and banks. Commonly cultivated.

Quality. The fruit aromatic, stimulant, diuretic: root sweet, nutritious.

Uses. Fruit in suppression of urine; poultice of the root sometimes applied to chapped nipples; but painful. Root a well-known esculent.

2. D. *gummifer* Lamarck.

Stem hispid; leaflets pinnatifid, with nearly smooth, thick, ovate incised segments; bracts of the involucre pinnatifid or trifid, shorter than the umbels.

Habitat. South of Europe on the sea-coast.

Quality. The roots yield a bitter balsamic substance, formerly called Sicilian Bdellium.

b

Œnanthe. *Linnæus.*

Umbels compound. *Involucres* variable. *Calyx* stiff, leafy. *Fruit* oblong, crowned by the stiffened styles. *Ridges* 5, very convex. *Vittæ* 1 to each furrow. *Albumen* subterete.

1. Œ. *crocata* Linnæus.—(Hemlock Dropwort.) Fig. 340.

Root large, branched, fusiform; stem-leaves pinnate, with lozenge-shaped leaflets cut long the upper sides; fruit cylindrical, in close hard heads.

Fig. 339.—Daucus Carota; *a*, a flower of the ray; *b*, an umbel of fruit; *c*, a transverse section of half a fruit magnified.

Habitat. Wet places and swampy meadows.
Quality. Poisonous ; narcotico-acrid ; loses its virulence in northern latitudes. Roots, resembling small parsnips, a frequent cause of fatal accidents.

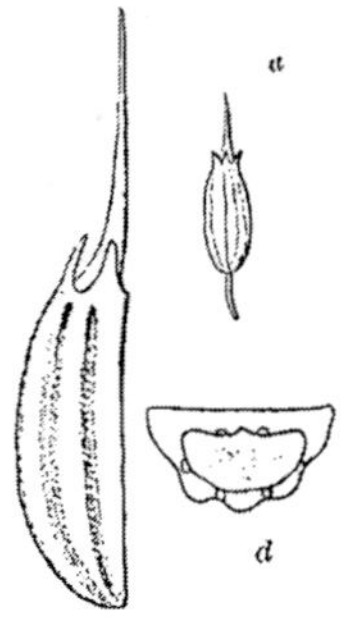

340

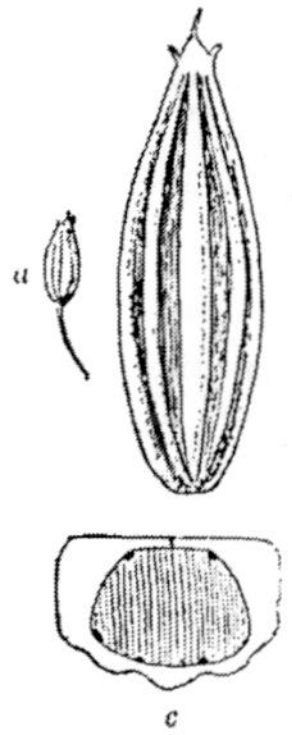

341

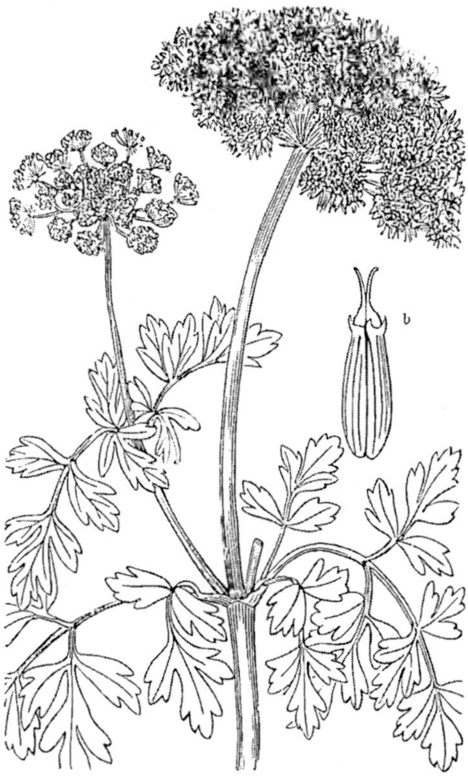

340

2. Œ. *Phellandrium* Sprengel. *Phellandrium aquaticum* Linnæus.—(Water Dropwort.) Fig. 34.

Rhizome jointed, with numerous whorled fibres ; leaves repeatedly pinnate, cut into innumerable fine dark-green segments.

Habitat. Ditches, ponds, and wet places.
Quality. As in the last, but less dangerous.

Anthriscus. *Hoffmann.*

Umbels compound. *Involucres :* general 0; partial of many bracts. *Calyx* obsolete. *Fruit* contracted at the side, beaked. *Ridges* 0, except on the beak, which has 5. *Vittæ* 0. *Albumen* half-terete, furrowed next the commissure.

1. A. *vulgaris* Persoon.

Stem smooth ; umbels lateral, stalked ; fruit ovate, hispid, about twice as long as the beak, which is smooth.

Fig. 340.—Œnanthe crocata; *a*, its fruit of the natural size ; *b*, ditto, magnified ; *c*, one of the halves ; *d*, its transverse section ; 341. *a*, Fruit of Œnanthe Phellandrium, natural size ; *b*, half of it magnified and seen from the back ; *c*, a transverse section of it.

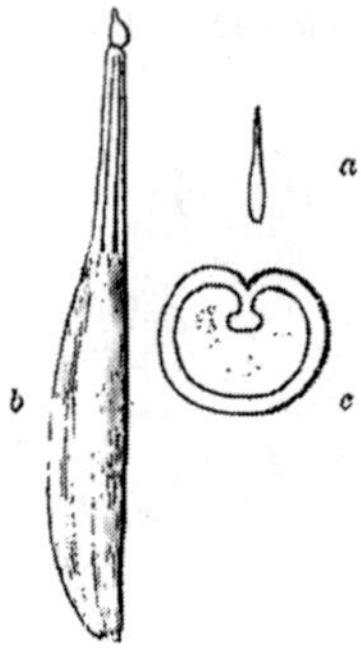

342

Habitat. Waste places, a common weed.
Quality. Deleterious; has been the cause of accidents in consequence of being mistaken for the following. Some Dutch soldiers, who gathered it by mistake for common Chervil, were poisoned by the soup into which it was put.—*Burnett.* Anthriscus sylvestris is reputed to be similar in its effects to Hemlock, only rather less narcotic. (Herba Cicutariæ *Officin.*)

2. A. *Cerefolium* Hoffmann.—(CHERVIL.) Fig. 342.
Stems hairy above the joints; umbels lateral, sessile; fruit smooth, about twice as long as the beak.

Habitat. Waste ground of Europe. Occasionally cultivated in gardens.
Quality. Leaves agreeably aromatic.
Uses. Grown merely for soups, and salads.

CUMINUM. *Linnæus.*

Umbels compound. *Involucre:* general of 2-4 bracts, partial halved, finally reflexed. *Calyx* of 5 lanceolate setaceous teeth. *Fruit* slightly contracted at the side. *Ridges* blunt, filiform, the lateral on the edge. *Vittæ* 1 beneath each prominent hairy furrow. *Albumen* nearly flat.

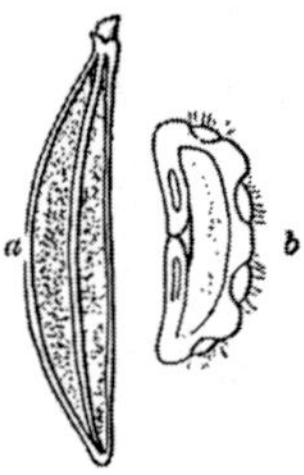

343

1. C. *Cyminum* Linnæus.—(CUMMIN.) Fig. 343.
Leaves multifid, setaceous; umbels 3-5-cleft; involucres longer than the hairy fruit.

Habitat. Egypt and the Mediterranean.
Quality. Mildly stimulant and carminative; discutient.
Uses. In the preparation of plaisters in veterinary practice; and in liqueurs.

FŒNICULUM. *Hoffmann.*

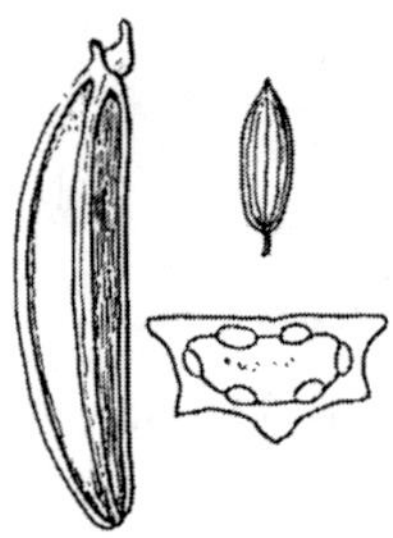

344

Umbels compound. *Involucre* 0. *Calyx* a tumid margin. *Fruit* taper. *Ridges* 5, prominent, bluntly keeled. *Vittæ* 1, large, to each furrow. *Albumen* plano-convex.

1. F. *officinale* Allioni.—(FENNEL.) Fig. 344.
Leaves multifid, with setaceous segments.

Habitat. Coast of the Mediterranean. Common in gardens.
Quality. An aromatic stimulant, carminative.
Uses. Chiefly as a potherb for flavouring sauces.

OPOIDIA. *Lindley.*

Umbels compound. *Involucres:* general obsolete, partial of many bracts. *Calyx* obscurely 5-toothed. *Fruit* terete, oval. *Ridges* = 3 dorsal angles. *Vittæ* one large one beneath each furrow, and one small one beneath each ridge. *Albumen* furrowed next the commissure.

345

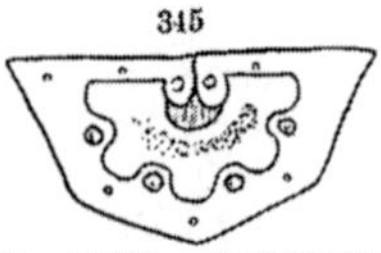

1. O. *galbanifera* Lindley. Fig. 345.
Leaflets oblong, obtuse, serrulate, decurrent.

Fig. 342.—*a*, Fruit of Anthriscus Cerefolium, natural size; *b*, one of its halves, magnified; *c*, its transverse section; 343. *a*, half the fruit of Cuminum Cyminum, magnified; *b*, its transverse section; 344. *a*, Fruit of Fœniculum officinale, natural size; *b*, one half magnified; *c*, its transverse section; 345. Transverse section of half a fruit of Opoidia galbanifera, magnified.

Habitat. Persia.
Quality. Supposed to yield the fetid gum-resin called Galbanum; but Dr. Pereira now thinks that the substance obtained from it is different from either Galbanum or Sagapenum.

ARCHANGELICA. *Hoffmann.*

Umbels large, compound. *Involucres:* general 0, partial of many bracts. *Calyx* 5-toothed. *Fruit* compressed from the back. *Ridges* 5, winged, the lateral short of the edge and broader than the dorsal. *Vittæ* 00, covering the plano-convex albumen, which is loose.

1. A. *officinalis* Hoffmann.—(ANGELICA.) Fig. 346.
Stem smooth, furrowed; leaves bipinnate, with ovate, or somewhat cordate broad serrated leaflets; upper petioles ventricose.

a

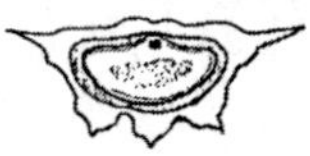

346

Habitat. Watery places.
Quality. Root and fruit pungent, aromatic, stimulant, tonic.
Uses. The candied stalks are stomachic. Chiefly employed in the preparation of gin.

ANETHUM. *Linnæus.*

Umbels compound. *Involucres* none. *Calyx* obsolete. *Fruit* compressed from the back, with a broad dilated edge. *Ridges:* 3 dorsal, filiform, equidistant; lateral lost in the margin. *Vittæ* 1 to each furrow. *Albumen* thin, lenticular.

1. A. *graveolens* Linnæus.—(DILL.) Fig. 347.
Segments of leaves long, setaceous; fruit elliptical; border flat.

Habitat. South of Europe, near the coast.
Quality. Aromatic, stimulant, carminative.
Uses. As a condiment; to relieve the flatulence and griping of infants.

a b

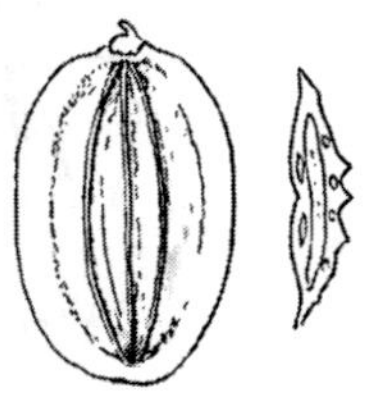

347

Fig. 346.—Archangelica officinalis; *a*, transverse section of half a fruit magnified. In this the vittæ are indistinctly shewn by the engraver; 347. *a*, Fruit of Anethum graveolens, magnified; *b*, a transverse section of one half of it.

PASTINACA. *Linnæus.*

Umbels compound. *Involucres:* general obsolete, partial 0. *Calyx* obsolete. *Fruit* thin, compressed from the back, surrounded by a broad border. *Ridges* 5, very fine, the 3 dorsal near each other, the 2 lateral distant and close to the edge. *Vittæ* 1 to each furrow. *Albumen* flat.

1. P. *sativa* Linnæus.—(PARSNEP.) Fig. 348.

Stem angular, furrowed; leaves pinnated, shining, hairy on the under side; leaflets oblong, blunt, crenate-serrate, the terminal 3-lobed.

Habitat. Chalky districts, especially on the coast.

Quality and *Uses.* Roots sweet, nutritious; a favourite article of food.

348

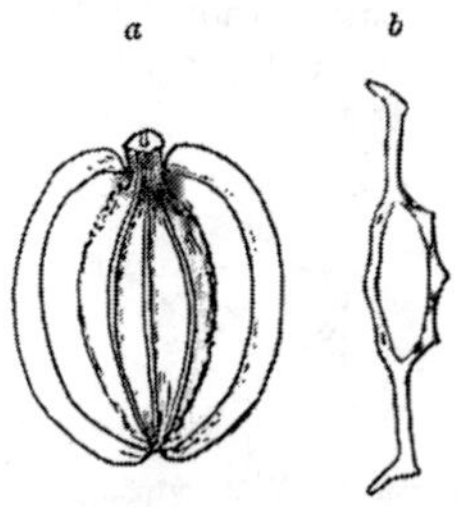

348

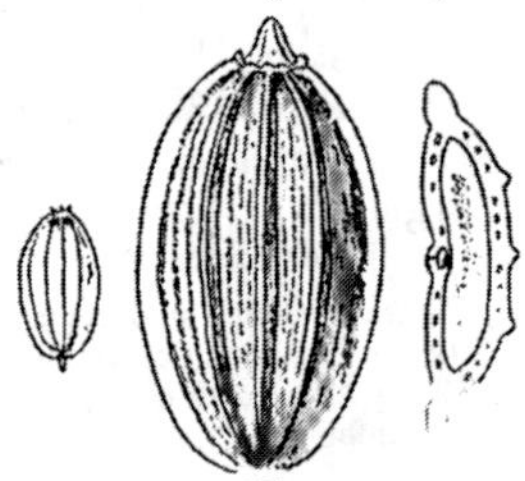

349

Fig. 348.—Pastinaca sativa; *a*, fruit magnified; *b*, its transverse section; 349. *a*, Fruit of Opopanax Chironium, natural size; *b*, the same magnified; *c*, transverse section of one-half.

OPOPANAX. *Koch.*

Umbels compound. *Involucres* general and partial, of few bracts. *Petals* roundish, entire, with an involute point. *Calyx* obsolete. *Fruit* thin, compressed from the back. *Ridges* 3, filiform. *Vittæ* 3 to each furrow, and 6-10 on the commissure. *Albumen* flat.

1. O. *Chironium* Koch.—(OPOPANAX.) Fig. 349.

Leaves bipinnate, with unequally cordate, crenate, obtuse segments.

Habitat. South of Europe.
Quality. One of the plants yielding a fetid antispasmodic gum-resin analogous in its effects to Ammoniacum.

NARTHEX. *Falconer.*

Umbels compound. *Involucres* 0. *Calyx* obsolete. *Fruit* thin, compressed at the back, with a dilated border. *Ridges* 3 only, dorsal. *Vittæ* 1 to each dorsal furrow, and 2 to the laterals. *Albumen* thin, flat.

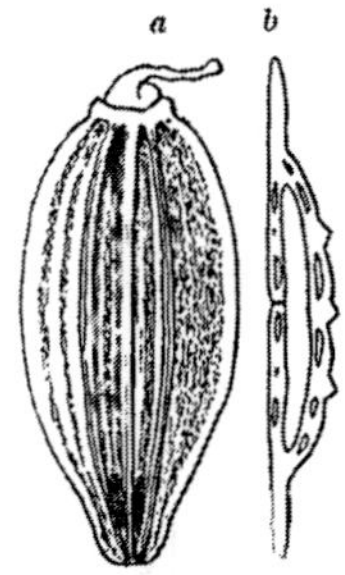
a b
350

1. N. *Asafœtida* Falconer. *Ferula Asafœtida* Linnæus.—(ASAFŒTIDA.) Fig. 350.

Radical leaves 3-parted; segments bipinnatifid, with oblong-lanceolate, obtuse, decurrent lobes.

Habitat. Laristan, Affghanistan, the Punjaub.
Quality. Gum-resin fetid, stimulating, antispasmodic.
Uses. In spasms and convulsions, chronic catarrh, flatulent colic, uterine obstructions.

N.B.—This is the genuine Asafœtida plant, but probably the substance is also yielded by even species of Ferula. It has been conjectured to have produced the Silphium or Laser of the ancients, but on unsatisfactory evidence. See Thapsia, p. 256.

FERULA. *Linnæus.*

Umbels compound. *Involucres* variable. *Petals* ovate, acuminate. *Calyx* slightly 5-toothed. *Fruit* thin, compressed from the back, surrounded by a broad border. *Ridges* 5, the dorsal filiform, the lateral disappearing in the border. *Vittæ* 3 or more in the dorsal furrows, 4 or more on the commissure. *Albumen* flat.

1. F. *persica* Willdenow. Fig. 351.

Stem dwarf, glaucous; leaves supradecompound, with distant segments, and linear-lanceolate cut lobes, dilated at the point; the first umbel sessile.

a b
351

Habitat. Persia.
Quality. There seems to be little doubt that this yields some part of the Asafœtida of commerce. Its fruit has been received from Persia as that of the Asafœtida plant; and an old plant at Chelsea yielded a gum-resin so like the drug that, until the re-discovery of Narthex, it was generally believed to be its real source.

Fig. 350.—*a*, Fruit of Narthex Asafœtida magnified; *b*, its transverse section; 351. Fruit of Ferula persica; *a*, natural size; *b*, magnified.

2. F. *orientalis* Linnæus.

Stem branched; leaves supradecompound; leaflets setaceous, minutely downy; upper petioles much inflated.

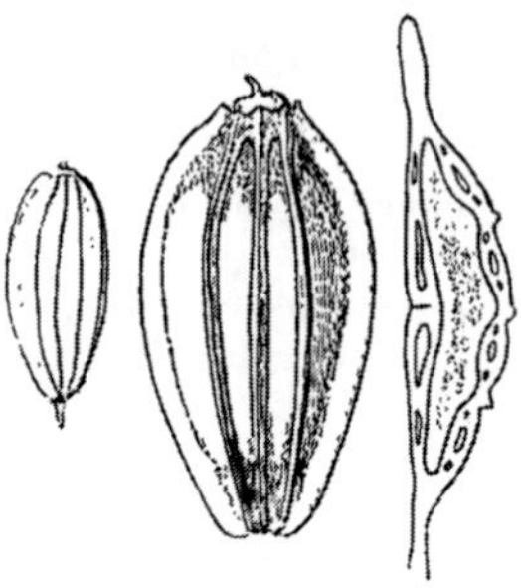

352

Habitat. Asia Minor, Greece.
Uses. Said to furnish African Ammoniacum; but if that drug is yielded by the Fashook of the Moors, then it is certain that its source is the next species.

3. F. *tingitana* Linnæus.—(AFRICAN AMMONIACUM.) Fig. 352.

Stem branched; leaves supradecompound, shining; segments oblong-lanceolate, much cut; upper petioles large and dilated; terminal umbel on a short stalk.

Habitat. Various places in the North of Africa.
Quality. The fetid gum-resin is analogous in its effects to common Ammoniacum.

HERACLEUM. *Linnæus.*

353

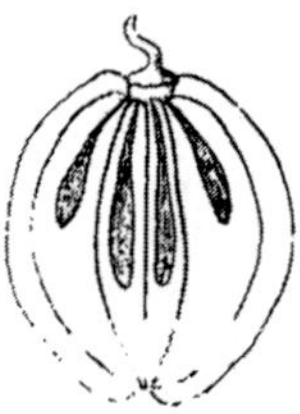

Umbels compound. *Involucres* obsolete. *Calyx* 5-toothed. *Fruit* compressed from the back, thin, with a broad border. *Ridges* 5, filiform, the 3 dorsal near each other, the lateral distant. *Vittæ* 1 to each furrow, short, clavate. *Albumen* thin, flat.

1. H. *Sphondylium* Linnæus.—(COW PARSNEP. HOGWEED.) Fig. 353.

Leaflets lobed or pinnatifid, cut and serrated; fruit smooth, emarginate; 2 vittæ on the commissure.

Habitat. Hedges and dry ditches; very common.
Quality. Acrid, vesicant. Rind of the root ulcerates the skin, if applied in a fresh state.

THAPSIA. *Linnæus.*

Umbels compound. *Involucres* variable. *Calyx* 5-toothed. *Fruit* compressed from the back. *Ridges:* 5 primary dorsal, filiform, 2 secondary lateral winged. *Vittæ* 1 beneath each of the 2 intermediate ridges, and 2 on the commissure. *Albumen* nearly flat.

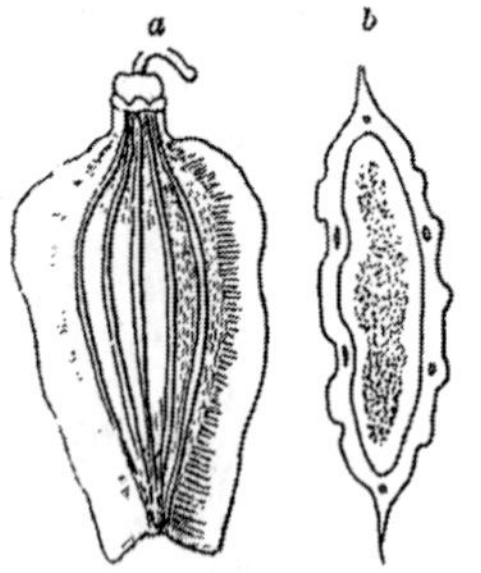

354

1. T. *garganica* Linnæus.—(ASA DULCIS.) Fig. 354.

Stem smooth; leaflets linear, acute, decurrent; fruit widely cordate.

Habitat. South of Europe and Barbary.
Uses. This, or a nearly allied species, called T. Silphium, yielded the Laser cyrenaicum or Asa dulcis, a drug enjoying the highest reputation among the ancients as an antispasmodic, deobstruent, and diuretic.

Fig. 352.—Fruit of Ferula tingitana, natural size; *b*, the same magnified; *c*, transverse section of the same; 353. Fruit of Heracleum Sphondylium, magnified; 354. *a*, Fruit of Thapsia garganica, magnified; *b*, its transverse section.

Dorema. *D. Don.*

Umbels proliferous, racemose. *Involucres* 0. *Calyx* obsolete. *Epigynous* disk cup-shaped, toothed. *Fruit* compressed from the back, surrounded by a broad border. *Ridges* 3, dorsal filiform, lateral very minute within the border. *Vittæ* of the furrows 0, of the commissure 4. *Albumen* flat.*

1. D. *ammoniacum* D. Don.—(Ammoniacum.) Fig. 355.

Leaves large, bipinnate, with pinnatifid segments, and oblong, obtuse lobes; petioles and fruit woolly.

Habitat. Persia.
Quality. The fetid gum-resin stimulating, discutient.
Uses. Deficient expectoration; as a plaister for glandular enlargements.

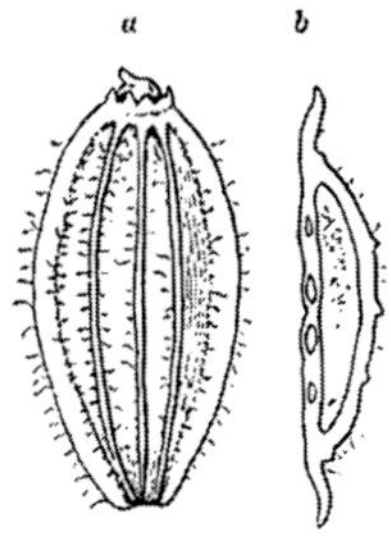

355

Eryngium. *Linnæus.*

Umbels simple, capitate. *Involucre* leafy. *Calyx* leafy. *Fruit* taper, obovate, scaly. *Ridges* 0. *Vittæ* 0.

1. E. *maritimum* Linnæus.

Leaves spiny-toothed, those next the root undivided, cordate, on the stem amplexicaul, palmate; bracts ovate, generally 3-lobed, spiny.

Habitat. Sandy coasts of the sea.
Quality. Root sweet, aromatic, tonic, diuretic.
Uses. The candied root used as an aperient, and in visceral obstructions. Reputed to be an aphrodisiac.

2. E. *campestre* Linnæus.

Leaves ternate, bipinnatifid, netted, those of the stem with lacerated auricles.

Habitat. Barren places. Europe.
Quality and *Uses.* As the last.

Natural Order, Ivyworts; *Araliaceæ* (V. K., p. 780.)

Prevailing Quality. Tonic? Acrid?

Aralia. *Linnæus.*

Styles diverging, 5.

1. A. *nudicaulis* Linnæus.

Leaf solitary, radical; petiole trifid: segments ovate, acute, serrated; scape shorter than the leaf.

Habitat. The United States.
Quality. Alterative and tonic.
Uses. Root said to equal Sarsaparilla in value.

Fig. 355.—*a*, Fruit of Dorema Ammoniacum; *b*, its transverse section, both magnified.

* In the "Flora Medica" the fruit is misdescribed in consequence of wrong specimens having been furnished by Mr. Don.

HEDERA. *Linnæus.*

Styles converging, or connate, 5-10.

1. H. *Helix* Linnæus.—(COMMON IVY.) Fig. 356.

Stem creeping by fibrous rootlets; leaves coriaceous, smooth, angular; umbels simple, downy.

Habitat. Bark of trees, and walls, everywhere.
Quality. Leaves bitter; fruit bitter, aperient, emetic.
Uses. Has had some reputation as a sudorific; leaves applied to cauterised surfaces. The gum-resin called Hederine, used by varnish makers, and said to be depilatory and emmenagogue.

Natural Order, Cornels; *Cornaceæ* (V. K., p. 782.)

Prevailing Quality. Astringent, tonic.

CORNUS. *Linnæus.*

Calyx nearly obsolete, 4-toothed. *Petals* 4, sessile. *Stamens* 4. *Style* 1. *Fruit* a 2-3-celled drupe.

Fig. 356.—Hedera Helix; *a*, magnified view of a perpendicular section of its seed.

1. C. *florida* Linnæus.

Flowers appearing with the leaves in close heads, surrounded by large white roundish bracts ; a tree.

Habitat. The United States.
Quality. Tonic, astringent, bitter.
Uses. Bark employed advantageously in intermittent fevers in the United States. It approaches Cinchona in its general effects, and is not inferior to it in the cure of intermittents. The young branches stripped of their bark, and rubbed with their ends against the teeth, render them white.

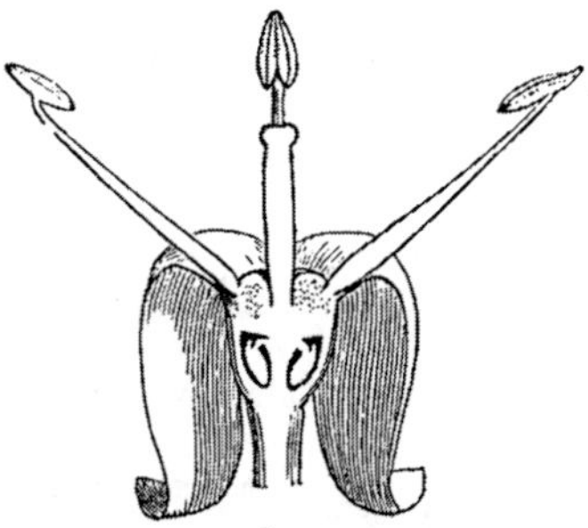

a

357

2. C. *sanguinea* Linnæus.—(Common Dogwood.) Fig. 357.

Leaves ovate, whole-coloured, downy beneath ; corymbs flat-headed.

Habitat. Hedges and woods, common.
Quality. Fruit contains a great quantity of oil, which is said to equal that of the olive.

3. C. *mas* Linnæus.—(The Cornel Tree. Cornelian Cherry.)

Flowers appearing before the leaves in small yellow heads with 4 inconspicuous bracts ; a tree.

Habitat. Woods and plantations from England to Japan.
Quality and *Uses.* Wood very hard and tough. Fruit austere, becoming subacid when perfectly ripe ; formerly fermented for a beverage. Bark said to have power in intermittents.

4. C. *suecica* Linnæus.

Flowers appearing with the leaves in close heads, surrounded by white bracts; leaves sessile, ovate ; herbaceous.

Habitat. Northern parts of Europe.
Quality. Berries tonic ; said to increase the appetite in a very remarkable manner.

Fig. 357.—Cornus sanguinea ; *a*, perpendicular section of its flower, magnified.

THE ASARAL ALLIANCE (V. K., p. 786.)

Natural Orders of Asarals.

Loranths (*Loranthaceæ.*) Ovary 1-celled.

Birthworts (*Aristolochiaceæ.*) Ovary 3-6-celled.

Natural Order, Loranths; *Loranthaceæ* (V. K., p. 789.)

Prevailing Quality. Unknown.

VISCUM. *Linnæus.*

Calyx obsolete. *Petals* 4. *Anthers* adnate to the petals, honeycombed.

1. V. *album* Linnæus.—(MISSELTO.)

Stem dichotomous, much branched ; leaves lanceolate, blunt, veinless.

Habitat. Europe, parasitical on trees.
Quality. Bark astringent. Berries yield a viscid matter of the same nature as birdlime.

Natural Order, Birthworts; *Aristolochiaceæ* (V. K., p. 792.)

Prevailing Quality. Stimulant, aromatic.

ARISTOLOCHIA. *Linnæus.*

Calyx tubular, oblique, inflated at the base. *Stamens* adnate to the style.

1. A. *Clematitis* Linnæus.—(COMMON BIRTHWORT.) Fig. 358.

Leaves roundish-cordate, stalked ; stem erect, striated ; flowers axillary, crowded, erect, with an ovate obtuse lip.

Habitat. Many parts of Europe, among rubbish, and in waste places.
Quality. Roots strong scented, powerfully stimulating.
Uses. Once in great repute as an aid in difficult parturition.

2. A. *rotunda* Linnæus. Fig. 359.

Leaves cordate, amplexicaul, obtuse ; stem nearly erect ; flowers solitary, sessile, erect, with an oblong lip.

Habitat. South of Europe.
Quality. Bitter, acrid roots stimulant and tonic.
Uses. In amenorrhœa as an emmenagogue ; in gout. Said to stupify snakes.

3. A. *longa* Linnæus.

Leaves cordate, ovate, retuse ; stem prostrate ; flowers erect, with a lanceolate acute lip ; root oblong.

Habitat. South of Europe.
Quality and *Uses.* As in A. rotunda.

4. A. *Serpentaria* Linnæus.—(SERPENTARY.) Fig. 360.

Leaves cordate, oblong, acuminate ; stem zigzag, ascending ; peduncles growing from the root, scaly ; flowers with a triangular mouth.

358

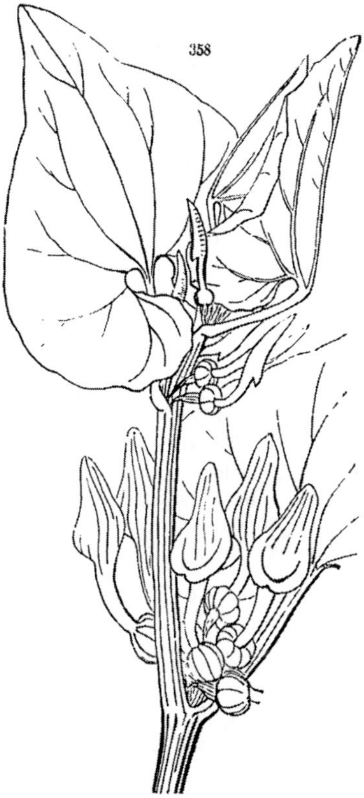

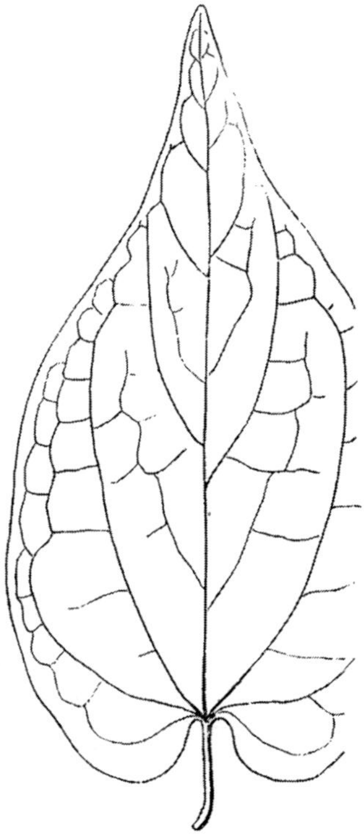

360

a

359

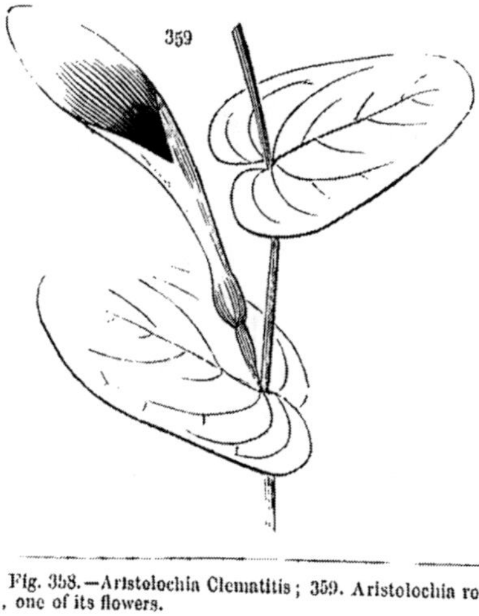

Habitat. United States.

Quality. Stimulant, nauseous, purgative; increases the pulse; diaphoretic.

Uses. In continued and intermittent fevers; low typhus; in the throat distemper. The root has a pene-

Fig. 358.—Aristolochia Clematitis; 359. Aristolochia rotunda; 360. Leaf of Aristolochia Serpentaria, *a*, one of its flowers.

trating resinous smell, and a pungent bitter taste. It acts as a tonic, and in certain cases as an antispasmodic and anodyne. It is peculiarly useful in supporting the strength and in allaying the irregular action which attends great febrile debility. Dr. Chapman considers it "admirably suited to check vomiting and to tranquillise the stomach, more particularly in bilious cases."

5. A. *anguicida* Linnæus. Fig. 361.
Leaves cordate, acuminate; stipules cordate, solitary, amplexicaul; calyx erect, with a lanceolate lip.

Habitat. Carthagena and Mexico.
Quality. The juice of the root chewed and introduced into the mouth of a serpent so stupifies it, that it may for a long time be handled with impunity; if the reptile is compelled to swallow a few drops, it perishes in convulsions. The root is also reputed to be an antidote to serpent-bites. *Jacquin*, l. c.

6. A. *Guaco*.—(The GUACO.) Fig. 362.
Leaves thin, oblong, almost parallel-edged, somewhat cordate, shortly acuminate, obtuse, smooth, 3-nerved at the base.

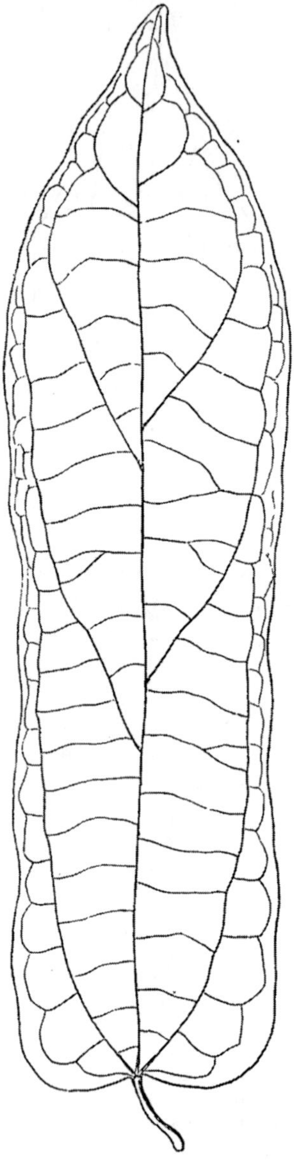

362

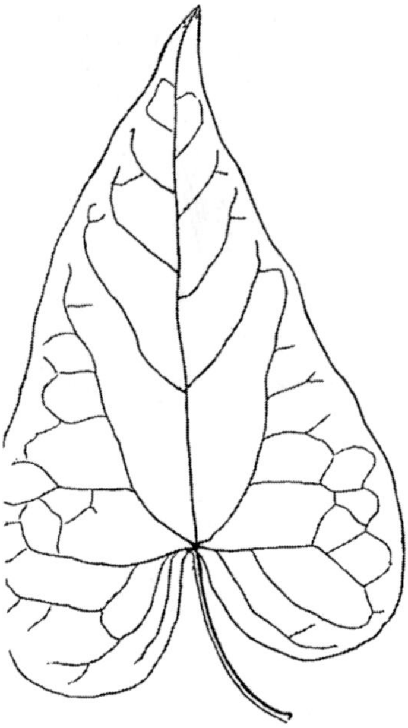

361

Fig. 361.—Leaf of Aristolochia anguicida; 362. Leaf of Aristolochia Guaco.

Habitat. Equinoctial America.

Quality and *Uses.* This appears, from the testimony of Dr. Hancock, to be the real Guaco, to which, as an alexipharmic, so much interest has attached by the relation of Humboldt. What is sold as Guaco in Colombia is certainly an Aristolochia of somo kiud. The accompanying figuro is taken from oue of Dr. Hancock's specimens.

ASARUM. *Linnæus.*

Stamens 12, horned, distinct from each other, and from the style. *Calyx* campanulate, 3-lobed.

1. A. *canadense* Linnæus.

Leaves reniform, mucronate, pubescent, in pairs.

Habitat. Canada.

Quality. A warm aromatic, stimulant and diaphoretic, aud as a substitute for Serpentary.

2. A. *europæum* Linnæus.—(ASARABACCA.) Fig. 363.

Leaves reniform, obtuse, hairy, in pairs.

Habitat. Woods of Europe.

Quality. Acrid, emetic, purgative, diuretic ? diaphoretic ; excites sneezing.

Uses. As a substitute for Ipecacuauha; as a counter irritaut in affections of the eyes, brain, &c., headache, toothache, ophthalmia. Drunkards in the south of France use it to sober themselves by emptying the stomach. Powder said to form the base of Cephalic snuff.

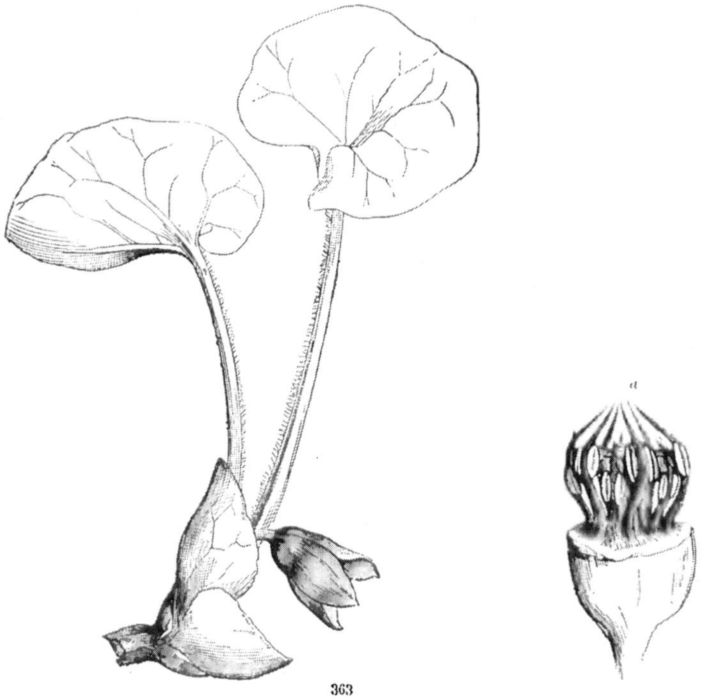

363

Fig. 363.—Asarum europæum ; *a*, the ovary and stamens magnified.

INDEX.

(THE NUMBERS REFER TO THE PAGES.)

ERRATA.

Page 71, line 16, *for* "GENTIANALS. Monopetalous. Placentæ parietal," *read* "GENTIANALS. Monopetalous. Placentæ parietal or axile. Embryo minute."

Page 71, line 18, *add* "Embryo large."